Approaches in Integrative Bioinformatics

Ming Chen • Ralf Hofestädt
Editors

Approaches in Integrative Bioinformatics

Towards the Virtual Cell

Editors
Ming Chen
College of Life Sciences
Zhejiang University
Hangzhou
People's Republic of China

Ralf Hofestädt
Department of Bioinformatics
and Medical Informatics
Bielefeld University
Bielefeld, Germany

Additional material to this book can be downloaded from http://extras.springer.com

ISBN 978-3-642-41280-6   ISBN 978-3-642-41281-3 (eBook)
DOI 10.1007/978-3-642-41281-3
Springer Heidelberg New York Dordrecht London

Library of Congress Control Number: 2013956814

© Springer-Verlag Berlin Heidelberg 2014
This work is subject to copyright. All rights are reserved by the Publisher, whether the whole or part of the material is concerned, specifically the rights of translation, reprinting, reuse of illustrations, recitation, broadcasting, reproduction on microfilms or in any other physical way, and transmission or information storage and retrieval, electronic adaptation, computer software, or by similar or dissimilar methodology now known or hereafter developed. Exempted from this legal reservation are brief excerpts in connection with reviews or scholarly analysis or material supplied specifically for the purpose of being entered and executed on a computer system, for exclusive use by the purchaser of the work. Duplication of this publication or parts thereof is permitted only under the provisions of the Copyright Law of the Publisher's location, in its current version, and permission for use must always be obtained from Springer. Permissions for use may be obtained through RightsLink at the Copyright Clearance Center. Violations are liable to prosecution under the respective Copyright Law.
The use of general descriptive names, registered names, trademarks, service marks, etc. in this publication does not imply, even in the absence of a specific statement, that such names are exempt from the relevant protective laws and regulations and therefore free for general use.
While the advice and information in this book are believed to be true and accurate at the date of publication, neither the authors nor the editors nor the publisher can accept any legal responsibility for any errors or omissions that may be made. The publisher makes no warranty, express or implied, with respect to the material contained herein.

Printed on acid-free paper

Springer is part of Springer Science+Business Media (www.springer.com)

Preface

The unprecedented accumulation of high-throughput data from genomics, transcriptomics, proteomics, metabolomics, phenomics, etc., has resulted not only in new attempts to answer traditional biological questions and solve longstanding issues in biology but also in the formulation of novel hypotheses that arise precisely from this wealth of data. At the present, with thousands of biological data resources and information systems inside the Internet, an unknown number of analysis tools, and exponential growths of molecular data (especially high-throughput data), the storage, processing, description, transmission, connection, and integrative analysis of this data becomes a great challenge for bioinformatics. Thus, the so-called Big Data becomes the new keyword describing the actual situation for which new software tools are needed to analyze this exponentially increasing data.

Important applications of *Big Data* are systems biology and systems medicine. For instance, hospital information systems represent complex patient data. The diagnosis process is now supported by new methods of biotechnology using, for example, high-throughput sequencing approaches. Therefore, we have complex patient data inside the hospital information system which needs to be stored, transported, and analyzed. New software tools are needed to allow the user-specific data access and analysis of this data. Overall, to develop and implement new tools for automatic data integration and analysis will help implement better diagnostic methods in practice. In the future, the entire genomes of patients will be stored within hospital information systems. Furthermore, it will be necessary to share the genome sequences inside the hospital computer network and analyze the genome data to detect, for example, cancer genes. With the availability of Internet, the automatic integration and analysis of data are of the most relevant research topics in computer science. In biology, such tools have become more and more important. Methods like high-throughput sequencing and omics analysis are responsible for the exponential data generation process.

This book will focus on the integration and analysis of omics data. The *Introduction* will present relevant biological background and an overview of these actual methods. When the Internet merged, methods such as data fusion and federated database systems became relevant. The initial tools were implemented and gave birth to a new field of research: Integrative Bioinformatics, which strives to implement user-specific integration and analysis of complex data. The *Introduction* of this book will give a definition and overview of this pertinent field of research. Since then, complex information systems have been developed and implemented. Finally, the data warehouse concept became more relevant. Today the data warehouse concept is still the best construction for the implementation of integrative information systems. The *Information Fusion and Retrieval* section will focus on the said data warehouse concept. Furthermore, this part of the book will give an overview of information retrieval and data mining tools, which allow the user-specific identification and integration of data. Based on the methods described here, we are able to implement user-specific integration tools. The analysis of this data can be done using statistic, visualization, or animation tools. Furthermore, modeling and simulation are important analysis methods. The *Network Visualization, Modeling, and Analysis* section will focus on methods for network prediction, network modeling, and simulation. In the case of network simulation, we prefer the Petri net method, which allows the parallel simulation of complex metabolic pathways. Our application section is divided into two parts. First, we focus on methods of *BioData Mapping*. One interesting aspect is the possibility of molecular disease mapping which allows the pathway prediction of any disease and the semiautomatic mapping of this pathway into a virtual 3D cell. The genotype-phenotype map enables us to uncover the casual networks inside the "black box" that lies between genotypes and phenotypes with advances in high-throughput and high-dimensional genotyping and phenotyping technologies. Another important and actual topic is presented by the *Biocomputation* section. After the reconstruction of a biological disease network, the identification of biomarkers or hubs for further analysis is important. To realize such tasks, the implementation of parallel algorithms is fundamental.

Important research topics for the next few years will be Big Data and Systems Medicine. Integrative Bioinformatics will be fundamental in developments for both fields and this book attempts to present an overview of relevant and actual research activities.

We are very grateful to all the authors for sharing their time, wisdom, and expertise. Finally, we want to thank Ms. Na Xu, the editor of Springer Beijing Office, for her continuous advice.

Hangzhou, People's Republic of China	Ming Chen
Bielefeld, Germany	Ralf Hofestädt
June 2013	

Contents

Part I Introduction

1 **Integrative Bioinformatics** 3
   Ming Chen and Ralf Hofestädt

2 **An Overview of Gene Regulation** 21
   Andrew Harrison and Hugh Shanahan

Part II Information Fusion and Retrieval

3 **Information Retrieval in Life Sciences: A Programmatic Survey** 73
   Matthias Lange, Ron Henkel, Wolfgang Müller,
   Dagmar Waltemath, and Stephan Weise

4 **Data Warehouses in Bioinformatics** 111
   Benjamin Kormeier

5 **Molecular Information Fusion in Ondex** 131
   Jan Taubert and Jacob Köhler

6 **Text Mining on PubMed** 161
   Timofey V. Ivanisenko, Pavel S. Demenkov,
   and Vladimir A. Ivanisenko

Part III Network Visualization, Modeling and Analysis

7 **Network Visualization for Integrative Bioinformatics** 173
   Andreas Kerren and Falk Schreiber

8 **Biological Network Modeling and Analysis** 203
   Sebastian Jan Janowski, Barbara Kaltschmidt,
   and Christian Kaltschmidt

**9 Petri Nets for Modeling and Analyzing Biochemical
    Reaction Networks** 245
    Fei Liu and Monika Heiner

Part IV BioData Mapping

10 Network Analysis and Integration in a Virtual Cell Environment 275
    Björn Sommer

11 Bridging Genomics and Phenomics 299
    Dijun Chen, Ming Chen, Thomas Altmann,
    and Christian Klukas

Part V BiocomputION

12 Parallel Computing for Gene Networks Reverse Engineering 337
    Jaroslaw Zola

13 Computational Biomarker Discovery 355
    Fan Zhang, Xiaogang Wu, and Jake Y. Chen

Part I
Introduction

Chapter 1
Integrative Bioinformatics

Ming Chen and Ralf Hofestädt

1.1 Introduction

Integrative Bioinformatics deals with the development of methods and tools to solve biological problems as well as providing a better understanding or new knowledge about biochemical phenomena by means of data integration and computational experiments [7]. Current high-throughput technologies such as NMR, mass spectrometry, protein/DNA chips, gel electrophoresis data, Yeast Two-Hybrid, QTL mapping, and NGS generate large quantities of high-throughput data. The challenge of Integrative Bioinformatics is to capture, model, simulate, integrate, and analyze this huge amount of data in addition to the data represented by hundreds of biological databases and thousands of scientific journals. The data needs to be integrated and made available in a consistent way to provide new and deeper insights into complex biological systems. Molecular biology produces this volume of data based on high-throughput technologies. One characteristic of this data is exponential growing. Therefore, storing and analysis of this molecular and cellular data essentially uses methods and concepts of Bioinformatics. Currently, there are more than 2,000 database and information systems available via the Internet, which represent this molecular data. Every year new molecular databases and information systems which can be used via the Internet crop up. The classical definition of an information system is based on a database system which represents the data and tools for the user-specific analysis of this data. Today an information system is or can be embedded into the Internet as shown in Fig. 1.1.

M. Chen
College of Life Sciences, Zhejiang University, Hangzhou, People's Republic of China
e-mail: mchen@zju.edu.cn

R. Hofestädt (✉)
Department of Bioinformatics and Medical Informatics, Bielefeld University,
Bielefeld, Germany
e-mail: hofestae@techfak.uni-bielefeld.de

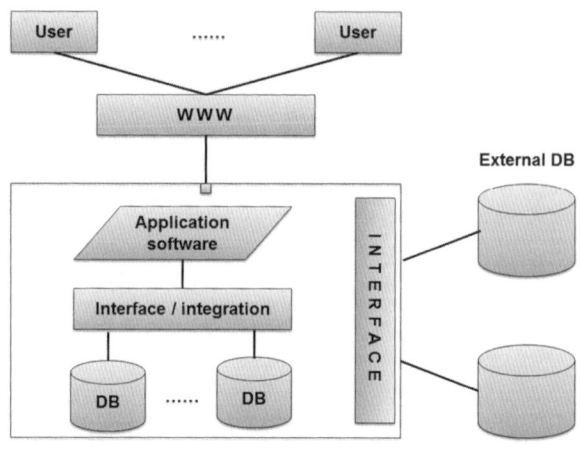

Fig. 1.1 Structure of a complex information system embedded into the Internet. *DB* denotes a database system

One characteristic of all these bio-information systems is that updates of implemented and running systems are constant. That is the reason that we have to handle a very dynamic collection of different molecular database and information systems. Most of these systems will lose their support sooner or later because financial support of most of such projects is usually temporary. This creates a complicated situation in molecular data today. All systems which are not adequately maintained must be identified and checked before using their data for further analysis. At that point it is important to note that the quality of molecular data presented by these systems via the Internet must be guaranteed by each owner of the database or information system. Until now, no quality standards have been defined for the practical use of this data. Taking a look to all these systems, we can safely say that most of them do not discuss this present data quality problem. Therefore, regarding any molecular database or information system, we have to be careful when using this molecular data for further analysis. Considering the actual molecular database and information systems which are now available, it would be good news for potential users to have a (semantic) overview of all these systems. A decade ago Nucleic Acids Research began to support this task by publishing an annual report of all molecular database systems. Recently we have developed an integrated database of the published biological databases and tools, named Da&To (http://bis.zju.edu.cn/DaTo), presenting helpful bio-web links including relevant database systems. Beyond the discussion of data quality, it is also important to mention that these systems are extremely heterogeneous in regard to the data structures, data representations, data access, etc. Therefore, it is not easy to implement the (semi-)automatic access to such bio-database and information systems. In terms of the represented data and database systems, we can differ between public data, open source data, and private data, which can only be used by special contracts. Overall we would like to call this situation the *molecular database problem*. One reason for this problem is the scientific foundation structure. In countries like Germany and most other countries, scientists have access to financial support via projects paid for by private foundations or government-supported foundations or paid directly by industry or the government. This kind of support is only for a limited time period. The normal time period for supporting a project is 3–5 years. After that time such projects and therefore most of the implemented systems will lose their support. This is why most of the database systems have a short life span even if they are available in the Internet years later. The only survival chance for a new and important system is to award new grants or to start a company. This is a very bad situation and Europe is presently trying to change it. The idea is to support national database centers in the future. However, this kind of solution will only help to solve this problem for the most relevant systems – so only for a subset of all systems. Another important problem is caused by the different formats and data storage techniques which are in use for all these systems. If there are no standards and no rules on how to prepare a molecular database, nothing will be changed in the future. Overall this was and is the main reason that the automatic process of data access continues to be difficult even if the data is available via the Internet. From the outset, the development and implementation of tools, which allow the user-specific data access based on that distributed and heterogeneous molecular data, was an important part of Bioinformatics. Therefore, new concepts and methods had to be developed and implemented to solve this task. At the beginning federated database systems seemed useful. The main disadvantage of this concept was and is that the data access process, which connects such autonomous running database systems, is time consuming. Behind these activities, specific integration tools like SRS [15] became popular, but did not allow complete user-specific integration efforts until now. In addition, most publications using integrative methods still use specific workflows. That means they extend their own data by integrating relevant external data, which has to be identified and extracted from specific database systems. Finally, they identify existing analysis tools or implement new tools for further analysis of this molecular data. This is the reason for the development of the data warehouse concept which has become successful for bio and medical applications during the past few years. The key idea of the data warehouse approach is to construct a new database system based on user-specific data (lab data) including user-specific external data (coming from relevant molecular databases) in combination with the user-relevant analysis tools. Therefore, a data warehouse can be interpreted as a complex web-based information system. Today more and more such *bio-data warehouses* are available.

When considering all these database activities, we can say that we presently have a wide area of databases and integration tools available. On the other hand, the analysis of this data is the key task of any user. Thousands of tools are published each year for the analysis of molecular data. The actual problem is that no one has a complete overview of existing algorithms and running tools for that kind of molecular data analysis. We would like to call this the *Bioinformatics Analysis Gap*. Different software tools could be implemented and new techniques and algorithms are appearing every day. In our case, analysis of molecular data can be a simple statistical approach or extend to complex simulations. Behind the database activities which are listed by the Nucleic Acids Research, we can see thousands of analysis tools which are available via the Internet. However, here we see the same situation: these tools are often no longer supported and the documentation of most of these tools is also poor. Furthermore, quality standards are not defined. From the beginning it was the idea of the international Journal of Integrative Bioinformatics (http://journal.imbio.de/) to focus on exactly these kinds of tasks: databases, integration, and analysis of integrated data [11]. After nearly 10 years of running this journal, we can see that the integration aspect is increasing in importance. Topics like Systems Biology and now Systems Medicine concentrate on this kind of data integration and analysis. In addition to the data storage and integration problem, the adequate analysis of this data is a key problem today.

1.2 Databases and Integration

Biological and biomedical data have been systematically stored in hundreds of public databases and information systems. A huge number of genes, enzymes, and biological pathways have already been identified, isolated, sequenced, and collected in these databases. For example, EMBL (http://www.ebi.ac.uk/embl/) and GenBank (http://www.ncbi.nlm.nih.gov/Genbank/) contain DNA sequences and databases like TRANSFAC/TRANSPATH (http://www.biobase.de/) bear the knowledge about gene expression. Metabolic pathways and their single biochemical reactions are stored in KEGG (http://www.genome.ad.jp/kegg/) and ExPASy (http://www.expasy.org/). BRENDA (http://www.brenda-enzymes.info) provides the kinetics of enzymatic-driven processes. Based on Da&To, we conducted a survey of all published biological databases and tools (a total of 14,117 till July 15, 2012) present in PubMed abstracts (over 3 million, since 1994). Undoubtedly, the USA, Germany, and the UK are the top three countries that published, respectively, 40.21, 8.64, and 7.54 % of all databases and tools. China (5.08 %) ranks the fourth, followed by France (4.53 %) and Japan (3.99 %). Most of these databases and tools were published in Bioinformatics-related journals. The top three journals for such publications are *Bioinformatics* (27.41 %), *Nucleic Acids Research* (20.54 %), and *BMC Bioinformatics* (8.06 %), which all together accounts for more than one-half of publications. The content of the publications was analyzed using MeSH terms that are the tags for the topics of articles (http://www.ncbi.nlm.nih.gov/mesh). We have found that the top 15 MeSHs rank differently over the years. By clustering them, we found that one category, containing the five following MeSHs, "Software," "Internet," "Animals," "Human," and "Algorithms," almost ranked nearly each year at the top of the top 15 MeSHs terms. Some hot spots switch over years. For instance, before 2002, the category of "Information Storage and Retrieval," "Computer Communication, Network," "Amino Acid Sequence, Data," and "Database, Factual" was a hot topic; while after 2003, "Database, Protein," "Database, Genetic," "User Computer Interface," and "Computational Biology, methods" overwhelmed over the formers. Other MeSH terms, occurring at lower frequencies, can indicate hot topic specific to some years. Further investigation shows that the correlation network of the MeSHs can be divided into 41 modules,

1 Integrative Bioinformatics 7

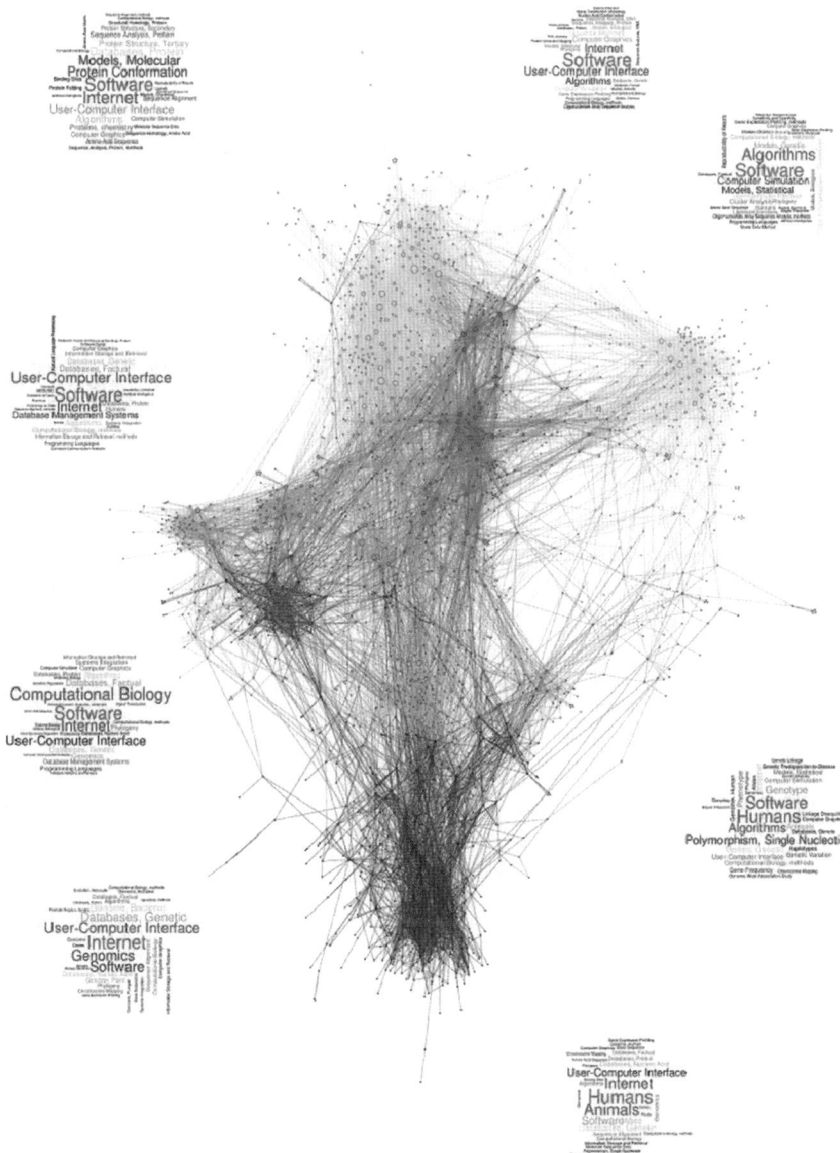

Fig. 1.2 The eight modules with their top 30 MeSHs

among which are 7 major modules (M40, M38, M31, M32, M28, M37, M39) counting nearly 96 % of all databases and tools. The filtered network with these seven modules is shown in Fig. 1.2 and their top 30 MeSHs are annotated.

As nearly all databases and tools were peer reviewed, giving them a quality assurance, also most of them are products of short-term research projects or

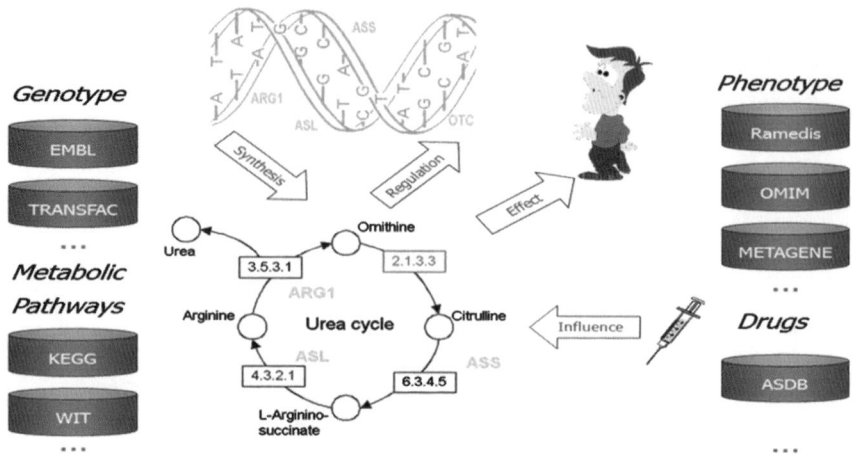

Fig. 1.3 Bioinformatics infrastructure for the analysis of metabolic diseases

PhD works, so they have to be freely accessible, but they are poorly maintained afterwards. We express the wish that some international nonprofit group could take care of them. Much more, we call upon to form a scientific society to maintain them under a kind of copyright agreement.

There is a special interest in supporting the Systems Medicine tasks today. Diseases are caused by gene defects and gene defects are responsible for defective metabolic pathways. The focus of molecular medicine is on using exactly this kind of data and analysis to understand the molecular behavior of any disease. Most of this biomedical data is collected and presented by OMIM (http://www.ncbi.nlm.nih.gov/Omim), which is a catalogue of medically important human traits, genes, and disorders thought to have a genetic basis. MedlinePlus is a premier source of health information for patients, families, and friends. Developed by the US National Library of Medicine, part of the National Institutes of Health, MedlinePlus contains web links to information on over 900 health topics. Other specific databases on inborn defects are Metagene (http://www.metagene.de), which is designed to support the diagnosis of inborn metabolism defects. RAMEDIS (http://www.Ramedis.de/) is a patient database of rare metabolic diseases. It develops a Bioinformatics system for representing, modeling, and simulating genetic effects on gene regulation and metabolic processes in human cells. This electronically available knowledge of genes, enzymes, metabolic pathways, and metabolic diseases increases rapidly. These databases are highly heterogeneous both in structure and in semantics and give only highly specialized views of the biological systems.

In Fig. 1.3 we can see that tools are needed which deal with the development of methods to facilitate the integration of data originating from multiple biological resources. To study and understand a disease, we have to identify the relevant biological networks which represent the molecular knowledge of the disease, as demonstrated in Fig. 1.3. Therefore, the first step is to identify the relevant molecular database systems. The second step is to identify the project-specific data inside any system. The last step is to extract this user-specific data and include this data in the user-specific information system. The implementation of integrative software methods will allow the identification, extraction, and prediction of these networks for any disease [17]. The diversity of interfaces offered by the different data resources requires the definition of interfaces and semantic tagging of the different types of data. There are, for example, databases offering their information by means of web services, others provide XML files or flat files, and there are several offering HTML-based interfaces. Actual research focuses on the development of methods for the automatic generation of interfaces to a diverse number of resources based on the researchers' needs and for the integration of the data contained in these resources. Therefore, different systems employ ontologies and controlled vocabularies to classify the information offered by the different data resources. Overall, development of a user interface and web services is an important task in realizing user-specific data integration.

Until now *flat file systems* (http://en.wikipedia.org/wiki/Flat_file_database) dominate the visible biological database systems within the Internet. A flat file is a data set which represents an implicit data structure. If a computer represents such a flat file, we can define this as a data resource. A flat file consists of different lines which represent data using the ASCII format. A simple example representing information about an enzyme shows this kind of data structure:

ENTRY	EC 2.1.3.3
NAME	Ornithine carbamoyltransferase
	Citrulline phosphorylase
	Ornithine transcarbamylase
CLASS	Transferases
	Transferring one-carbon groups
	Carboxyl- and carbamoyltransferases
SYSNAME	Carbamoyl-phosphate
	L-ornithine carbamoyltransferase

The end of each line is determined by a specific character or character list (often enter). Furthermore, special separators are often used to identify different data within one line. Keywords are in capital letters on the left side of each line such as ENTRY, NAME, and SYSNAME. The keywords are important signals for the so-called parser systems which realize the automatic identification and extraction of data regarding such flat file systems.

The so-called data-based information systems are often systems where data is represented as HTML data sets or other structured data. These systems represent organized data. The main reason using data-based information systems is the user-specific data access and data analysis via the Internet. Data modification inside the system can be done only by the owner or administrator of the system. The data access is based on the workload of the representing network and Internet. Based on the URL of the data-based information system, relevant HTML pages representing the data can be identified. Overall that means that complex parser systems must be developed and implemented to identify the user-specific data. After this identification process, the parser is able to extract the data using simple copy and paste functions.

More comfortable are Internet interfaces. Therefore, CGIs and mechanisms out of the JavaWorld such as servlets or applets are useful. Based on these mechanisms, a complex computation and representation of data is possible. The best situation for implementation of automatic data access procedures is that data is represented by a database system. In that case, data can be identified and extracted directly using the database query language (e.g., SQL for relational database systems).

The presence of numerous informational and data resources on biological data described above raises the acute problem of data integration and suitable access. From the beginning of the Internet, more and more tools were developed for user-specific data integration and analysis. Today, a lot of integration tools for biological data sources are available and in use. These systems are based on different data integration techniques, e.g., federated database systems (ISYS [16] and DiscoveryLink [10]), multi-database systems (TAMBIS [18]), and data warehouses (SRS [15] and Entrez [19]).

ISYS stands for Integrated SYStem and can be characterized as a component-based implementation. The main goal of ISYS is to provide a dynamic and flexible platform for integrating molecular biological data sources. This system was developed as a Java application. The system must be installed based on a local computer system. Different platforms like MS Windows or Solaris are supported. The locally installed system accesses the distributed data sources on the Internet. One main feature is the global view of integrated data sources with the help of a global scheme. Materialization of the integrated sources is not required. ISYS provides a JDBC (Java Database Connectivity) driver.

DiscoveryLink system was developed by IBM. It is also based on federated database techniques. A federated system requires the development of a global scheme. Thereby, the degree of integration must be rated as tight. DiscoveryLink accesses its original data sources through views. Read-only SQL is supported as query language. A JDBC and an ODBC (Open DataBase Connectivity) driver are also provided, and different output formats can be generated as well. The TAMBIS integration system is based on multi-database techniques. It can be used through a Java applet. Due to the use of a multi-database query language, it is not necessary to construct an integrated global scheme. Therefore, the degree of integration can be described as loose. As a query language in TAMBIS, a variety of the Collection Programming Language (CPL) [20] is implemented. CPL is hardwired into the system architecture. This is why it is not so easy to use this query language from outside of the system. Other disadvantages of TAMBIS are the absence of an API or other public interfaces. The number of input formats, which is limited to one, generated by the Java applet, proves also to be disadvantageous.

SRS is based on local copies of each integrated data source with a special format that is described in the ICARUS language specification. ICARUS can help represent the structure of the integrated data source. Through the use of these local copies, SRS is completely materialized. But during this transfer into the new format, no scheme integration is realized. Therefore, the degree of integration can be characterized as loose. SRS runs on a web server and is accessible via any web browser. An HTML interface for data queries is provided. Furthermore, the system can be queried by constructing special URLs. But no query languages like SQL or OQL (Object Query Language) are supported. SRS offers also a C-API. Various output formats are possible (HTML or ASCII text). One problem with the result presentation in SRS is the necessity to parse the outputs for further computer-based processing. The absence of any scheme integration is also disadvantageous for the use of the SRS. Similar to SRS is the Entrez system. This system integrates only data sources of NCBI. No materialization of the integrated sources is realized. Entrez uses views of the original sources. Consequently, scheme integration cannot be established. Therefore, the degree of integration can be classified as loose. The statements about SRS to query the system are completely transferable to Entrez. There are no standard query languages, no standardized API, or no other interface standards like JDBC. HTML is the only interface provided. Another Entrez feature is the manual construction of special URLs. Various output formats prove to be useful. These include HTML or ASCII text as well as XML and ASN.1 files. The greatest disadvantage of the Entrez approach is the restricted number of integrated data sources (only NCBI internal data sources) and the missing support of query languages.

Furthermore, many more integration tools are available, most of them implemented based on research projects. In our book Chaps. 4 and 5 will focus to this topic.

1.3 BioWeb

Besides the structured data deposited in the molecular databases, biomedical literature is published on the scale of over 500,000 documents per year and hosts unstructured knowledge. Besides the databases discussed in Chap. 2, information systems like PubMed (http://www.ncbi.nlm.nih.gov/pubmed) are bibliographic databases that access bibliographic information and abstracts of published articles in biomedical journals. Another archive of biomedical and life science literature is PubMed Central (PMC). One of the values of PMC is the collection of full-text articles, each of which complies with a common format. Due to the daily routine of checking these bibliographic databases and very often the overwhelmingly long list of search results, several web-based PubMed derivatives have been developed to help users quickly and efficiently search and retrieve relevant publications utilizing the services provided by PubMed and PMC databases. BioText (http://biosearch.berkeley.edu/) differs from the other PubMed derivatives in searching the full text and figure/table captions beyond the abstracts. Based on this kind of literature data, different, so-called text and data mining tools are available today. These tools

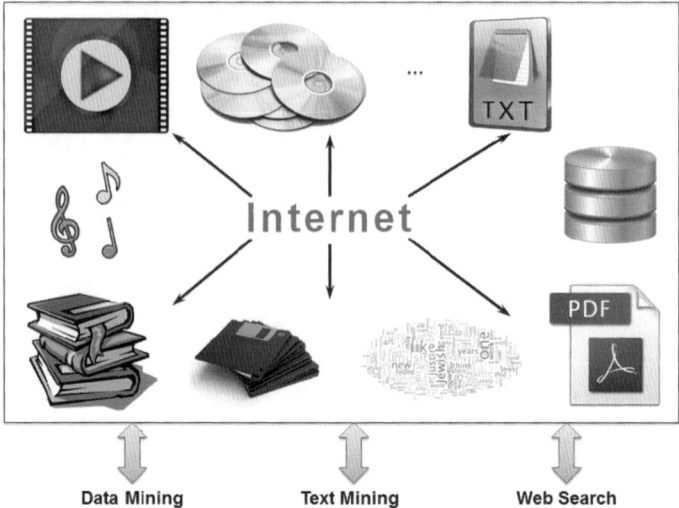

Fig. 1.4 Biosearch and Biomining are tools for user-specific search via the Internet (icons from the web)

realize user-specific access to global data inside the web. To realize user-specific access, text and data mining tools as well as search engines are necessary for implementation (Fig. 1.4).

Today, we are more or less at the beginning of developing and implementing complex mining and search tools which will allow identification and extraction of user-specific web data. At present, this subject seems to be a new research topic of information systems and one key issue is web semantics influencing mining and searching. In our book, different chapters will focus on the important topic of new mining and search tools. For web search, Chap. 3 will present an overview and a "biogoogle" web search tool. Chapter 6 will present an overview of text mining and a text mining tool. All in all, this topic is more complicated than simple database integration, and the key problem is to handle the semantics of the web data, which is still not solved.

1.4 Analysis and Simulation of Biological Networks

Based on the molecular database and information systems which are available via the Internet, analysis of bio-data is the second important step of Integrative Bioinformatics. Focusing on this actual research topic, the analysis of biological networks is a central issue for the future of biotechnology and molecular medicine as already seen in Chap. 2. Besides approaches of genome sequence comparison, genome annotation, and enzyme assignment, Bansal [4] describes a framework of automated reconstruction of metabolic pathways using the information on orthologous and homologous gene groups archived in the GenBank. Allen [2] presents a reconstruction method using the exploration of gene expression data with factor analysis. Factor analysis is shown to identify and to group genes according to membership within independent metabolic pathways for steady-state microarray gene expression data. Boyer and Viari [5] propose a new formulation for the problem of metabolic pathway reconstruction. They use an idea similar to that of Arita [3] to consider chemical compounds as sets of individual atoms and reactions as transfers (partial injections) of atoms between compounds.

Moreover, several software tools have been developed to assist reconstruction of pathways. For instance, PathoLogic [13] is used by PathMiner by McShan et al. [12]. However, they have a number of limitations. Predicting each gene function based solely on sequence similarity often fails to reconstruct cellular functions with all the necessary components. They do not contain comprehensive information about metabolic pathways, such as physical and chemical properties of the enzymes that are involved. Some approaches are not fully computer aided. The individual database search process requires too much human intervention, and the quality of annotation largely depends on the knowledge and work behavior of human experts. The future of metabolic pathway analysis may depend upon its ability to capitalize on the wealth of genetic and biochemical information currently being generated from genomic and proteomic technologies.

An ideal system for metabolic pathway reconstruction would at least include a web-based architecture to allow remote and local access to the different biological databases. It would offer a proven approach that can perform complex queries, data transformations, and data integration in one powerful biological tool, without requiring extensive programming. An automated primary and secondary database update and report system would enable the internal data to remain consistent, accurate, and reliable, with the ability to incorporate information flowing from experimental validation, such as gene expression, enzyme catalyzation, protein interaction, and pathways. An essential feature would include a quality assurance process to allow quick distribution of queries and retrieve primary results. In light of these desirable features, Sebastian Janowski designed and implemented the VANESA information system (Chap. 8) which has a single common data representation to handle the diverse range of rudimentary data such as enzymes, proteins, and metabolites as well as incomplete or fragments of gene sequences of metabolic pathways. VANESA is able to edit, extend, visualize, and analyze biological networks.

Nucleic acid and protein sequence comparison is an important tool in genome informatics. Initial clues to understand the structure or function of a macromolecular sequence arise from homologies to other macromolecules that have been previously studied. Many applications and tools, such as BLAST (http://www.ncbi.nlm.nih.gov/BLAST) and FASTA (http://www.ebi.ac.uk/fasta3), were developed to further understand the biological homology and estimate evolutionary distance. Recently the emphasis of research efforts has begun to turn away from gene sequences to metabolic pathways. It is therefore not surprising that the development of computational algorithms to predict metabolism function from gene, amino acid sequences, and metabolic networks is now a core aim of Bioinformatics. As more genomes are sequenced and the metabolic pathways reconstructed, it becomes possible to perform biological comparison from a biochemical-physiological perspective. Alignments represent one of the most powerful tools for comparative analysis of metabolism. Metabolic pathway alignment is of importance to study biology evolution, pharmacological targets, and other biotechnological applications, such as metabolic engineering and metabolism computation. A metabolic pathway alignment is a mapping of the coordinates of one pathway onto the coordinates of one or more other pathways. For example, the same metabolic pathway from two organisms may have diverged if the organisms evolved from a common ancestor, where individual metabolites and enzymes may have been changed, added, or lost in one pathway. This alignment involves recognition of metabolites that are common to a set of function-related metabolic pathways, interpretation of biological evolution processes, and determination of alternative metabolic pathways. Moreover, it aids in function prediction and metabolism modeling. Although researches on genomic sequence alignment have been intensively conducted, until now the metabolic pathway alignment has been less studied. Several approaches of metabolic pathway alignment have already been made by Dandekar et al. [8], Forst and Schulten [9], and Pinter et al. [14]. However, their definitions of pathways are the traditional biochemical pathways such as glycolysis, the pentose phosphate pathway, and the citric acid cycle. Less effort is made on analysis of gene regulatory networks as well as signaling pathways. Sebastian Janowski is handling biological networks which include metabolic pathways and signal pathways. His VANESA tool also includes different alignment algorithms.

In this book we present different chapters which represent new tools and methods for reconstruction, visualization, and analysis of biological networks. The aforementioned system, VANESA, is attending to of all these research topics and offers tools for this type of bionetwork analysis.

1.5 Bio-data Warehouse

Having the Internet and hundreds of molecular database and information systems which represent an exponential data-growing process, we can identify this continually increasing molecular data collection as the backbone of the virtual cell. That means that the information which represents the virtual cell is on the increase within the Internet every day. Furthermore, access and analysis of this data is fundamental for the development of bio-research. The practical situation is that any bio-research group tries to discuss a fundamental question (hypothesis). Based on their own data (in-house or lab data), they need access to literature and database systems to construct their specific model or working hypothesis. To discuss this model (hypothesis) or test the quality of this model, user-specific data access has to first be implemented. Having the complete biological knowledge (e.g., the representative biological

1 Integrative Bioinformatics

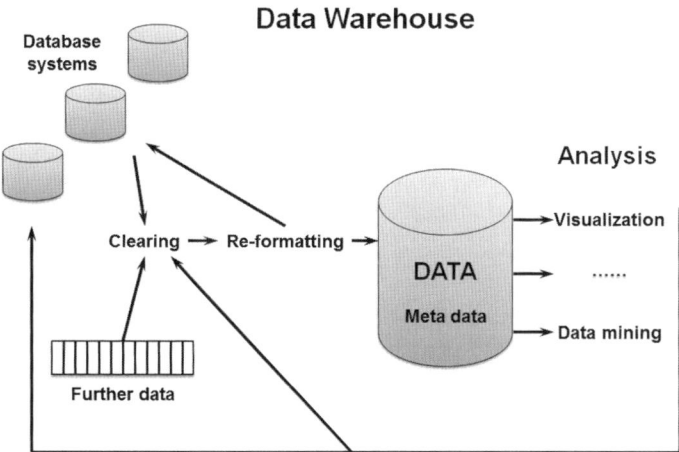

Fig. 1.5 Architecture of a data-warehouse system

network of a disease), analysis tools can be used to discuss the open question or hypothesis. In the case of Biology, today this working strategy is called Systems Biology. Therefore, data extension will support the model development process which is the first step taken for that kind of work. The next step is represented by data analysis processes. The final step is to use this model as a specification for the implementation of the simulation tool. Based on the simulation tool, experiments can be simulated in hypothetical or virtual worlds. To realize such an analysis scenario, we need to implement complex software systems. At present it seems that bio-data warehouses are the best solution for data integration and analysis. The idea behind the bio-data warehouse was to create a new database system based on the relevant distributed data which is available in the Internet and relevant for the project. Furthermore, the integration and access to the user-specific data analysis tools is the second part of such a system. In contrast to integration by data linking methods, data sets are identified and extracted from the original data resources. Furthermore, the extracted data will be cleaned and finally transformed into the new database system (called metadata system). This kind of integration can be called true data integration as many resources verge into one new database system.

Figure 1.5 shows the architecture of the data warehouse. The data for the new database system, which is in the center of the bio-data warehouse, comes from in-house data (lab data) and from different data sources and database systems which are available via the Internet. The second characteristic component of such a structure is the analysis part of the system. User-specific analysis tools have to be implemented or identified and integrated so that the user-specific analysis will be supported. Chapter 4 shows the detailed description of the data integration component of a data warehouse. Furthermore, this chapter presents an actual warehouse concept which is currently in use in different projects. One of these projects is the VANESA system presented by Sebastian Janowski. As mentioned previously, VANESA is a complex information system for the reconstruction and analysis of biological networks. Such complex networks can be analyzed (tools of VANESA) or systematically visualized 3D, which is discussed in Chap. 7. This kind of data and representation is the backbone of the representation of a virtual cell, which is discussed in the Chap. 10. The other important analysis is the simulation of this type of data. As regards the literature, we can identify different methods and concepts for simulation of biological networks or biochemical knowledge. Overall the Petri net representation became well known during the last 10 years [1, 6].

In this book, the Chap. 9 from Fei Liu and Monika Heiner will present an overview of Petri net modeling and simulation of biological networks.

1.6 Problems and Future

Molecular data is available via the Internet. Therefore, around 2,000 database systems are available today. One problem using this kind of data is that most of these systems are only supported for a few years. Another fundamental problem is the data quality of these systems which can never be guaranteed. Most Bioinformatics projects will receive financial support for the implementation of new database and data analysis systems. The data quality is still not the focus of all these projects until now. Even if it would be the main focus, we could not solve this problem, because a high rate of error is already within the technologies which are presently in use. Furthermore, the preparation of the experiments, the analysis technology, and the data interpretation represents a high error rate. Another problem is finding financial support so that relevant database systems which are already implemented can be serviced, supplemented when necessary, and implemented in the future. To realize the high goals, we need standards of data representation so that access and data quality checks can be easily implemented in the future. The data quality problem has not been at the center of discussion until now. However, taking a look at the practical situation of bio-data handling, we can say that this problem will come to focus very soon as most of the represented data seems to be dirty. Based on this dirty bio-data, Systems Biology and Systems Medicine cannot be successful in the future. One solution of the bio-data quality problem could be the wiki approach. This is more or less the only chance to guarantee high quality and the actualization of a dynamic bio-data for the future. But this approach will be difficult to implement because the relevant database systems are already in the hand of private companies and they will take to this idea of data representation. Also the industry will have problems with these kinds of solutions. However, in the future all foundations and government research centers can require all supported cooperation partners to follow this rule of data representation. For the user-specific data integration, it appears that the bio-data warehouse concept is already the solution for most applications. Therefore, we need bio-data warehouse shells for potential users and open software standards for the interfaces so that the updating problem of such bio-data systems can be more easily solved. The integration of data is one important task which is more or less solved today. The other important task is the identification of relevant analysis tools and the integration of these tools within user-specific bio-data warehouses. This was the main reason publishing this book because in the near future this problem needs to be solved using new methods of Integrative Bioinformatics. The main application of Integrative Bioinformatics is to support diagnosis and therapy methods and concepts. Based on complex bio-data warehouses, we are able to realize the molecular disease mapping problem [17]. In this case, we are able to identify the relevant biological networks of any disease. Based on this knowledge, we are able to study the biochemical and genetic behavior of these networks, which will allow the identification of targets, mutations, and molecular defects which can be responsible for the disease. Molecular medicine is already working in that direction, and the new scientific topic which joins together Medicine, Molecular Medicine, and Integrative Bioinformatics is called Systems Medicine. To understand the behavior of life, we have to implement the virtual cell. The vision of the implementation of the virtual cell is the key connection between Molecular Medicine, Systems Biology, Systems Medicine, and Integrative Bioinformatics.

WWW-List of Selected Molecular Information Systems

Genes

EMBL – http://www.ebi.ac.uk/
The "EMBL Nucleotide Sequence Database" represents all known DNA and RNA sequences.
GenBank – http://www.ncbi.nih.gov/Genbank/
NIH genetic sequence database.
HGMD – http://www.uwcm.ac.uk/medical_genetics/
HGMD represents the Human Gene Mutation Database.

Proteins and Enzymes

ENZYME – http://www.expasy.org/enzyme/
ENZYME is a repository of information relative to the nomenclature of enzymes.
LIGAND – http://www.genome.ad.jp/ligand/
The Ligand Chemical Database for Enzyme Reactions is linking chemical and biological aspects of life in the light of enzymatic reactions.
PDB – http://www.rcsb.org/pdb/
PDB presents 3D macromolecular structure data primarily determined experimentally by X-ray crystallography and NMR.
PIR – http://pir.georgetown.edu/
This database is a comprehensive, annotated, and nonredundant set of protein sequence databases in which entries are classified into family groups and where alignments of each group are available.

PROSITE – http://www.expasy.ch/prosite/
PROSITE is a database of protein families and domains.
REBASE – http://rebase.neb.com/
Restriction Enzyme data BASE is a collection of information about restriction enzymes.
SWISSPROT – http://www.expasy.org/sprot/
Protein sequence database.

Pathways

CSNDB – http://geo.nihs.go.jp/
The Cell Signaling Networks DataBase (CSNDB) is a data and knowledge base for signaling pathways of human cells.
KEGG – http://www.genome.ad.jp/
The Kyoto Encyclopedia of Genes and Genomes represents information of pathways that consist of interacting molecules or genes.

Gene Regulation

TRANSFAC – http://www.biobase.de
This database presents data about gene regulatory DNA sequences.
TRRD – http://www.bionet.nsc.ru/
Transcription Regulatory Regions Database.

Metabolic Diseases

OMIM – http://www3.ncbi.nlm.nih.gov/
The Online Mendelian Inheritance in Man database is a catalogue of human genes and genetic disorders authored and edited by Dr. Victor A. McKusick and his colleagues.
PATHWAY – http://oxmedinfo.jr2.ox.ac.uk/
PATHWAY is a database of inherited metabolic diseases. The database is divided into two sections: substances and diseases.
PEDBASE – http://www.icondata.com/health/pedbase/
PEDBASE is a database of pediatric disorders. Entries are listed alphabetically by disease or condition name.
RDB – http://www.rarediseases.org/
The Rare Disease Database is a delivery system for understandable medical information to the public, including patients, families, physicians, medical institutions, and support organizations.

References

1. Alla H, David R (1998) Continuous and hybrid Petri nets. J Circuit Syst Comput 8(1):159–188
2. Allen HD (2001) Reconstruction of metabolic pathways by the exploration of gene expression data with factor analysis. Dissertation, Virginia Polytechnic Institute and State University, Blacksburg, VA

3. Arita M (2000) Metabolic reconstruction using shortest paths. Simulat Pract Theory 8:109–125
4. Bansal AK (2000) A framework of automated reconstruction of microbial metabolic pathways. In: Proceedings of the IEEE international symposium on bio-informatics and biomedical engineering, Arlington, VA, 8–11 November, pp 184–190
5. Boyer F, Viari A (2003) An initio reconstruction of metabolic pathways. In: ECCB'2003 (European conference on computational biology), 27–30 September, Paris
6. Chen M, Hofestädt R (2003) Quantitative Petri net model of gene regulated metabolic networks in the cell. In Silico Biol 3(3):347–365
7. Collado-Vides J, Hofestädt R (2002) Gene regulation and metabolism – post genomic computational approaches. MIT Press, Cambridge, MA
8. Dandekar T, Schuster S, Snel B (1999) Pathway alignment: application to the comparative analysis of glycolytic enzymes. Biochem J 1:115–124
9. Forst CV, Schulten K (1999) Evolution of metabolisms: a new method for the comparison of metabolic pathways using genomics information. J Comput Biol 6:343–360
10. Haas LM, Schwarz PM, Kodali P, Kotlar E, Rice JE, Swope WC (2001) DiscoveryLink: a system for integrated access to life sciences data sources. IBM Syst J 40:489–511
11. Hofestädt R (ed) (2005) Yearbook bioinformatics 2004. IMBio, Informations management in der Biotechnologie e.V, Magdeburg
12. McShan DC, Rao S, Shah I (2003) PathMiner: predicting metabolic pathways by heuristic search. Bioinformatics 19(13):1692–1698
13. Paley S, Karp PD (2002) Evaluation of computational metabolic pathway predictions for *Helicobacter pylori*. Bioinformatics 18(5):715–724
14. Pinter RY, Rokhlenko O, Yeger-Lotem E et al (2005) Alignment of metabolic pathways. Bioinformatics 21(16):3401–3408
15. Schaftenaar G, Cuelenaere K, Noordik JH, Etzold T (1996) A Tcl-based SRS v. 4 interface. Comput Appl Biosci 12(2):151–155
16. Siepel A, Farmer A, Tolopko A, Zhuang M, Mendes P, Beavis W, Sobral B (2001) ISYS: a decentralized, component-based approach to the integration of heterogeneous bioinformatics resources. Bioinformatics 17:83–94
17. Sommer B, Ivanisenko V, Arrigo P, Hofestädt R (2012) Prediction and 3D visualization of biological networks using cytological disease mapping. EMBnet J 18(Suppl B):115–116
18. Stevens R, Baker P, Bechhofer S, Ng G, Jacoby A, Paton NW, Goble CA, Brass A (2000) TAMBIS: transparent access to multiple bioinformatics information sources. Bioinformatics 16:184–185
19. Tatusova TA, Karsch-Mizrachi L, Ostell JA (1999) Complete genomes in WWW Entrez: data representation and analysis. Bioinformatics 15:536–543
20. Davidson SB, Overton C, Tannen V, Wong L (1997) BioKleisli: a digital library for biomedical researchers. Int J Digit Libr 1:36–53

Chapter 2
An Overview of Gene Regulation

Andrew Harrison and Hugh Shanahan

Abstract It is not unreasonable to assume that in the near future next-generation sequencing techniques will allow the sequencing of all the DNA and expressed types of RNA involved in a given response or process. Such a range of data will be necessary to unravel the complexities of the multiple layers involved in the regulation of gene expression.

In this article we discuss a broad range of studies about gene regulation. These involve studies of processes such as transcription and splicing, the production of a variety of transcripts, and the involvement of protein–nucleic acid composites such as chromatin. We seek to shed light on common themes that are beginning to develop in these rapidly evolving, but intimately related, fields.

Keywords Gene expression • Post-transcriptional processing • Epigenetics • Non-coding RNA • Genome tertiary structure

Acronyms

CPSF	cleavage and polyadenylation specificity factor
CstF	cleavage stimulation factor
CTCF	CCCTC-binding factor
CTD	carboxy-terminal domain
dsDNA	double-stranded DNA
dsRNA	double-stranded RNA

A. Harrison (✉)
Department of Mathematical Sciences and School of Biological Sciences,
Essex University, Essex, UK
e-mail: harry@essex.ac.uk

H. Shanahan
Department of Computer Science, Royal Holloway, University of London, Surrey, UK

EJC	exon junction complex
NDR	nucleosome-depleted region
ncRNA	non-coding RNA
miRNA	microRNA
Poly(dA:dT)	double-stranded sequence of DNA composed of AT pairs
PAP	poly(A) polymerase
PASR	promoter-associated sRNA
PROMPTS	promoter-associated transcripts
PTM	posttranslational modification
RNAi	RNA interference
RNP	ribonucleoprotein
hnRNP	heterogeneous nuclear ribonucleoprotein
mRNP	messenger ribonucleoprotein
RNAPII	RNA polymerase II
ssDNA	single-stranded DNA
ssRNA	single-stranded RNA
sRNA	short RNA
siRNA	small interfering RNA
SR Protein	serine-rich protein
TSS	transcription start site
TSSa-RNA	transcription start-site-associated RNAs

2.1 Introduction

The development of sequencing technologies has resulted in dramatic reductions in sequencing costs over the last decade [1]. There are already a broad range of high-throughput sequencing technologies [2], with others, such as nanopore technology [3], expected to arrive in the very near future. Our increasing ability to sequence nucleic acids quickly and cheaply will transform many biological areas of research [4]. This includes medicine, and the sequencing, and resequencing, of individuals is already helping to illuminate the genetic changes responsible for cancer progression [5]. The new sequencing technologies are also being used increasingly in fields previously dominated by microarrays. Deep sequencing of RNA and recording its abundance in the sample, referred to as RNA-Seq [6–8], has generated much excitement and it has been claimed that it represents a revolutionary tool for transcriptomics [9]. We are still in the early days of the revolution and many of the RNA-Seq studies to date have been of a descriptive nature with basic data analysis [10]. However, there is a rapid growth in techniques and software to analyse next-generation RNA-Seq datasets [11] and increasingly sophisticated analyses are likely to become the norm. Even in the absence of sophisticated analysis techniques, there have been some fascinating results; for example, such experiments suggest that, for humans, approximately 75 % of the total mRNAs within a cell are common to all tissues, with about 8,000 protein-coding genes ubiquitously expressed [12].

However, transcriptome complexity is observed to vary between tissues, with areas such as the brain, kidney and testis expressing a greater diversity of mRNA than tissues such as the muscle and liver. Other techniques whether sequencing is being utilised include the measurement of protein–DNA interactions via ChIP-Seq [13].

Nonetheless, current next-generation sequencing presents challenges in assembly and sequence accuracy due to short read lengths and method-specific sequencing errors [14]. Understanding the physical causes impacting upon the fidelity of sequencing is important in establishing the error composition of any sequence. For example, a limitation of the 454 technology relates to sequences containing consecutive instances of the same base, such as AAA or GGG [2]. With this technology, the length of homopolymers is inferred from the signal intensity because there is no terminal molecule preventing multiple contiguous additions at a particular cycle. This results in a greater error rate than results from discriminating between incorporation and nonincorporation. The major error type for the 454 platform is insertion-deletion rather than substitution, whereas the dominant error for Illumina/Solexa is substitution, rather than insertions or deletions [2]. There are also other biases in RNA-Seq data which may limit its adoption for large-scale systems biology experiments [15]. For some applications, microarrays are more sensitive than the current sequencing technologies. This is leading to many groups using hybrids of sequencing and microarrays together, utilising the advantage of both approaches whilst minimising the disadvantages of each technology's limitations [14, 16].

The meta-analysis of large datasets of gene expression is now helping to underpin systems biology models, increasingly pointing to how the interactions between groups of closely coupled proteins underpin gene expression in humans and other higher eukaryotes [17, 18]. The implicit assumption of many current models in systems biology is that regulation is for the most part mediated by transcriptional regulatory networks [19]. However, this view has faced significant difficulties and blind studies to perform the high-throughput identification of transcription factor targets have provided very poor results [20]. Part of the problem is that the regulation of gene expression in eukaryotes is very complex and strongly modulated by a number of mechanisms beyond simple transcription factor complex formation. Our understanding of the components involved in gene regulation, their complexity as well as the interplay between different layers of regulation utilised within cells has expanded rapidly in parallel with our ability to utilise high-throughput sequencing.

Systems analysis of gene expression is identifying coordination and coupling in transcription, coordination among transcription factors, coupling among transcription factors and chromatin remodelers, a nuclear organisation coupled to transcription, interwoven layers of mRNA processing involving the coupling of transcription and splicing, coupling of transcription and export with quality control processes, dynamic messenger ribonucleoprotein complexes, regulation of cytoplasmic events from within the nucleus, coupling between transcription and ribosomal synthesis and links between protein synthesis and degradation [21]. Many of the molecular interactions responsible for coordination are being mapped out biochemically [22], detailing the lines of feedforward and feedback between chromatin, RNA, multifunctional proteins and ribonucleoprotein (RNP) complexes.

These complexities are leading to designs of next-generation sequencing experiments increasingly requiring integrative approaches to bring together knowledge of the multiple layers of regulation of gene expression. There are many threads linking these layers and in this review we give a broad overview about the rapid progress in understanding the regulation of genes via several mechanisms. We will begin with transcriptional and post-transcriptional events. We will then discuss how the structure of chromatin radically affects transcription. Following this, the major role that non-coding RNA plays in regulation will be discussed. We will also highlight a number of common themes that are emerging across these different layers of regulation. Finally, we discuss how next-generation sequencing is poised to play a significant role in the systems biology of the future, the huge data management problem we face and how it will likely transform how we work together to better understand gene regulation.

2.2 Transcription and Beyond

The transcription cycle begins with preinitiation complex formation, RNA polymerase II (RNAPII) recruitment, a transition to an initiating and then an elongating RNAPII, and progressing to termination [23]. RNAPII will do work as it progresses through transcription and the amount of energy required to break and make bonds depends upon tertiary interactions between RNAPII, chromatin, nascent RNA and ribonucleoproteins (RNPs).

2.2.1 The Dynamic Nature of RNAPII

It is increasingly clear that subtle changes in the structure of RNAPII occur as it progresses through the transcription cycle. In particular, a relatively unstructured protein domain lies below the RNA exit channel [22], the carboxy-terminal domain (CTD) of RNAPII, and this serves as a binding pad for many nuclear factors, playing a key role during transcriptional and co-transcriptional processing, including terminating transcription.

The CTD has a simple heptad repeat structure, Tyr-Ser-Pro-Thr-Ser-Pro-Ser ($Y_1S_2P_3T_4S_5P_6S_7$), with 52 repeats in mammals [24]. The last repeat of the CTD in vertebrates is followed by a conserved ten amino acid extension. Thirty-one of the fifty-two repeats in the human CTD differ from the consensus heptad in at least one position, with most of the nonconsensus repeats towards the carboxy terminal of the CTD [24]. The presence of these divergent repeats enables additional functionality. As shown in Fig. 2.1, dynamic and reversible modifications to CTD

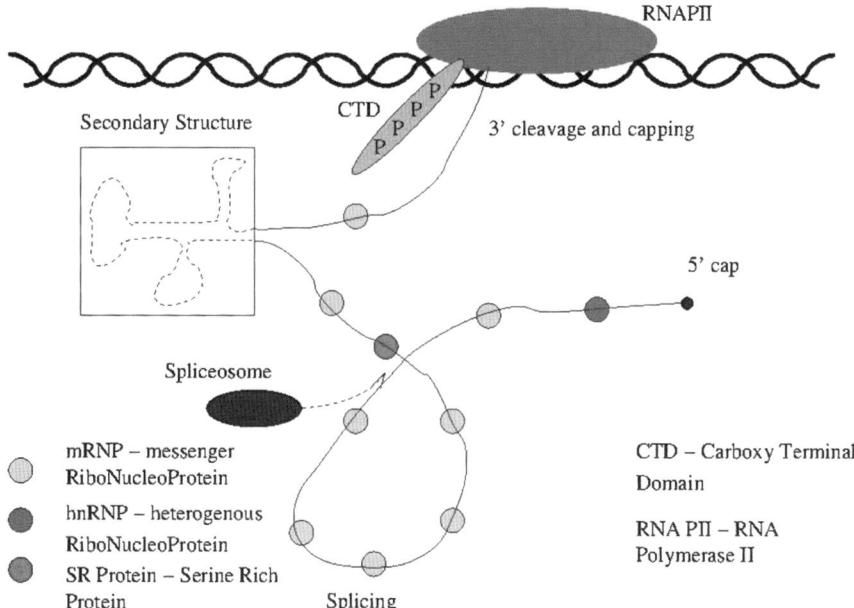

Fig. 2.1 A schematic diagram of the processes and interactions that occur in pre-mRNA and how splicing is implemented (described in Sects. 2.2.1, 2.2.2, and 2.2.3). The formation of bonds between the pre-mRNA and ssDNA is carried out by the formation of RNA secondary structure or binding with mRNPs. Splicing is enhanced by SR proteins or inhibited by hnRNPs. Initiation and termination of the transcript is aided by complex formation triggered by PTMs in the CTD region of RNA PII

occur during the transcription cycle, including phosphorylation, glycosylation as well as changes to the isomeric state of prolines. The appropriate recruitment of factors at different stages of the cycle is closely related to these modifications and a CTD code describing these coordinated changes is being actively sought [24].

All three serines of the CTD consensus repeat can undergo phosphorylation [24]. Ser2 and Ser5 are dynamically phosphorylated and dephosphorylated during the transcription cycle. Phosphorylation of Ser2 residues plays a major role in enabling RNAPII to progress into an elongating form, as well as being involved in splicing and polyadenylation events. Phosphorylation of Ser5 residues is greatest near the 5' end of genes, with Ser5 phosphorylation helping in the addition of a methylguanosine cap to the 5' end of the newly synthesised RNA. The CTD loses most of its Ser5 phosphorylation before RNAPII reaches the polyadenylation signals at the 3' ends of protein-coding genes. The dephosphorylation of Ser2 and Ser5 during the transcription cycle is required for recycling RNAPII. Dynamic phosphorylation of Ser7 has a role in some protein-coding genes at their 3' termini, involved in either terminating transcription or 3' processing. Tyrosine and threonine can also undergo phosphorylation, but it is presently unclear what functions these play. Experiments to unravel the role of threonine are complicated by it being found in 15 positions in the nonconsensus repeats, as well as in its canonical position 4 in the consensus repeats. Serines and threonines can also be glycosylated, with phosphorylation and glycosylation appearing to be mutually exclusive [24]. Isomerisation of the two peptide-prolyl bonds, at positions 3 and 6, also occurs, resulting in four possible configurations in each repeat.

2.2.2 The Folding and Binding of Proteins to Nascent Pre-mRNA

Alterations in RNA structures represent a regulatory mechanism for many cellular processes [25]. There is an intimate relationship between the binding of messenger ribonucleoproteins (mRNPs) and RNA secondary structure, with some proteins binding to single-stranded RNA (ssRNA) sequences [26] and others to double-stranded RNA (dsRNA) sequences [27]. Heterogeneous RNPs (hnRNPs) are very abundant in the cell and RNA–protein interactions act to modify the form of RNA secondary structures and may act to inhibit the existence of structures in some cases [28].

Pre-mRNA is free to fold only within a limited period after transcription, with an upper limit of $\sim$100 nucleotides [29]. It is likely that co-transcriptional wrapping up of RNA by folding, or through binding by mRNPs, occurs rapidly in order to minimise the possibility of genomic mutations induced by the formation of R-loops during transcription [30]. As shown in Fig. 2.1, an R-loop is a structure in which an RNA molecule is partially or completely hybridised with one strand of a double-stranded DNA, leaving the other strand unpaired [31]. Transcriptional R-loop formation in higher eukaryotes is highly correlated with chromosome instability. Little is known about the molecular mechanisms responsible for R-loops influencing genome stability, but single-stranded DNA is more vulnerable to mutations than double-stranded DNA [32]. Thus, extensive R-loop formation will result in these transcribed regions being more susceptible to DNA-damaging agents by increasing the frequency of single-stranded regions. R-loops will also act to slow down elongation of RNAPII [22].

2.2.3 Post-transcriptional Splicing

RNA-Seq results suggest that almost all human protein-coding genes undergo alternative splicing [33]. Furthermore, over 80 % of genes produce a minor isoform with a relative abundance of 15 % or more of the major isoform. It has been recently proposed that most alternative splicing is a consequence of noise in the splicing machinery [34]. However, alternative splicing and polyadenylation are observed to vary significantly between tissues, with coordinated changes in alternative splicing and polyadenylation between many genes being observed, suggesting that alternative splicing provides a central contribution to the evolution of phenotypic complexity in mammals [33].

Pre-mRNA splicing occurs co-transcriptionally in all eukaryotes [22]. However, there is little overlap between groups of genes that are differentially spliced and those that are differentially expressed [35]. As shown in Fig. 2.1, a small number of RNA-binding proteins, usually members of the serine-rich protein (SR protein) and hnRNP families, are involved in splicing regulation and the interplay of these positive and negative factors acts to modulate the inclusion, or otherwise, of exons [36]. SR proteins help to activate splicing by binding to exons and recruiting the spliceosome. Most members of the SR protein family have their binding to RNA affected by the conformation of the target RNA [37]. SR proteins exert some of their stimulatory effect through stabilising RNA–RNA interactions during spliceosome assembly and splicing catalysis [38]. HnRNP proteins, in contrast, usually repress splicing by interfering with the spliceosome's interactions with splice sites. In particular hnRNP proteins may disrupt RNA–RNA interactions through sequestering sequences [38].

The binding of these positive [39] and negative [40] regulators of splicing has been shown to depend on RNA secondary structures. There seem to be two mechanisms involved in how RNA secondary structure affects the choice of 5′ and 3′ splice site and branch point elements. The most common process results from the presence of structural elements which may hinder the accessibility of selected sequences by splicing factors [37] – depending on the system analysed, this inhibition has been observed to target only the acceptor site, the donor site or both. The second mechanism occurs when RNA secondary structures that do not involve the conserved splicing sequences can vary the relative distance between these elements – these changes then result in considerable variation in splice site usage or efficiency [37]. Structural constraints also affect less-defined cis-acting sequences such as exonic/intronic splicing enhancers or silencer elements [41]. Furthermore, RNA secondary structure has been proposed to influence splicing. For example, secondary structural elements involving both exonic and intronic sequences have been found in the *dystrophin* gene [42].

The advent of high-throughput sequencing experiments, in conjunction with exon arrays, enables observations of co-regulated splicing events in groups of genes, as well as the determination of sequence motifs associated with these events [35, 43]. Some of the sequence motifs now being associated with tissue-specific alternative splicing are consistent with the binding patterns previously identified for known splicing regulators, such as NOVA and FOX [35]. This suggests that as catalogues of isoform expression profiles increase, they will provide sufficient sensitivity to enable the discovery of weaker motifs indicative of novel splicing regulators. There is a need for such analysis as there are more than 300 RNA-binding proteins in mammalian genomes that may act as splicing regulators, yet little is presently known about their binding specificity or their involvement in particular splicing events [35]. Mining of the existing datasets is already highlighting positional dependencies in the

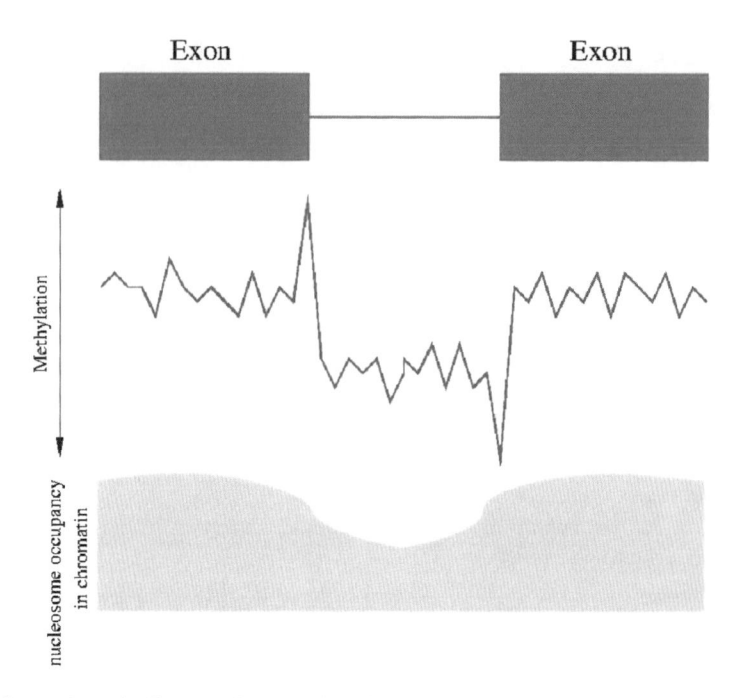

Fig. 2.2 A schematic diagram of the relationship between exon–intron boundaries, methylation and nucleosome occupancy as described in Sect. 2.2.3. As noted in Sect. 2.3.6, there is a noticeable peak in methylation (specifically CpG) at exon–intron boundaries and a trough at intron–exon boundaries [48]

binding of regulators, with both NOVA and FOX binding as enhancers when they are downstream of an alternative splicing exon, whereas they act as repressors when they bind on the upstream side [35]. Combinations of hundreds of RNA features are being assembled as part of large data-mining efforts to identify the principal components of the splicing code [44].

Unravelling evidence of co-regulated splicing events in several genes is nontrivial as it is very likely that splicing regulation can occur at every possible step of the spliceosome assembly and catalysis pathway. Furthermore, there are large numbers of factors involved in the splicing of each transcript and stochastic events may be important during splicing because simple binding kinetics determines the assembly pathways for a given pre-mRNA substrate [38]. Spliceosome assembly is also modulated in response to transcriptional events and chromatin structure [38]. The rate of elongation affects splice site selection and exon skipping and, thereby, the nature of the information expressed from a gene [45, 46]. Post-transcriptional processing also involves a close relationship with how DNA is modified in its accessibility during transcription [36, 47]. As shown in Fig. 2.2, chromatin organisation marks exon–intron structure and chromatin structure, via histone modifications, modulates exon selection [48, 49]. Transitions in DNA methylation across junctions of exons and introns may also be involved in splicing [50]. The differences in transcription rates that result from these chromatin modifications, as well changes in nucleosome density [47], may be the principal cause for a large proportion of tissue-specific, or development-specific, alternative splicing events [36]. It is now possible to use high-throughput sequencing technologies to map histone methylation states across the human genome [51]. We are therefore likely to see high-throughput sequencing used in further integrative studies of the dynamic interplay between proteins modifying chromatin, interacting with RNA, and their resulting impact on alternative splicing. Efforts to crack the splicing code, e.g. [44], are likely to be enhanced by knowledge about the tissue-specific modifications that chromatin undergoes within particular genes of interest.

2.2.4 RNA Editing

RNA editing can provide a source of sequence variation between transcripts from the same gene. The most common form of editing in eukaryotes is A-to-I, in which adenosine is converted into inosine within double-stranded RNA and the inosine is subsequently treated as guanosine by the spliceosome and ribosome [52]. Such editing is apparent because of differences in the RNA sequence and the DNA sequence. A-to-I editing is essential for the maintenance of normal life in mammals [53]. Editing can undergo spatiotemporal regulation [52]. Furthermore, RNA editing and alternative splicing are coupled, as modifying the RNA sequence can result in novel splice sites [54]. Moreover, as shown in Fig. 2.3, multiple editing sites within the same transcript are weakly correlated and so results in the production of diverse transcriptomes, eclipsing the variety resulting from alternative splicing but with less impact on the protein composition within cells [53]. The diversity resulting from RNA editing may be a principal contributor to the adaptive evolution of phenotypic complexity in mammals and be a dominant source of transcript diversity in the brain [55]. However, editing has also been associated with a number of human pathologies [56]. In particular, alterations in RNA editing impact upon a number of psychiatric disorders [57], in particular upon an individual's responsiveness to serotonergic drugs. Polymorphisms in editing genes have also been recently associated with extreme old age in humans [58]. High-throughput sequencing has already been used to identify RNA-editing sites [59], and we are likely to see many further studies in this area.

Meta-analysis of sequence differences in the small RNA component of rice and Arabidopsis [60] indicates that sequences of many transcripts are likely modified in vivo. These include N1-methyl modified purine nucleotides in tRNA, potential deamination or base substitutions in microRNAs, $3'$ microRNA uridine extensions and $5'$ microRNA deletions. However, the impact of editing, and other post-transcriptional modifications, can mimic RNA-sequencing errors and a number of sequence variations previously classed as sequencing errors may in fact result from editing and other modification [60].

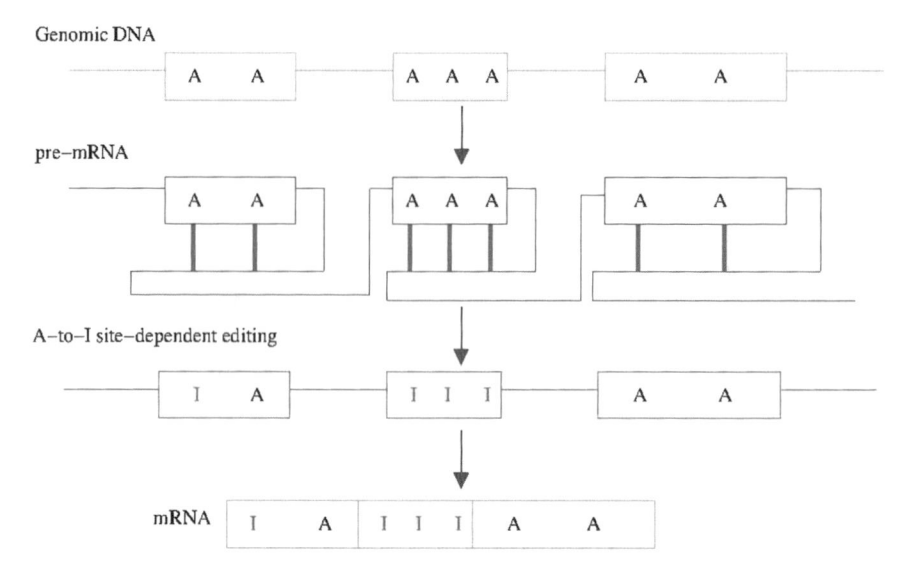

Fig. 2.3 A summary of RNA editing as described in Sect. 2.2.4. The actual editing occurs in double-stranded pre-mRNA which can then be edited by ADAR. We note that A-to-I editing is site dependent, i.e. not every A is edited to an I and the editing depends on the site and condition

2.2.5 The Processing of the 3′ Ends of Transcripts

There are a number of molecular mechanisms involved in processing the 3′ ends of pre-mRNAs in metazoans [61]. Transcripts are cleaved before acquiring a polyadenylation (poly(A)) tail and the efficiency and specificity of this 3′ processing is regulated by large protein complexes, involving many factors. Transcription factors and activators affect 3′ processing and there is also crosstalk between factors involved in transcription, splicing and this processing machinery. Furthermore, the CTD of RNAPII helps to couple this regulatory network through acting as a site for gathering and delivering polyadenylation factors [61].

The molecular machinery involved in 3′ processing has a complex architecture, containing over 80 proteins. As outlined in Fig. 2.4, there are several sub-complexes, including cleavage and polyadenylation specificity factor (CPSF), cleavage stimulation factor (CstF) and poly(A) polymerase (PAP) [61]. The poly(A) signal consists of two sequence elements: an AAUAAA hexamer, or a variant such as AUUAAA, is found 10–30 nucleotides upstream of the cleavage site that binds CPSF; a U/GU-rich region is located approximately 30nt downstream of the cleavage site and associates with CstF. The majority of transcriptional units contain more than one poly(A) signal and the alternative choices act to change the coding sequence or the sequences of the 3′ untranslated region. This results in alternative protein isoforms or transcripts that differ in their stability, localisation, transport and translation properties [61]. Tissue-specific regulation of alternative polyadenylation has a higher frequency than other types of alternative splicing [33].

2 An Overview of Gene Regulation

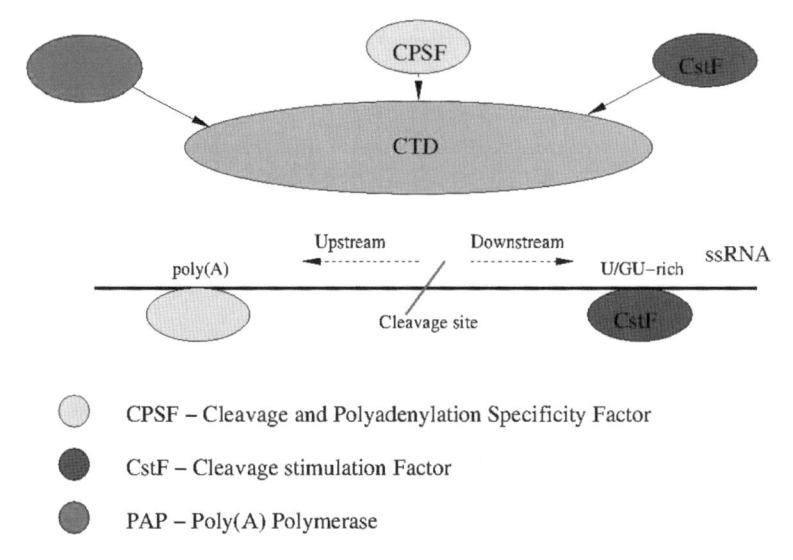

Fig. 2.4 A summary of the processing that occurs during cleavage of the 3' end of a transcript as described in Sect. 2.2.5. The sub-complexes PAP, CPSF and CstF are not an exhaustive list of the sub-complexes required for 3' processing. The CTD of RNA PII gathers and delivers polyadenylation factors. The cleavage site lies between an upstream poly(A) region (10–30 nucleotides of the cleavage site), which CPSF binds to, and a downstream, U/GU-rich region ($\sim$30 nucleotides from the cleavage site) that CstF associates to

The interplay between several mechanisms involved in regulating 3' end processing determines which of the transcriptional unit's sites are chosen to be polyadenylated [61]. Regulatory factors can compete with CPSF and CstF binding to their sequence elements. There can also be cooperative interactions, resulting from proteins bound to the transcript increasing the rate at which CPSF and CstF are able to bind their respective elements. Factors bound to the pre-mRNA can inactivate PAP. The rate of transcriptional elongation can shift the kinetic competition between processes, resulting in not enough time for upstream sites to be chosen and therefore the subsequent polyadenylation of downstream sites. Differential expression of individual proteins which make up part of the large 3' processing complexes will act to preferentially select suboptimal cis-elements. Factors involved in polyadenylation can also be sequestered to the cytoplasm. The factors can also become bound into other complexes in the nucleus, which can result in different choices of site. The factors can also be posttranslationally modified, again altering which of the several sites are chosen to be polyadenylated.

Chromatin structure also impacts upon the regulation of alternative polyadenylation [62]. The canonical polyadenylation signal 6-mer, AATAAA, is a poly(dA:dT) tract, and such tracts act to stiffen DNA and deplete nucleosomes. Indeed, [62] find that human polyadenylation sites (PAS) have strong nucleosome depletion in conjunction with downstream nucleosome enrichment. Moreover, the downstream nucleosome affinity is associated with increased usage of the PAS when there are multiple sites available.

2.2.6 Post-transcriptional Modifications and Folds Used in Quality Control and Regulation

A number of post-transcriptional modifications are used by the cell to check the fidelity of transcripts as they are produced. The addition of a cap to the 5' of the nascent transcript is likely a switch that enables RNAPII to move from an abortive state into a fully elongating state [22]. The poly(A) tail acts to enable transport of mRNAs from the nucleus to the cytoplasm and affects both their stability and the rate at which they are translated [61].

A key quality control process is the nonsense-mediated mRNA decay (NMD) pathway [22]. This involves the exon junction complex (EJC), a group of proteins which are deposited on spliced transcripts about 20 nucleotides upstream of exon–exon junctions [22]. During the first round of translation for a newly synthesised transcript, the presence of at least one EJC which is 50 or more nucleotides downstream of a stop codon results in the transcript and recently translated peptide being rapidly degraded. This targets those transcripts in which the first in-frame stop codon is poorly placed for transcript termination, resulting in the constitutive stop codon being either in the last exon or within 50 nucleotides of the final exon–exon junction [22].

EJC deposition possibly evolved to enhance protein production and mRNA surveillance [22]. However, NMD is used to play several regulatory roles in the cell, other than just simply removing aberrant transcripts. For example, a number of splicing factors appear to alter the production of their own isoforms in order to target their transcripts to the NMD pathway whenever their intracellular concentrations become too high. Moreover, splicing makes for better translation resulting from the interactions between EJCs and complexes associated with ribosomes [22]. Furthermore, the EJC also interacts with proteins involved in directing mRNA localisation.

It is not just the EJC that acts to modulate the efficiency of an mRNA's localisation and translation efficiency. A number of features of the untranslated regions of mRNAs control their metabolism [63], the regulation of which is likely to depend on the tertiary structure of RNA as well as trans-acting factors. For example, *cis*-acting elements in the mRNA, usually in the 3' untranslated regions (UTR), mediate the subcellular region to which the transcript is localised. Whereas, other elements in the 3' UTR, such as AU-rich elements, regulate mRNA decay. Translation efficiency also depends on structures in the 5' UTRs as well as the length of the 5' UTR.

2.2.7 Sequence Variations

Genome resequencing of individuals will identify the differences with other genome sequences, and identify single-nucleotide variations, and whether the individual is homozygous or heterozygous for such variations. Individual alleles may contain distinctive sequences and heterozygous individuals may produce expression of different RNA sequences. RNA-Seq has now been used to detect single-nucleotide variations in expressed exons of the human genome [64].

2.3 The Structure of Chromatin Impacts upon Gene Regulation

Transcription and post-transcriptional processing occurs whilst RNAPII is progressing through chromatin. Rather than just being a naked strand of DNA, instead chromatin is a complex mixture of nucleic acid, proteins and covalently bound modifications.

2.3.1 Nucleosomes

Constraints on DNA arise from its interactions with group of eight basic histone proteins, collectively known as nucleosomes [65]. DNA and nucleosomes are arranged as beads on a string, with a linker of naked DNA sequence bridging two neighbouring DNA-wrapped nucleosomes. The nucleosomes act to neutralise the self-repulsion of DNA resulting from the negatively charged phosphates in its backbone, enabling DNA to be packaged efficiently and fit into the confined space of the nucleus. As shown in Fig. 2.5, the histone core is composed of two copies of four histone proteins (H2A, H2B, H3 and H4). Each octamer consists of two H3–H4 histone dimers bridged together as a stable tetramer that is flanked by two separate H2A–H2B dimers [66]. DNA coils through a left-hand toroid around the histone core, with approximately 147 bases looping 1.65 times around each nucleosome, with each histone core anchoring 34–36 DNA base pairs through electrostatic, hydrogen and nonpolar interactions [66]. A further linker histone, H1, protects internucleosomal linker DNA near the nucleosome entry-exit point [66]. DNA and nucleosomes may undergo further compacting into transcriptionally inactive 30 nm fibres [65], as well as other high-order compactions.

A short basic stretch flanking lysine around position 16 of the histone H4 N-terminal domain directs internucleosomal contacts, which modifies high-order chromatin structures [66]. The interaction between residues 16 and 20 of histone H4 and two acidic patches on the C-terminal α-helices of histone H2A present on an adjacent nucleosome mediates in salt-dependent folding of chromatin. Acetylation of lysine residues relieves positive charges, perturbing histone-DNA contacts and affecting nucleosome stability [66]. Indeed, acetylation of lysine 16 on H4 (H4K16ac) prevents the compaction of nucleosome arrays in vitro, likely via electrostatic repulsion and hindering H2A contacts [66]. The acetylation of H4K16 also repels ATP-dependent chromatin-remodelling complexes, such as ACF, which will only interact with histones in the absence of H4K16ac [66].

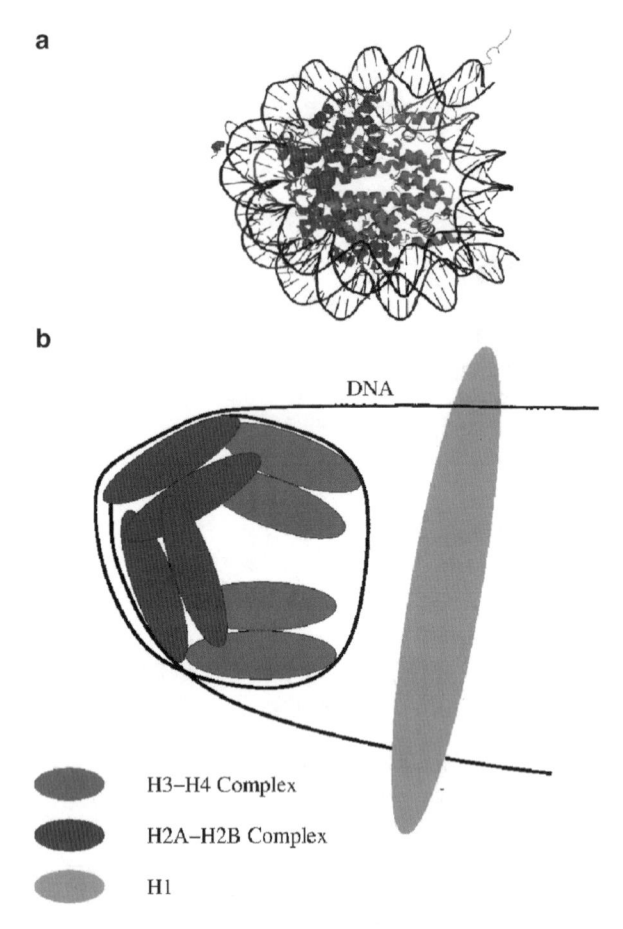

Fig. 2.5 The nucleosome as described in Sect. 2.3.1. In (**a**) a cartoon representation of a nucleosome structure determined from x-ray crystallography is shown (PDB code 1aoi) [223]. The histone structure H1 was not determined in the structure. In (**b**) a schematic diagram to represent the entire nucleosome, including the histone H1 structure, is shown. The core nucleosome structure is composed of eight domains which are composed of four dimers of H2, H3 and H4 histones. The DNA loops around this structure following this order of dimers: H2A–H2B, H3–H4, H3–H4 and H2A–H2B. The H1 histone binds to the entry and exit DNA giving the structure stability. The DNA turns 1.65 times and is comprised of 147 bases

2.3.2 Nucleosome Variants

Variants of histone combinations contribute to the properties of the nucleosomal core particle and its role in building specialised structures as well as altering transcriptional activity [66]. Histones H4 and H2B are largely invariant, whereas there is more variety with H3 and H2A.

The non-canonical H2A.Z is conserved from lower to higher eukaryotes. Nucleosomes can only incorporate one type of H2A variant because of steric clashes between loops in H2A and H2A.Z. H2A.Z impacts upon nucleosome stability and chromatin folding, resulting from a small destabilisation within H2A.Z-H3 interactions and a longer H2A.Z acidic patch, relative to H2A, used in H4 NTD binding. Despite its conservation, there remains uncertainty about the function of H2A.Z resulting from the rapid turnover rates of H2A.Z-containing nucleosomes [66]. Another H2A variant unique to mammals is H2A.Bbd. H2A.Bbd–H2B dimers dock on the (H3–H4)$_2$ tetramer, producing nucleosome core particles that organise about 118 base pairs of DNA but which are considerably less stable than the canonical nucleosomes. The variant H2A.Bbd lacks the ubiquitinatable C-terminal domain as well as the acidic patch that contacts the H4 N-terminal domain, making nucleosomes containing H2A.Bbd resistant to salt-induced chromatin folding. H2A.Bbd may reside within active chromatin [66].

Mammals have evolved a replication-dependent H3 variant, H3.1, that only differs from the non-canonical variant, H3.2, by the substitution of a single amino acid [66]. H3.2-containing nucleosomes are probably associated with heterochromatin. Whereas, the H3.3 histone variant differs from H3.1 by five amino acids and is associated with euchromatin. H3.3-containing nucleosomes are unstable, with the H3.3 histone undergoing rapid turnover. The displacement of nucleosomes during transcription appears to be the primary role for H3.3 [66]. Cysteines that are found in H3 variants may act to stabilise H3–H4 tetramers through disulphide bridges, particularly under oxidative conditions [67]. A further cysteine in H3.1 variants may also result in stabilising disulphide bridges between neighbouring nucleosomes which both contain H3.1s, helping to compact higher-order structures of chromatin [67]. There is also an H3.CenH3 variant that is involved in chromatin structures associated with kinetochore assembly and function [66]. The different forms of chromatin resulting from H3 variants and posttranslational modifications may result in chromosomes having a "barcode structure" [67], influencing epigenetic states during cellular differentiation and development.

The linker histone H1 acts to seal the two turns of nucleosomal DNA and is required for changes in conformation between extended and compact chromatin [68]. H1 also plays a role in establishing the spacing between nucleosomes, maintaining the level of methylation in particular regions of the genome, regulating a subset of cellular genes and acting to control development [68]. There are 11 variants of the linker histone H1 which is more than twice greater than the variability of any core histone. H1 variants also show a greater degree of divergence from each other than do the variants for other histones [68]. The variants are distinctive about when they appear in the cell cycle, but there is presently considerable uncertainty in their functionality [68].

Fig. 2.6 A schematic diagram of the relationship between poly(dA:dT) density, substitution rates and its position between nucleosomes as described in Sect. 2.3.3

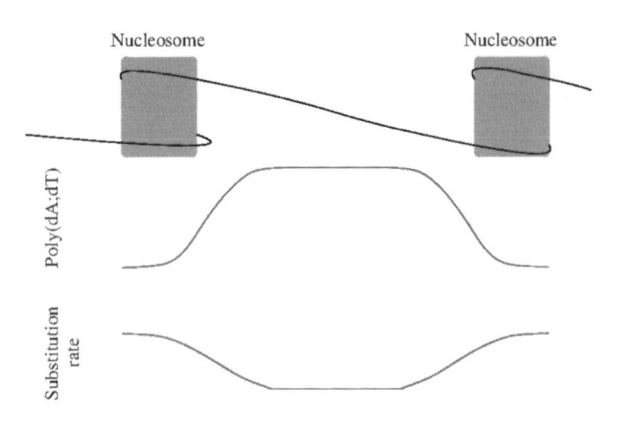

2.3.3 Nucleosome Positioning, Promoters and Gene Regulation

The code through which the genome sequence acts to position nucleosomes is increasingly understood [69–71] and our knowledge of the role nucleosomes play in gene regulation is rapidly evolving. DNA sequences have different abilities to bend and modify their helical twist and these differences are amplified when wrapping around the sharp bends of the nucleosome [70]. The bending around nucleosomes is facilitated through approximately 10 bp periodicity of specific dinucleotides. However, tracts of poly(dA:dT) are rigid and predicted to be unable to efficiently loop around histones. As shown in Fig. 2.6, such tracts are, as expected, observed to be free of nucleosomes [72] and play an important role in regulating nucleosome positioning within neighbouring genomic sequences. The nucleosome positioning code works in tandem with other regulatory codes in DNA [73], and amino acid content of proteins are likely to be modified as a function of nucleosome occupancy [74]. Moreover, [75] (see Fig. 2.6) have observed that substitution rates in linker regions are approximately 10–15 % lower than in nucleosomal DNA, which may be associated with higher DNA repair efficiencies in linker regions compared to nucleosomes. The roles that nucleosomes play in regulating transcriptional start sites, discussed below, in conjunction with differences in rates of insertions and deletions, and point mutations, between DNA wrapped around nucleosomes and that in linker regions, act to leave a nucleosome-associated periodic pattern in genome sequences, ultimately moulding the DNA sequence over evolutionary time scales [76].

The movement of nucleosomes by a few bases along DNA can dramatically alter the accessibility of the genomic sequence. Variations in genome sequences subsequently impact on nucleosome affinities and promoter structure, resulting in distinct modes of gene regulation [72, 77]. Functional promoters in eukaryotes must attract RNAPII and also evade the effects of nucleosomal repression. Cryptic transcription may occur if the suppression induced by nucleosomes does not function [78]. Typically, transcription start sites are found in nucleosome-free regions [47] as a major mechanism for suppressing transcription is to wrap potential transcription start sites around nucleosomes [79].

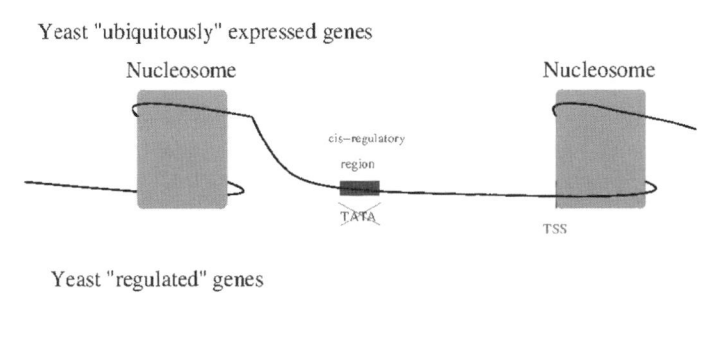

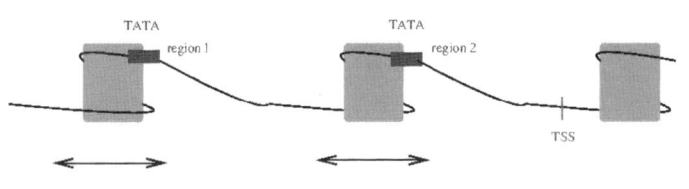

2 An Overview of Gene Regulation

Fig. 2.7 A schematic diagram relating the positioning of cis-regulatory regions, the transcription start site (TSS) and nucleosomes for ubiquitously expressed and regulated yeast genes as described in Sect. 2.3.3. In the ubiquitous case, cis-regulatory regions tend to lie in the linker regions with the TSS at the start of a nucleosome. In addition such cis-regulatory regions do not have a TATA box. Regulated genes, on the other hand, have their cis-regulatory regions lie in the exposed regions of the nucleosomes and can be exposed or hidden more as individual nucleosomes slightly shift their position. These regions tend to have a TATA box

The nucleosome positioning signals are used by eukaryotes to regulate gene expression with distinct noise and activation kinetics through altering the architecture of promoters [72]. As outlined in Fig. 2.7, "ubiquitously" expressed genes in yeast have open promoters [80], characterised by a poly(dA:dT) tract resulting in a large nucleosome-depleted region (NDR) close to the transcription start site, in conjunction with accurately positioned nucleosomes further upstream. Associated cis-regulatory sequences reside within the NDR and the lack of nucleosomes means that transcription factors can bind to the regulatory DNA without competition. TATA-binding boxes are typically not found within these promoters.

As outlined in Fig. 2.7, "regulated" genes in yeast have covered promoters [80], with a more evenly distributed nucleosome positioning, resulting in transcription factors and nucleosomes competing for access to DNA. Transcription factor binding sites tend to be exposed on the nucleosome surface, near the border with a linker [81]. The nucleosome positioning sequences for these promoters result in high nucleosome occupancy close to the transcription start site. The regulation of chromatin, via subtle changes in nucleosome positioning and accessibility of DNA to transcription factors, enables large dynamic changes in expression [72, 77, 82]. TATA elements are frequently associated with this group of promoters [82].

Mammalian genes that have broad CpG-enriched promoters tend to produce multiple transcription start sites and are typically ubiquitously expressed [83]. The regulation of which start site is chosen is associated with the methylation state of the promoters [84]. Whereas mammalian genes with promoters containing a TATA box tend to produce a sharp single transcriptional start site and typically produce tissue-specific expression [83, 85]. In mammals promoters containing a TATA box evolve slower than promoters containing CpG islands [86]. Furthermore, the sequence of DNA at human promoters, enhancers and transcription factor binding sites, in contrast to yeast, typically encodes high intrinsic nucleosome occupancy [87], with these regions depleted in nucleosome-excluding poly(dA:dT) tracts.

The structure of DNA wrapped around nucleosomes details the tertiary structure of a gene within a sequence and structural variations at the chromatin level are likely to play a role in the regulation of the co-transcriptional processing of RNA. Long poly(dA:dT) tracts, which exclude nucleosomes, are avoided in exonic sequences, enabling an increased density of nucleosomes in exons [73]. Furthermore, differences in linker lengths between nucleosomes in exons and introns may result in different chromatin-packing arrangements [73]. The positioning of nucleosomes is also involved in exon definition events during co-transcriptional processing [49, 88] and nucleosome depletion has been associated with the regulation of polyadenylation [62]. Furthermore, [89] have identified peaks in the density of H2A.Z-containing nucleosomes just downstream of start codons and just upstream of stop codons in human T-cells.

2.3.4 Dynamic Nucleosomes and Gene Regulation

The regulation of the dynamics through which DNA alters its binding around nucleosomes is intimately involved in controlling gene expression [90] and the different mechanisms are outlined in Fig. 2.8. Models of such regulation are founded on the idea that the regulation results from a competition between nucleosomes and other DNA-binding proteins [91]. The affinities that these molecules have for the sequence (binding affinity landscape) dictate their competitive and cooperative interactions [91]. High nucleosome occupancy tends to reinforce cooperative interactions between transcription factors in displacing nucleosomes [87].

DNA accessibility and nucleosome positioning are also regulated through the action of ATP-dependent chromatin-remodelling complexes. Chromatin-remodelling complexes are present at many promoters [92] and the dynamic repositioning of nucleosomes has been associated with selecting the transcriptional start site [65] as well as other aspects of the initial stages of transcription [93]. Also, histone-devoid transcriptional start sites, in conjunction with the active cycling of factors on and off a promoter, permit formation of preinitiation complexes that are poised for transcription to be initiated [66, 90], a different state from a gene that is fully repressed. The evolution of chromatin-remodelling complexes is likely associated with changes in chromatin regulation during the evolution of vertebrates

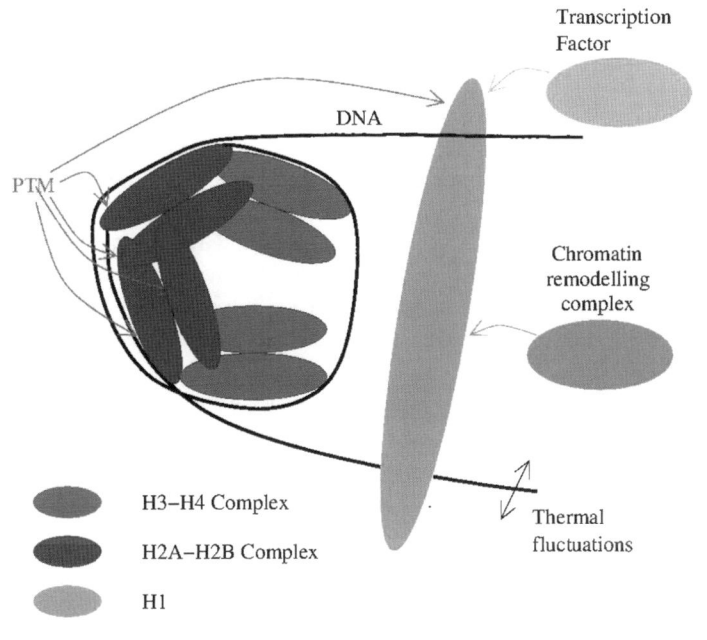

Fig. 2.8 A summary of the regulatory mechanisms that can be applied to the nucleosome as described in Sect. 2.3.4. PTMs can be applied to the H2 histones. The H1 histone can interact with transcription factors or chromatin-remodelling complexes. Furthermore, thermal fluctuations may result in the transient exposure of DNA regulatory sites to proteins

from unicellular eukaryotes [92]. Complexes, such as the ISWI (imitation switch; [94]) family, are involved in regulating higher-order chromatin structure [92], promoting regularly spaced nucleosomes and gene silencing [66]. Whereas the complex SWI/SNF (switching defective/sucrose non-fermentation; [94]) transiently exposes DNA regulatory sites through creating loops on a nucleosome's surface [65]. Some of these SWI/SNF complexes, such as BAF complexes in mammals, undergo progressive changes in subunit composition during the transition from a pluripotent stem cell to a multipotent progenitor cell [92]. At least four ATP-dependent remodelling complexes have nonredundant and specialised roles in maintaining pluripotent chromatin within stem cells. Tissue-specific complexes may enable matching between chromatin remodelling and transcription factors [92]. This can result in co-regulation of many genes or be restricted to the activation or repression of a single gene.

The disruption or displacement of nucleosomes will modify the rate at which polymerases pass over the DNA or the rate at which transcriptional factors will bind [65]. There are transcription-coupled changes in DNA topology or local chromatin structure, with histone and nucleosome removal during elongation of RNAPII [47]. The transit of RNAPII across the transcription unit is preceded by a leading wave of histone posttranslational modifications that open the chromatin and transiently displace nucleosomes [95]. There are at least two processes by which this happens. The first results from the nucleosome within transcriptionally active genes having two components, a fluid H2A–H2B dimer and a stable H3–H4 tetramer – H3–H4 tetramers are ~20 times more stable than H2A–H2B dimers [90]. The stability of the H2A–H2B dimer within the nucleosome will be further decreased by posttranslational modifications such as ubiquitylation, phosphorylation and acetylation. H2A–H2B dimers can also be exchanged through the actions of ATP-dependent chromatin-remodelling complexes. The movement of the H2A–H2B dimer could enable transcription factors and polymerases to access binding sites on DNA. The second process results from the linker histone H1 and its subtypes, associated with greater than 80 % of the nucleosomes in a mammalian nucleus, having residence times of a few minutes in interphase, consistent with dynamic interactions [90]. However, these residence times are variable and governed by the phase of the cell cycle, posttranslational modifications to H1, the subtype of H1 and competition for binding sites with other competing factors, such as transcription factors and chromatin-remodelling complexes, each of which themselves show dynamic interactions with chromatin [90]. Thus, alterations in residence times of H1 can result in changes to the residence time of a transcription factor, changing the balance between repression and activity. Thermal fluctuations of DNA wrapped around nucleosomes may also result in transient exposure of DNA regulatory sites to proteins. Such exposure is most energetically favourable towards the entrance and exit points of the DNA around the nucleosome and, indeed, transcription factor binding sites tend to be exposed near the border with a linker [81].

2.3.5 Histone Tails

Each histone has a tail which is targeted by a broad range of chemical moieties at multiple sites. Virtually all exposed polar residues (and some of the prolines) within the tails of histones are subject to covalent posttranslational modifications (PTMs). These include acetylation, methylation, phosphorylation, ubiquitylation, ADP-ribosylation, glycosylation [90] and SUMOylation [95], with lysine residues modified by up to three methyl groups. Acetylation and methylation results in only a small chemical group being added to the tail. However, ubiquitylation and SUMOylation are large appendages, almost the same size as the histone proteins, and their bulk could lead to more prominent changes in chromatin structure [95]. There is strong purifying selection among histone proteins and these targeted residues [66]. There has been considerable effort in establishing whether, and how, combinations of moieties on groups of histone tails act to produce a histone code that is used to regulate chromatin compaction and transcription [66].

A series of interlocking histone PTMs occurs during initiation, early elongation and mature elongation [95]. The transcriptional state of chromatin is correlated with several histone PTMs [66]. For example, hypoacetylation of H4K30me3 (trimethylated lysine residue at position 30 of the H4 tail) and H3K27me3 is associated with silenced chromatin, whereas hyperacetylation of H3K4me3 and H3K36me3 is associated with actively transcribed chromatin. Moreover, the distribution of these marks can be distinctive, with H3K4me3 present at the beginning of genes whereas H3K36me3 accumulating within the body and towards the downstream region of genes [90]. However, single histone marks do not fully prescribe chromatin structure and its impact upon transcription and different marks works in combination when interacting with histone-binding proteins [66, 90]. Furthermore, experiments on transcriptionally synchronised genes are beginning to unravel a transcriptional clock controlled by dynamic nucleosomes [90]. Changes in the methylation and acetylation status of the histone pass through cycles, with particular combinations of histone modifications never coexisting on the same nucleosome at the same time. However, the sequence of events at a nucleosome appears to depend on many factors and there have been no simple rules describing the order of events [90]. In particular, different causes for why a gene is induced produce distinctive histone modifications [90]. Moreover, different sets of histone modifications act to regulate gene expression in high-CpG-content promoters and low-CpG-content promoters [23].

2.3.6 DNA Methylation

Cytosines within chromatin can be covalently modified so that they carry a methyl group at position 5 within their pyrimidine rings [96]. 5-Methylation of cytosine does not affect its base pairing with guanine, and cytosine is still replicated as cytosine. A consensus view has been that DNA methylation always appears in a CpG context (C followed by a phosphate and then a G, i.e. CG is on the same strand). Methylation of CpG has high mutagenic potential [96], as 5-methylcytosine can be deaminated to thymine. Such transitions accumulate over the course of evolution resulting in CpG dinucleotides being markedly unrepresented in genomes of vertebrates given the fraction of cytosines and guanines in the genome. However, there are islands of CpGs which are found at the expected frequency, and these tend to overlap with gene promoters [96].

Once a methyl is added to cytosine, it can be copied to newly synthesised DNA, resulting in an epigenetic memory that can be conserved during cell division. The DNA methylation pattern is maintained through mammalian development by DNMT1, a methyltransferase that is associated with the replication complex [97]. During cell division and DNA replication, DNMT1 is involved in recognising methylated CpG residues on hemimethylated DNA and methylates the opposite strand. However, epigenomic profiles also undergo targeted methylation and demethylation alterations during development, and these differential changes in methylation play a crucial role in cell lineage commitment [98]. For example, targeted repression and de novo methylation of genes responsible for pluripotency occur at gastrulation, whilst the embryo is beginning to separate into germ layers [97]. The importance of epigenetic alterations that affect tissue-specific differentiation is such that their dysregulation could be the principal mechanism through which epigenetic changes cause cancer [84]. Different tissues show marked differences in DNA methylation [99], such that tissue-matched profiles from adult patients of different ages have more in common with each other than do disparate tissues from the same individual. Indeed, broad methylation patterns show tissue-specific conservation from humans to mouse [84], such that the methylation profiles of human and mouse brain cells, or human and mouse heart cells, have more in common than do the profile of a human brain and human heart cell.

Methylation acts to change the properties of chromatin. For example, methylation of DNA acts to modify nucleosome formation and positioning [96]. Biophysical and structural studies of DNA indicate that CpG methylation reduces backbone flexibility and dynamics, decreasing local DNA deformability. The position of 5-methyl group in the major groove increases steric hindrances on DNA wrapping around the nucleosome [96]. Only altering the conformation of a few nucleosomes through methylation may result in a significant impact upon the regular spacing arrays of nucleosomes expected to be involved in producing higher-order chromatin structure [96].

The methylation of cytosines affects how chromatin can subsequently bind to trans-acting factors and RNAPII. The binding of methylation to gene promoters can act to suppress transcription, and so any methylation associated with genes was believed to be indicative of transcriptional repression. However, this view is undergoing revision following the results from whole-genome epigenomic observations [99]. For example, a key function for differential methylation during differentiation is associated with changes in alternative transcription start sites [84]. Hypomethylation of promoters in conjunction with higher levels of gene-body methylation is positively correlated with transcription [50]. There is also recent evidence that DNA methylation acts to mark out aspects of gene structure within chromatin but shows cell-type-specific differences. DNA methylation peaks are found at the transcriptional start site in human T-cells [89]. Whereas DNA methylation shows a trough at the transcriptional start site in human embryonic stem cells and fibroblasts [50]. Both [50] and [89] find a drop in DNA methylation at the transcriptional termination site. Exons typically show higher CpG methylation fractions than do introns [50]. Interestingly, there is a sharp peak in CpG methylation at the exon–intron junction and a sharp dip in CpG methylation at the intron–exon junction [50], suggesting that transitions in DNA methylation are involved in splicing regulation. DNA methylation also peaks just downstream of the start codon and just upstream of the stop codon, suggesting that DNA methylation may be used as a signal for the addition, or removal, of co-transcriptional modifications that will only be utilised during translation at the ribosome [89]. Gene-body methylation may also inhibit incorrect choice of start sites for transcription [99].

The view that methylation is restricted to CpG sites is being questioned due to results from the first DNA methylomes that are now being sequenced at base resolution [100]. Almost 100 % of the methylcytosines in fully differentiated fibroblast cells are indeed in a CpG context, whereas pluripotent embryonic stem cells show almost 25 % of the methylcytosines in a non-CpG context (C followed by a C, A or T; [100]). Moreover, 99 % of the methylation of CpG sites occurs on both strands (the opposite strand is also CpG), whereas methylation on CHG (where H = A, C or T) is highly asymmetrical, with 98 % of the cases being found only on one of the strands rather than both [100]. Moreover, within embryonic stem cells, non-CpG methylation is enriched within gene bodies but is depleted in DNA-protein-binding sites and enhancers [100].

At this time it is unclear whether gene-body methylation, and its marking out of gene structure, is restricted to subsets of genes in particular cell types. The biological implications of such methylation as well as methylation's interplay with transcriptional elongation and splicing are still uncertain. Indeed, the initial findings from whole-genome methylation profiles, from a small number of cell types, indicate that we are still someway from understanding the biology of gene-body DNA methylation. But the rate of discovery suggests that the next few years will lead to a significant illumination of the role of DNA methylation in gene regulation across the genome.

2.3.7 DNA Methylation Interactions with Histone Tails

There are regulatory interactions between enzymes involved in processing DNA methylation and histone modifications [97] and these interactions play a crucial role in mammalian development [101]. For example, G9a contains an SET domain which acts as a histone methyltransferase, and G9a also contains an ankyrin domain which recruits the DNA methyltransferases DNMT3A and DNMT3B. DNA methylation patterns are erased in the early embryo, resulting from passive demethylation caused by DNA (cytosine-5)-methyltransferase 1 (DNMT1) being excluded from the nucleus [98]. Methylation profiles across the genome are then re-established in each cell at approximately the time of implantation, through a wave of de novo methylation whilst ensuring the CpG islands remain unmethylated [97]. As shown in Fig. 2.9, the de novo DNA methylation template is written through histone modifications, with patterns of methylation of H3K4 across the genome being formed in the embryo before de novo DNA methylation. CpG islands in the early embryo have RNAPII bound to them and this acts to recruit H3K4 methyltransferases. Whereas the rest of the genome contains nucleosomes with unmethylated H3K4. Subsequently, de novo methylation occurs through the action of DNMT3A and DNMT3L (DNMT3-like, a paralogue of DNMT3A) complexed with DNMT3B. This recruits methyltransferases to DNA by binding to histone H3, whereas any form of methylation of H3K4 acts to inhibit this methylation. This results in de novo DNA methylation taking place at CpG sites throughout the genome but being prevented at CpG islands because of the presence of H3K4me. This model explains the strong anti-correlation between DNA methylation and H3K4me in a number of cell types [97]. Moreover, a DNMT3A–DNMT3L tetramer may oligomerise on DNA-containing histones without H3K4me and lead to the nearly global methylation of the mammalian genome [101].

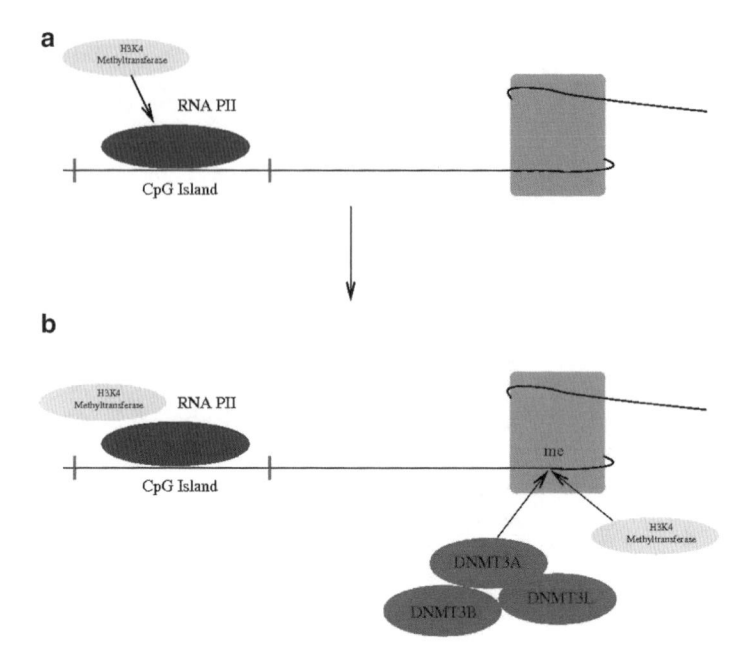

Fig. 2.9 A schematic diagram explaining the mechanism how non-CpG island sites are methylated in embryos as described in Sect. 2.3.7. In (**a**) CpG island sites are first bound by RNA PII sites which then recruit the H3K4 methyltransferase, sites that will be methylated are bound to nucleosomes. In (**b**) H3K4 methyltransferase in conjunction with DNMT3B, DNMT3A and DNMT3L to methylate the relevant CpG site. The CpG island sites are protected from methylation by the previously bound H3K4 methyltransferase

During post-implantation development, further epigenetic reprogramming occurs in primordial germ cells [101]. DNA methylation patterns are re-established by DNMT3A and the DNMT3B–DNMT3L complex at imprinted loci and transposable elements during gametogenesis. Targeting to transposable elements may involve Piwi-interacting RNAs, whilst targeting to imprinted genes involves the interactions of DNMT3L with unmethylated H3K4 tails [101].

DNA methylation also helps to maintain patterns of histone modifications through cell division [97]. During replication and cell division, regions that are methylated tend to be reassembled in a closed conformation, containing histones that are non-acetylated. Whereas unmethylated DNA gets repackaged in a conformation that is more open, containing nucleosomes whose histone tails are acetylated [97]. The mediation between DNA methylation and histone modifications likely results from methylcytosine-binding proteins such as MECP2 and MBD2, which are able to recruit histone deacetylases to methylated DNA. Enzymes such as G9a and DNMT1 may also interact with DNA methylation sites and direct H3K9 dimethylation [97].

2.4 The Spatial Organisation of the Genome in the Nucleus Acts to Regulate the Expression of Genes

2.4.1 Gene Expression Is Localised

It is increasingly clear that genomes and gene regulation are organised non-randomly in the nucleus [102]. Most nuclear events occur in spatially defined sites and in dedicated nuclear bodies, rather than occurring ubiquitously throughout the nucleus [103]. The formation of structures in the nucleus, such as Cajal bodies [104], results from stochastic assembly and self-organisation. Similarly, biological processes such as the formation of the DNA damage response may result from self-organising events [105]. Indeed, whole genomes may be considered as self-organising entities during mitosis, with networks of co-regulated gene expression and chromosomal association that are mutually related during differentiation resulting in self-organising lineage-specific chromosomal topologies [106]. The density of RNAPII may also act to regulate the colocalisation of gene expression [107].

Heterochromatin regions of the genome are usually found at the periphery of the nuclear membrane and are usually silent, whereas more open chromatin associated with active genes is typically found towards the centre. This is outlined in Fig. 2.10. Such a situation is consistent with biophysical models of the entropic organisation of self-avoiding polymers which suggest that long flexible polymers (associated with gene-rich chromosomes) will move to the centre of a confining sphere, whereas compact polymers (heterochromatin) will move to the periphery [108]. There is increasing evidence that gene activation or silencing is frequently associated with repositioning of the locus relative to nuclear compartments [109]. Active genes dynamically colocalise to shared sites of ongoing transcription [110] and genes such as Myc have been observed to preferentially relocate to regions in the nucleus at the same time as other genes with which they are co-regulated [111]. The movement of DNA into loops can result in proximal associations between co-regulated genes which are separated along the genome sequence [112]. The initiation step of transcription is required to tether genes to the same foci [113], but even in the absence of transcription, there are still localised concentrations of RNAPII [113]. A model consistent with much of the experimental data is that there are transcription zones within the nucleus in which RNAPII is concentrated locally through self-assembly processes [114]. These dense regions of nuclear RNAPII concentration have been termed factories, and it is possible that a transcriptional factory model may describe an aspect of the architecture of all genomes [115].

The synthesis of mRNA in mammalian cells is observed to be stochastic [116], with developmental genes exhibiting pulses of activity [117]. The stochastic nature of gene expression results from dynamic passage of genes through transcription factories [112]. Selection pressures will act upon groups of genes undergoing coordinated stochastic transcriptional regulation, and chromosome organisation shows the signature of selection for reduced gene expression noise [118]. Other

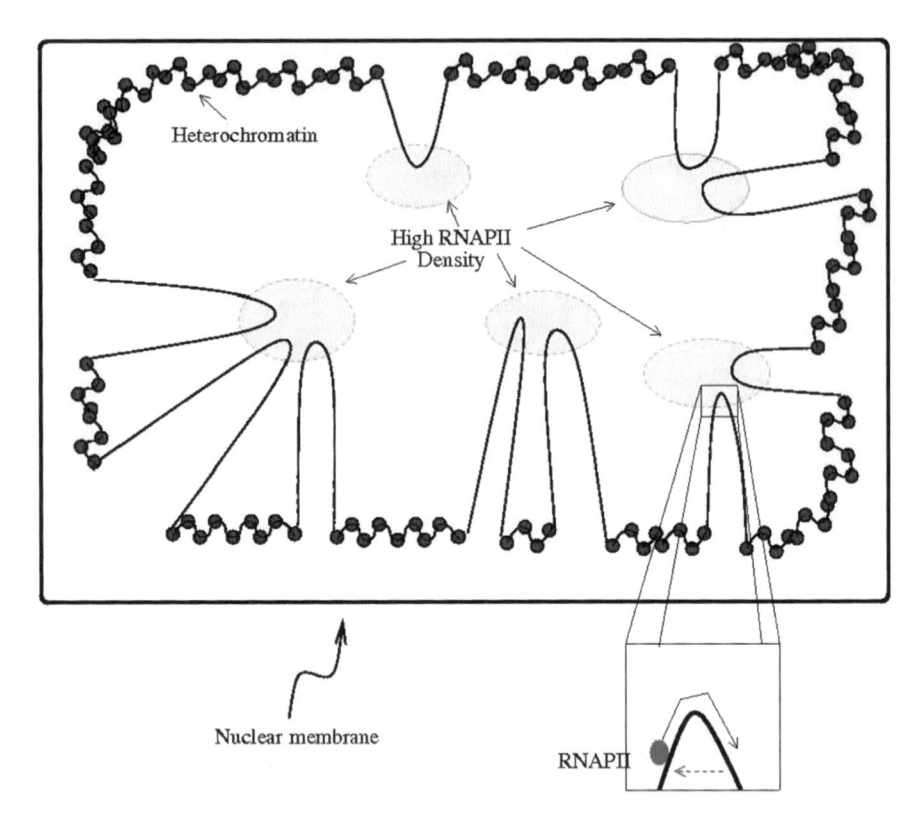

Fig. 2.10 A schematic diagram of chromatin organisation and transcription factories as outlined in Sects. 2.4.1 and 2.4.2. Heterochromatin (*blue* and *black circles* and *lines*) generally lies close to the nuclear membrane. Free DNA loops extend into the centre of the nucleus where it passes through regions of high RNA PII density referred to as transcription factories. Loops colocalise and hence are co-regulated exhibiting a similar noise structure in their expression. This is consistent with the stochastic nature of expression. The expanded region to the bottom of the diagram posits a hypothesis that as the start and end regions of the gene are physically close to each other, RNA PII can be re-recruited for transcription or be used for surveillance (Color figure online)

aspects of transcriptional regulation constrain the organisation of genes on eukaryotic chromosomes [119–121], with the 3D regulation of gene expression directly impacting upon genome evolution [122].

2.4.2 Loops and Networks of DNA Interactions Regulate Gene Expression

Genomes show tissue-specific spatial organisation [123] and cell nuclei frequently contain chromosome territories [124]. There is increasing evidence for three-dimensional networks of chromosomal interactions [125].

The topology of DNA around individual genes modifies gene regulation. As shown in the blown up region of Fig. 2.10, loop structures in which the promoter and terminator of a gene are in close proximity are associated with gene activity [22]. The role of the loop may be to increase the efficiency of recycling RNAP II back to a promoter after it has reached the end of a gene. The loop may also be involved in surveillance, with the results of an initial round of transcription being checked to ensure authentic signals are in place.

There are also cell-type-specific long-range looping interactions between enhancers and promoters which establish three-dimensional chromatin structures, such as for the CFTR gene [126, 127].

The conformational contacts between separate regions of chromatin change during cellular differentiation [128]. For example, extensive spatial chromatin remodelling accompanies gene repression during cellular differentiation [128], with repression of Hox A9, 10, 11 and 13 expression associated with the formation of distinct higher-order chromatin contacts between genes. Whereas, different chromatin conformations are associated with transcriptional activity. Major changes in higher-order structures of chromatin interactions are being associated with the regulation of transcriptional activity in increasing numbers of gene clusters, including the Bithorax complex in Drosophila [129] and the human apolipoprotein [130] and Hox A [128] gene clusters. The chromatin conformations may act as an epigenetic memory [129]. The conformation changes during differentiation may also be evolutionarily conserved [128].

DNA regulatory elements known as insulators mediate chromatin interactions, resulting in the formation of chromatin loops [131]. The name arises from the insulator's role in preventing inappropriate interactions between groups of enhancers and promoters. CCCTC-binding factor (CTCF) is one such insulator protein. CTCF contains three domains, one of which is a DNA-binding domain with 11 zinc fingers. It is evolutionarily conserved from insects to mammals, and over 80 % amino acid residues are identical between human, chicken and frog and up to 100 % conservation within the zinc finger-containing region [132]. CTCF binds in tens of thousands of places across the genome, with the binding sites grouping into different classes [133], each of which exhibits distinct evolutionary, genomic, epigenomic and transcriptomic features. The chromatin architecture and form at CTCF-binding sites can result in cell-type-specific changes [134]. Understanding the code by which CTCF acts to coordinate the three-dimensional position and regulation of genes within a cell's nucleus is being actively sought [135]. CTCF is believed to fit tightly into the linker region between nucleosomes [135], which results in positioning of a nucleosome over a site acting to occlude the binding of CTCF [136]. Furthermore, CTCF is sensitive to the presence of a 5-methyl group in the major groove of DNA [96] and CTCF can only bind to unmethylated DNA [135]. Moreover, the binding of CTCF, possibly through the action of chromatin remodelers, acts to accurately position 20 nucleosomes, enriched in the histone variant H2A.Z, both upstream and downstream of the site [137]. CTCF also acts to demarcate cell-type-specific chromatin domains associated with active (H2AK5ac) and repressive (H3K27me3) histone modifications [136]. CTCF has also been found

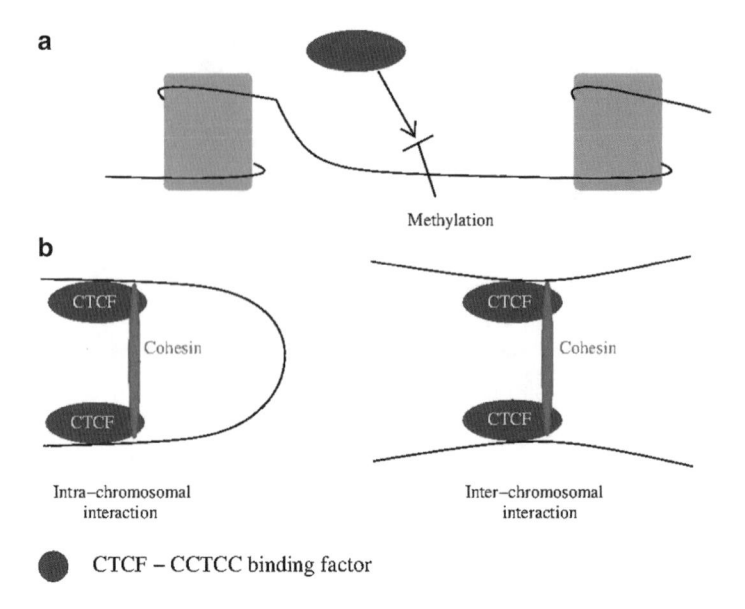

Fig. 2.11 A schematic diagram of how CTCF and cohesin can bind chromosomes as explained in Sect. 2.4.2. In (**a**) CTCF can bind between nucleosomes but cannot bind if its site is methylated. In (**b**) CTCF in conjunction with cohesin can bind between the same chromosome or between different chromosomes

to have a close relationship with the borders of lamina-associated domains [138], 0.1–10 megabase domains that are believed to anchor chromosomes to the nuclear envelope.

The important role that CTCF plays in establishing patterns of nuclear architecture and transcriptional control in vertebrates [139] is likely related to CTCF binding to cohesion [140–143], which creates intrachromosomal and interchromosomal links (shown in Fig. 2.11), resulting in a cell-type-specific network [134] that determines the three-dimensional structure of the genome [144]. Cohesin and CTCF are also involved in the maintenance of imprinting of loci such as the IGF2 (insulin-like growth factor 2)/H19 (H19 fetal liver mRNA) genes. A set of enhancers downstream of H19 play a role in regulating expression of both IFG2 and H19 – within developing embryos IGF2 is paternally expressed and H19 is maternally expressed [141]. On the maternal locus there are two unmethylated regions between IGF2 and H19 and where CTCF and a ring of cohesion can associate. The interaction between CTCF cohesion from this pair of regions results in a loop of chromatin-containing IGF2 which is then insulted from the action of the downstream enhancers, and only H19 is subsequently expressed. However, on the paternal locus, the region between IGF2 and H19 is methylated resulting in CTCF being unable to bind, leading to the H19 locus being bypassed and IGF2 being able to interact with the enhancers downstream of H19 [141].

Other elements may be involved in forming higher-order chromatin structures in the nucleus. For example, Polycomb response elements mediate the formation of chromosome higher-order structures in the Bithorax complex [129].

2.5 Regulation of Gene Expression by Non-coding RNA

2.5.1 Short Non-coding RNAs Associated with the Start, End and Enhancers of Genes

Short RNAs cluster at the 5' and 3' ends of genes [145]. A class of short transcripts close to transcription start sites of genes have been observed to be present at low abundance [79]. They have been named by several groups (promoter-associated sRNAs (PASRs, [145]), transcription start-site-associated RNAs (TSSa-RNAs, [146]) hereafter PASRs). PASRs are mostly derived from nucleosome-free DNA [79]. As shown in Fig. 2.12, they flank active promoters, with a peak in the abundance of short RNA antisense transcripts found ~250 nucleotides upstream of a gene's transcription start site [146, 147] and a peak in the abundance of short sense transcripts found between approximately 50 nucleotides [146] and 2.5 kilobases [148] downstream of the transcription start site [147]. Such divergent transcription appears common for active promoters as most of them have engaged polymerases upstream, in an orientation opposite to the proximal gene [147]. There is a correlation in expression between PASRs and their proximal gene, suggesting they are both responding to a common inducement of expression, even though the transcripts are in opposite directions. The density of antisense termini-associated sRNAs (TASRs), found towards the 3' ends of genes, is similarly correlated with the expression of the proximal gene [145].

A further source of gene-associated short RNAs is enhancers [149, 150]. Enhancer RNAs (eRNAs) have already been found in macrophages [149] and neurons [150] and it is likely that they will be identified in many, if not all, mammalian cell types. Enhancers overlap a sizeable fraction of extragenic transcription sites in higher eukaryotes [149]. In the studies of [149, 150], only a fraction of all enhancers were found to be associated with RNAPII and eRNA synthesis, suggesting that

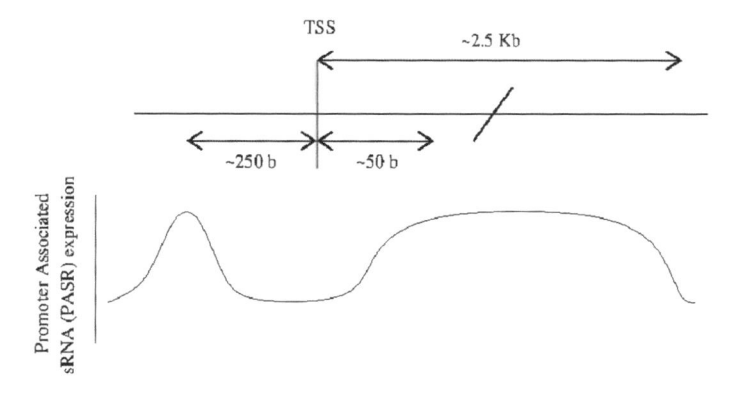

Fig. 2.12 A schematic diagram of the relationship between PASR expression levels and a TSS as explained in Sect. 2.5.1

there are a number of regulatory components involved with each enhancer. Changes in eRNA synthesis are correlated strongly with changes of mRNA expression at nearby genes [150], suggesting that eRNA synthesis may require a dynamic interaction between an enhancer and a promoter. Furthermore, upstream extragenic transcription frequently precedes the induction of an adjacent coding gene [149]. Transcripts from enhancers are not polyadenylated and they show little bias in transcribing both strands [150] as well as being very unstable [149].

The level of H3K4me at the enhancer and eRNA synthesis are tightly correlated, and so the process of eRNA synthesis may be to establish and maintain chromatin in a state required for enhancer function [149]. Indeed, it is likely that the function of many of these gene-associated short RNAs, including PASRs and TASRs, is to mediate transcription-coupled changes in chromatin structure [79]. Such changes may involve the prevention of nucleosomes obstructing transcription factor binding sites [147] or facilitating initiation through the impact of negative supercoiling [146] behind the passage of RNAP II. These will help promoter regions maintain a state poised for subsequent regulation [146]. Polymerase resides on approximately 30 % of human genes, with RNAP II observed to be pausing, appearing to wait for a signal to begin elongating [147]. Genes that are developmentally regulated or that respond to extracellular triggers are those that are likely to have pre-engaged RNAP II [22], so as to speed up the rate at which the gene is ready for transcription. It is likely that there is a rate-limiting step that stops RNAPII fully escaping into elongating [79]. It is presently unclear what this trigger is, but it is likely to be associated with pre-mRNA processing [22].

2.5.2 Long Non-coding RNAs

Significant numbers of long ncRNAs are regulated during development [151]. In particular, the binding of transcription factors, along with evidence of selection, conserved secondary structure, splicing patterns and subcellular localisation, suggests the explicit regulation of non-coding transcription [152]. Long ncRNAs can act as coactivators of transcription factors [153]. They can also act as "ligands" for RNA-binding proteins, causing an allosteric change from an inactive to active conformation, which in turn can inhibit transcription through modifying transcription factor and histone acetyltransferases [154]. Non-coding RNAs also modulate the subcellular localisation of some transcription factors [151]. Non-coding RNAs can also bind to, and regulate the action of, RNA polymerase II during heat shock [155]. Also, some of the transcripts labelled as non-coding may in fact be the source of functional small peptides [156].

The wide variety of regulatory roles ncRNAs can play are shown in Fig. 2.13. A number of chromatin-modifying enzymes contain RNA-binding motifs [157] and long non-coding RNAs recruit chromatin-remodelling complexes to genomic loci [152, 158–162]. Long non-coding RNAs act to direct genomic methylation [163]. They also provide a scaffold for histone-modifying enzyme recruitment

2 An Overview of Gene Regulation

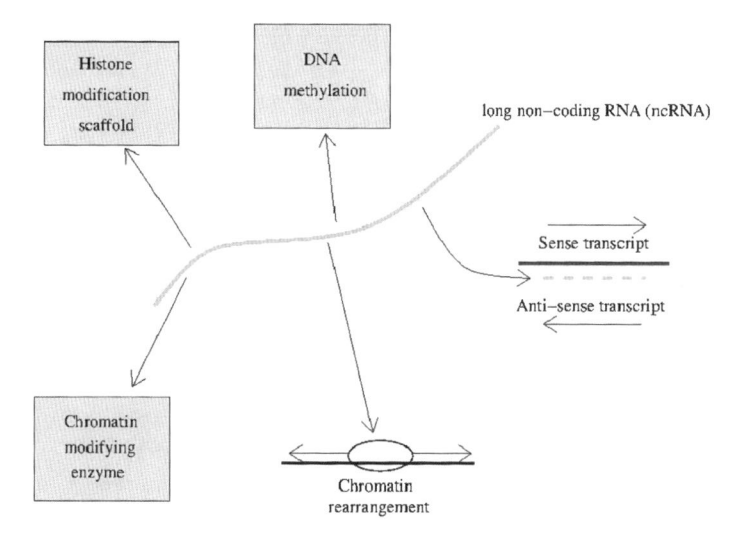

Fig. 2.13 An outline of the various roles long ncRNAs can play in regulation as explained in Sect. 2.5.2

[164], leading to heterochromatin formation [165]. The non-coding transcripts act as local modulators of chromatin structure, triggering chromatin modifications which then expand along the chromosome, even though the neighbouring regions are not complementary to the original transcript [164]. The expansion of the induced chromatin changes may just be restricted locally, or they can expand further and may underpin genomic imprinting [166] and X chromosome inactivation [167]. Another example is the expression of hundreds of long ncRNAs that are sequentially expressed along the Hox loci, defining chromatin domains of differential histone methylation and accessibility [168]. One of the ncRNAs in the Hox loci recruits the Polycomb chromatin-remodelling complex and silences transcription across 40 kb in trans through inducing chromatin to enter a repressive state.

Natural antisense transcripts can overlap part or all of another transcript [164] and many protein-coding genes can be regulated by their antisense transcript partners. The antisense transcripts can bind to their sense partners and enhance their stability, through modifying the binding of an HuR protein and suppressing deadenylation and decapping [169]. The binding of an antisense transcript can also induce changes in RNA secondary structure which act to expose AU-rich elements and make the sense transcript prone to degradation [170]. Interactions between sense and antisense transcripts can also block the binding sites of other regulatory factors such as microRNAs. This appears to be the case for β-secretase, a transcript regulated by its antisense partner and likely related to the pathogenesis of Alzheimer's disease [171].

Antisense RNAs typically undergo fewer splicing events than sense transcripts [172]. However, natural antisense transcripts can modify the alternative splicing isoforms of their sense partners [172, 173] and may also impact upon alternative polyadenylation [164]. Furthermore, long ncRNAs can be processed to yield small

RNAs and they can also modulate the efficiency by which other transcripts are cut into small RNAs and interact with the RNAi pathway [174]. Endogenous siRNAs have been observed to map to overlapping regions between sense and antisense RNAs, and the RNAi pathway could regulate both the sense and antisense transcripts in these cases [175]. However, the RNAi pathway is not responsible for antisense-mediated regulation of the expression of some genes [175]. Duplex formation of sense and antisense partners may also interact with the RNA-editing pathway [176].

In a number of cases, it appears to be the act of transcribing a non-coding transcript, rather than the transcript itself, which acts to regulate a nearby protein-coding gene. Transcriptional interference resulting from collisions between RNA polymerases producing the sense and overlapping antisense expression may occur [177], but this is likely not to be the predominant regulatory pathway mediated by antisense transcripts [164]. Transcription of an ncRNA can pass across the promoter of the protein-coding gene and interfere with transcription factor binding, preventing the expression of the protein-coding gene [151]. Transcriptional elongation induces the addition of histone marks that act to prevent transcription initiation from locations within the body of the transcript [151]. ncRNA transcription can induce histone modifications that repress the transcription of an overlapping protein-coding gene. Furthermore, continuous transcription of ncRNA can prevent silencing of genes by proteins such as Polycomb group proteins [178]. Non-coding RNAs can also help to recruit Trithorax group proteins [179] which help to main active transcription states by counteracting the effects of the Polycomb proteins.

2.5.3 Regulation by MicroRNAs

MicroRNAs (miRNAs) are short ($\sim$22nt) non-coding single-stranded RNAs [180]. They function by usually repressing mRNAs post-transcriptionally through complementary binding to partial overlaps in target mRNAs. They play a central role in coordinating the activities of many thousands of transcripts and they play an integral role in the development and regulation of different cell types and tissues [181].

There are several good reviews of miRNA biogenesis, e.g. [182]. RNA polymerase II mediates the transcription of most miRNAs and is summarised here and Fig. 2.14. Pri-miRNAs are long primary transcripts which typically form a stem hairpin structure, a terminal loop and ssRNA flanking segments. The nuclear enzyme Drosha, assisted by DGCR8 (DiGeorge syndrome critical region gene 8), cleaves the RNA near the stem of the hairpin, about 11 bp away from the ssRNA–dsRNA junction [182]. This releases a pre-miRNA which is then exported from the nucleus by the protein exportin-5. In the cytoplasm, the enzyme Dicer further cleaves the pre-miRNA near the terminal loop to yield a duplex of $\sim$22nt. One of the strands is loaded into an Argonaute (AGO) protein, and this is used to guide complementary target mRNA sequences for repression.

miRNAs have played a significant role in the phenotypic evolution of metazoans [183] and there is a close coupling between miRNA evolution and the establishment

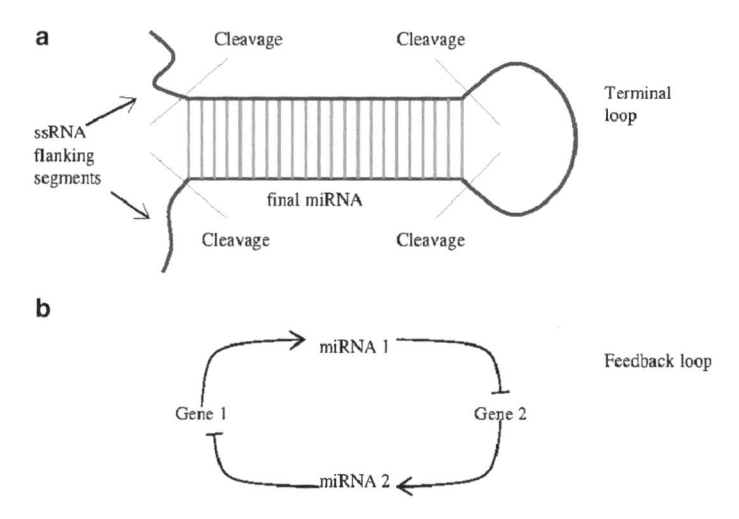

Fig. 2.14 (a) A schematic diagram of a pri-miRNA and the region that eventually forms the miRNA. In (b) we list one regulatory mechanism miRNAs can play in regulating genes explained in Sect. 2.5.3. This case ensures that either gene 1 or gene 2 is expressed

of tissue identities early in bilaterian evolution [184]. An expansion in the number of miRNAs has also been hypothesised to lie behind the origin of vertebrate complexity [185]. The increase in new miRNA families is likely due to the ease in which they are formed along with the wide impact they have on gene networks. The formation of a new miRNA is likely related to pervasive transcription of sequences containing hairpin loops, each of which is only a few mutations away from being a new miRNA [180, 186]. Once a miRNA is operational, and modifying the regulation of many genes, it undergoes very strong purifying selection meaning that their sequences are extremely well conserved [180, 186], making miRNAs excellent phylogenetic markers [187]. However, the targets to which miRNAs bind show little conservation in animals, indicating that miRNA regulatory networks have undergone extensive rewiring during metazoan evolution [180]. Unlike the continuous formation of new miRNA families, there has been a much smaller expansion and evolution of transcription factors during metazoan evolution [187]. Gene duplication is the dominant source of new transcription factors. There is a greater chance of evolutionary advantage for a duplicated transcription factor to undergo a few mutations and bind to new targets of DNA than it is for a non-transcription factor family member to mutate enough to be able to bind to DNA [187].

There appear to be two broad classes of miRNAs [188]. Class I miRNAs are regulated by large numbers of transcription factors and are likely to function within developmental programmes. Whereas class II miRNAs are regulated by small numbers of transcription factors and are likely to function in maintaining tissue identity in adults. The widespread regulation of genes by miRNAs leads to many pathologies resulting from disruptions in the regulation of miRNAs, and they are being increasingly identified as being involved in a range of diseases [189], including many neurodegenerative diseases [190]. Because of concerns about off-target effects of new drugs, it is also being recognised increasingly that development in pharmacogenomics will require greater knowledge of miRNAs [191].

The expression of many transcription factors is subject to miRNA regulation. Feedback motifs are rare in pure transcription factor networks [192], and miRNAs provide the necessary post-transcriptional feedback [193, 194]. miRNAs usually repress gene expression, but not always [195]. One of the roles microRNAs might play is to tune expression at threshold points [183], such that stochastic gene expression programmes will have less noisy outcomes [196]. This type of regulation is required as noise can induce bimodality in positive transcriptional feedback loops [197]. The resulting robustness leads to stabilised developmental pathways, increasing phenotypic reproducibility [198]. The networks through which microRNAs act to regulate self-renewal in stem cells, as well as the transformation of stem cells into differentiated cells, are beginning to be mapped out [199].

There are differences between transcription factor and miRNA regulation related to biological processes in which they are involved. In animals, the repression of miRNAs is usually weak compared to TF-mediated repression [180] and it increasingly appears that miRNAs act to fine-tune the translational and transcriptional output of TFs [200]. miRNAs can act to quickly suppress or reactivate protein production at ribosomes [201], whereas changes in TF binding modifying transcription rates take longer before the information feeds through to protein production [195]. Furthermore, the actions of miRNAs, unlike TFs, can be localised to different parts of a cell. This compartmentalising can then be used in processes such as neurons requiring to regulate gene expression on a synapse-specific scale rather than across the cell [202].

2.6 Common Themes

2.6.1 Structural Considerations

2.6.1.1 The Shape of RNA Impact upon Gene Regulation

RNA molecules form stable secondary and tertiary structures in vitro and in vivo [203]. RNA secondary structures play an important role in binding splicing factors [38], and the search for novel RNA-binding targets for well-known proteins can be enhanced if secondary structure is taken into account [204]. Furthermore, the binding of microRNAs to target sequences depends upon the local tertiary structure of RNA [205]. Moreover, RNA editing also depends on the structure of the RNA, as ADAR converts adenosines to inosines (A to I) using double-stranded RNA substrates.

The reliable computational prediction of RNA structure would be very useful in understanding its underlying function; however, despite some progress in the area, it remains a highly challenging problem [206]. Buratti and Baralle [37] cautioned against the use of *in silico* predictions of pre-mRNA structure such as those obtained by Mfold [207] and Pfold [208]. Buratti and Baralle [37] noted that existing computer algorithms provide a folding prediction (and usually more than one) for virtually any RNA sequence and are strongly biased by the length of the RNA examined. Buratti and Baralle [37] discussed the example of NF-1 gene transcripts, which are implicated in the generation of human tumours. Correlations between the in silico changes in secondary structure and splicing in NF-1 are heavily dependent on the RNA window. This makes it difficult to assign significance to them.

2.6.1.2 The Shape of DNA Impacts upon Gene Regulation

Gene regulation is related to the properties of chromatin in the nucleus. This ranges from posttranslational modifications of histone tails which alter their propensity to bind to each other or to allow transcription, to the movement of histones affecting accessibility of binding sites for transcriptional factors, to the looping of DNA that bring the 5′ and 3′ ends of active genes into proximity, to CTCF acting to regulate networks of binding between different chromosomes and to the movement of co-regulated genes in and out of transcriptional factories.

2.6.2 Gene Structure Is Written Out in Chromatin

The density of nucleosomes and lengths of linkers between nucleosomes differ in exons and introns [73]. The positioning of nucleosomes, as well as histone modifications, is involved in co-transcriptional splicing decisions [48, 49]. Moreover, nucleosome depletion has also been associated with the regulation of polyadenylation [62].

Gene-body DNA methylation also likely plays a role in splicing as exons show higher methylation fractions than do introns and there are sharp transitions in methylation states at exon–intron junctions [50]. Differences in DNA methylation also occur at transcriptional start sites and termination sites [50, 89]. DNA methylation acts to make DNA more rigid [96], and so the regulation of co-transcriptional events may involve a feedback between nucleosome positioning and DNA methylation. Interactions between DNA methylation and histone tail PTMs may also play a role in regulating these events.

There are peaks in DNA methylation as well as in the density of H2A.Z-containing nucleosomes just downstream of start codons and just upstream of stop codons in human T-cells [89]. Given that the use of start and stop codons is not required till translation, an exciting possibility arising from the observations of [89] is that the chromatin markings may be indicative of co-transcriptional modification events which act to label where a protein starts and finishes.

2.6.3 Interacting Codes

Gene regulation appears to be intimately controlled through the actions of several codes – namely, the modulation of a regulatory mechanism by the DNA or protein sequence it encounters. Within the DNA sequence, there is a nucleosome positioning code and this is increasingly well understood. There is likely a CTCF code which helps to regulate the three-dimensional positioning of genes within a cell's nucleus. The heptad repeats in the CTD of RNAPII can undergo different types of posttranslational modifications and these are intimately involved in regulating the binding of factors required for many of the steps in transcription, post-transcriptional processing and termination of transcription [24].

The tails of histones can undergo many types of posttranslational modifications. However, cracking this histone code is proving challenging [66]. This is further complicated by the interactions that occur between the CTD code and the histone code [24]. Moreover, other chromatin-associated proteins, such as HP1, are also posttranslationally modified resulting in the possibility of subcodes with the histone code [209]. Furthermore, H3 histone variants modify the properties of chromatin and their distribution along chromosomes is analogous to a barcode [67]. In addition, much of the impact of H1 variants on the histone code remains to be determined [68].

2.6.4 Kinetics and Competition Between Processes Underpin Gene Regulation

Self-organisation and assembly of structures such as Cajal bodies [104] depends upon the time-dependent concentrations of subcomponents. Furthermore, the movement of genes in and out of transcription factories will also result in changes to the rate of expression [112]. The form of chromatin also causes differences in elongation rates which in turn affect splice site selection [47]. There are a number of other steps available for regulation in splicing [38], encompassing a large number of kinetic events. The kinetic parameters may have a determining role in splice-site choice [36]. It is clear that in order to model how the changing form of recently transcribed RNA impacts on post-transcriptional processes such as splicing, we need to consider the implications of RNA secondary structure, the binding of ribonuclear proteins, the speed of transcription, the form of chromatin and any histone modifications and the dynamic interplay between all these processes.

There is binding competition between a number of processes. Many transcription factors and chromatin-associated proteins have highly transient interactions with chromatin, undergoing rapid cycles of binding and unbinding [103]. The high levels of molecular crowding in the nucleus help to increase the efficiency of binding resulting in local changes in density dramatically altering the rate at which nuclear structures form [103]. Nucleosomes and transcription factors each have affinities for a DNA sequence and competitive and cooperative interactions between these proteins act to determine their occupancy [70]. The cycling of factors on and off promoters enables the formation of poised transcriptional complexes [90], which are typically observed in approximately 30 % of promoters [22].

There are also interactions between different types of regulators. miRNAs can bind to exon–exon junctions, suggesting that they can target splice isoforms [210]. An intron retention event can lead to transcripts containing miRNA-binding sites that they would not otherwise have [211]. Moreover, the biogenesis of miRNAs can result in crosstalk to pre-mRNA splicing [212]. The binding sites of RNA-binding proteins can overlap with microRNA target sites [213] and RNA structure also acts to modify microRNA binding [205]. RNA editing is also coordinated with splicing [54] and there is a close interplay between editing and miRNAs [52]. There is also crosstalk between RNA editing and RNA interference [214]. Next-generation sequencing will increasingly underpin experiments to map out these networks of interactions [43].

2.6.5 Gene Regulation Can Be Tissue Specific

Three-quarters of the mRNA in a cell are common across tissues, and about 8,000, or approximately one-third, of human protein-coding genes are ubiquitously expressed [12]. However, much of the rest of RNA appears to be tissue specific and likely underpins phenotypic complexity in mammals. Alternative splicing and alternative polyadenylation vary between tissues [33]. The majority of retrotransposon expression is tissue specific [215]. RNA editing is enhanced in the brain [52]. Long non-coding RNAs show developmental regulation [151]. miRNAs function in developmental programmes and maintain tissue identity [181]. The state of chromatin also changes as cells transform from pluripotent stem cells into multipotent progenitor cells and the composition of chromatin-remodelling complexes are tissue specific [92].

2.7 Putting It All Together: How Would You Cope if You Could Sequence Everything?

Despite the ferocious complexity of the different mechanisms involved in gene regulation, common themes are emerging as demonstrated in the previous section and summarised as a mind map in Fig. 2.15. It is not unreasonable to assume that in the near future next-generation sequencing techniques will allow the sequencing of all the DNA and expressed types of RNA involved in a given response or process [4]. Such a range of data will be necessary to unravel the complexity of the multilayered regulation of gene expression.

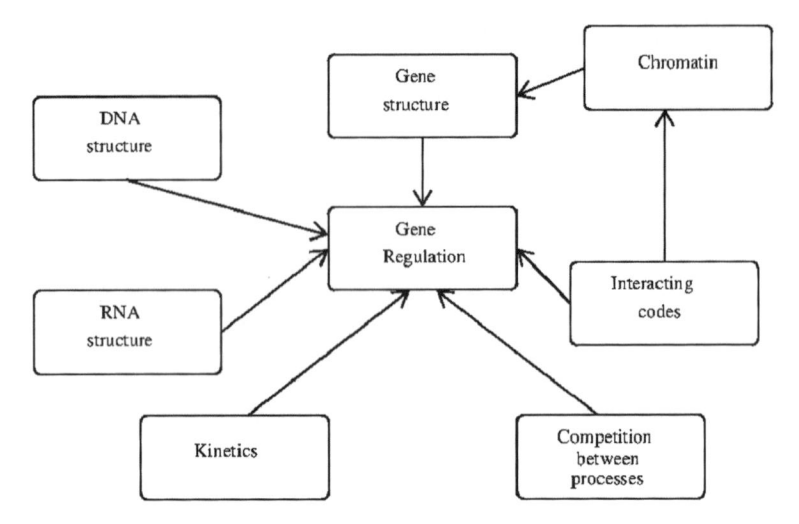

Fig. 2.15 Mind map of Sect. 2.6

A better understanding of chromatin and RNA biology will play a central role in how we use cross-species information reliably. For example, alternative splicing is likely to be one of the principal contributors to the evolution of phenotypic complexity in mammals [33]. The splicing patterns in mammalian model organisms, such as mice, are therefore likely to differ with humans in a number of ways, and so differing populations of isoforms may complicate the interpretation of the negative side effects of pharmaceuticals. RNA editing will also result in different transcript populations in humans compared to other mammals [55], again complicating studies to identify how drugs impact on gene expression systems. A further complication is that of miRNAs, which play a key role in regulating tissue-specific transcription. There are more than 100 extra miRNAs in humans compared to chimpanzees and more than 150 extra miRNAs in human compared to mouse [183], and these extra miRNAs are likely to result in gene expression patterns being found in humans that are not found in our nearest neighbours.

One of the fundamental goals of systems biology is to generate meaningful quantitative models of the regulation of gene expression. In order to do this, not only must there be a significant increase in the types of data being collated (as we have shown in this review), the amount of each type must also be increased considerably. This is necessary to circumvent the so-called curse of dimensionality where the output from all the genes is measured but only in a small number of conditions. It will be necessary to bring multiple studies together, so as to identify some of the subtle changes in gene expression that are biologically meaningful [17]. This indicates a huge increase in the amount of data being gathered, processed and analysed. Already, genomics is one of many fields facing a deluge of data [216]. Bioinformatics repositories are already at the petabyte scale [217] – the growth of sequencing data will result in the repositories transcending the exabyte scale within the decade. The archiving of next-generation sequencing data has well-established resources such as the Short Read Archive [218]. In order to cope with the flow of data, the Short Read Archive is adopting high-speed file transfer protocols, at present `fasp` (Aspera Inc.). However, the transfer of data between external bioinformatics laboratories is already leading to increasing problems in keeping up-to-date [219] and things will only get worse. Moreover, the management of next-generation sequencing data within institutions is already leading to a number of bottlenecks, requiring increasing resources to be spent on systems administration and computers [220] rather than on personnel to make use of the data. A further cost which is only likely to escalate is that of power to run the facility. The computational and staffing issues being faced by the genomics community are likely to limit the democratisation of sequencing.

Genomics is not alone in facing a need for processing very large datasets. The state of the art has arisen from commercial use [216], with organisations such as Google efficiently processing searches and data mining on enormous datasets. Virtually all of these organisations are rapidly implementing data centres to cope with their data-processing requirements. The economies of scales associated with centres mean that they can sell the resources to external users, through the cloud computing model. Bioinformaticians have now begun to look at cloud computing as one feasible solution to cope with the rapid growth of data [221]. There are now increasing needs for large-scale biological data and computational infrastructure to be developed on international scales, such as ELIXIR in Europe [222].

All of this will result in ever larger datasets requiring ever larger computational and experimental infrastructure, as well as larger-sized teams to cope with the data and use it to discover new biology. We can sequence everything, we can afford to do so, we can learn huge amounts, and we will have to likely change some of our working practises to be able to fully utilise the technology.

References

1. Schloss J (2008) How to get genomes at one ten-thousandth the cost. Nat Biotechnol 26:1113–1115
2. Shendure J, Ji H (2008) Next-generation DNA sequencing. Nat Biotechnol 26:1135–1145
3. Branton D, Deamer D, Marziali A, Bayley H, Benner S, Butler T, Di Ventra M, Garaj S, Hibbs A, Huang X, Jovanovich S, Krstic P, Lindsay S, Ling X, Mastrangelo C, Meller A, Oliver J, Pershin Y, Ramsey J, Riehn R, Soni G, Tabard-Cossa V, Wanunu M, Wiggin M, Schloss J (2008) The potential and challenges of nanopore sequencing. Nat Biotechnol 26:1146–1153
4. Kahvejian A, Quackenbush J, Thompson J (2008) What would you do if you could sequence everything? Nat Biotechnol 26:1125–1133
5. Pleasance E, Cheetham R, Stephens P, McBride D, Humphray S, Greenman C, Varela I, Lin M, Ordonez G, Bignell G, Ye K, Alipaz J, Bauer M, Beare D, Butler A, Carter R, Chen L, Cox A, Edkins S, Kokko-Gonzales P, Gormley N, Grocock R, Haudenschild C, Hims M, James T, Jia M, Kingsbury Z, Leroy C, Marshall J, Menzies A, Mudie L, Ning Z, Royce T, Schulz-Trieglaff O, Spiridou A, Stebbings L, Szajkowski L, Teague J, Williamson D, Chin L, Ross M, Campbell P, Bentley D, Futreal P, Stratton M (2010) A comprehensive catalogue of somatic mutations from a human cancer genome. Nature 463:191–196

6. Denoeud F, Aury J, Da Silva C, Noel B, Rogier O, Delledonne M, Morgante M, Valle G, Wincker P, Scarpelli C, Jaillon O, Artiguenave F (2008) Annotating genomes with massive-scale RNA sequencing. Genome Biol 9:R175
7. Mortazavi A, Williams B, McCue K, Schaeffer L, Wold B (2008) Mapping and quantifying mammalian transcriptomes by RNA-Seq. Nat Methods 5:621–628
8. Yassour M, Kaplan T, Fraser H, Levin J, Pfiffner J, Adiconis X, Schroth G, Luo S, Khrebtukova I, Gnirke A, Nusbaum C, Thompson D, Friedman N, Regev A (2009) Ab initio construction of a eukaryotic transcriptome by massively parallel mRNA sequencing. Proc Natl Acad Sci U S A 106:3264–3269
9. Wang Z, Gerstein M, Snyder M (2009) RNA-Seq: a revolutionary tool for transcriptomics. Nat Rev Genet 10:57–63
10. Morozova O, Hirst M, Marra M (2009) Applications of new sequencing technologies for transcriptome analysis. Annu Rev Genomics Hum Genet 10:135–151
11. Pepke S, Wold B, Mortazavi A (2009) Computation for ChIP-seq and RNA-seq studies. Nat Methods Suppl 6:S22–S32
12. Ramsköld D, Wang E, Burge C, Sandberg R (2009) An abundance of ubiquitously expressed genes revealed by tissue transcriptome sequenced data. PLoS Comput Biol 5:e1000598
13. Ma W, Wong W (2011) The analysis of Chip-Seq data. Methods Enzymol 497:51–73
14. Wall P, Leebens-Mack J, Chanderbali A, Barakat A, Wolcott E, Liang H, Landherr L, Tomsho L, Hu Y, Carlson J, Ma H, Schuster S, Soltis D, Soltis P, Altman N, dePamphilis C (2009) Comparison of next-generation sequencing technologies for transcriptome characterization. BMC Genomics 10:347
15. Oshlack A, Wakefield M (2009) Transcript length bias in RNA-seq data confounds systems biology. Biol Direct 4:14
16. Coombs A (2008) The sequencing shakeup. Nat Biotechnol 26:1109–1112
17. Needham C, Manfield I, Bulpitt A, Gilmartin P, Westhead D (2009) From gene expression to gene regulatory networks in Arabidopsis thaliana. BMC Syst Biol 3:85
18. Tsankov A, Brown C, Yu M, Win M, Silver P, Casolari J (2006) Communication between levels of transcriptional control improves robustness and adaptivity. Mol Syst Biol 2:65
19. Tavazoie S, Hughes JD, Campbell MJ, Cho RJ, Church GM (1999) Systematic determination of genetic network architecture. Nat Genet 22(3):281–285
20. Scheinine A, Mentzen WI, Fotia G, Pieroni E, Maggio F, Mancosu G, De La Fuente A (2009) Inferring gene networks: dream or nightmare?: Part 2: challenges 4 and 5. Ann N Y Acad Sci 1158:287–301
21. Komili S, Silver P (2008) Coupling and coordination in gene expression processes: a systems biology view. Nat Rev Genet 9:38–48
22. Moore M, Proudfoot N (2009) Pre-mRNA processing reaches back to transcription and ahead to translation. Cell 136:688–700
23. Karlić R, Chung H-R, Lasserre J, Vlahoviček K, Vingron M (2010) Histone modification levels are predictive for gene expression. Proc Natl Acad Sci U S A 107:2926–2931
24. Egloff S, Murphy S (2008) Cracking the RNA polymerase II CTD code. Trends Genet 24:280–288
25. Klaff P, Riesner D, Steger G (1996) RNA structure and the regulation of gene expression. Plant Mol Biol 32:89–106
26. Antson A (2000) Single stranded-RNA binding proteins. Curr Opin Struct Biol 10:87
27. Carlson C, Stephens O, Beal P (2003) Recognition of double-stranded RNA by proteins and small molecules. Biopolymers 70:86–102
28. Dreyfuss G, Matunis M, Pinol-Roma S, Burd C (1993) hnRNP proteins and the biogenesis of mRNA. Annu Rev Biochem 62:289–321
29. Eperon L, Graham I, Griffiths A, Eperon I (1988) Effects of RNA secondary structure on alternative splicing of pre-mRNA: is folding limited to a region behind the transcribing RNA polymerase? Cell 54:393–401
30. Aguilera A (2005) Cotranscriptional mRNP assembly: from the DNA to the nuclear pore. Curr Opin Cell Biol 17:242–250

31. Li X, Manley J (2006) Cotranscriptional processes and their influence on genome stability. Genes Dev 20:1838–1847
32. Lindahl T, Nyberg B (1974) Heat-induced deamination of cytosine residues in deoxyribonucleic acid. Biochemistry 13:3405–3410
33. Wang E, Sandberg R, Luo S, Khrebtukova I, Zhang L, Mayr C, Kingsmore S, Schroth G, Burge C (2008) Alternative isoform regulation in human tissue transcriptomes. Nature 456:470–476
34. Melamud E, Moult J (2009) Stochastic noise in splicing machinery. Nucleic Acids Res 37:4873–4886
35. Hallegger M, Llorian M, Smith C (2010) Alternative splicing: global insights. FEBS J 277:856–866
36. Nilsen T, Graveley B (2010) Expansion of the eukaryotic proteome by alternative splicing. Nature 463:457–463
37. Buratti E, Baralle F (2004) Influence of RNA secondary structure on the pre-mRNA splicing process. Mol Cell Biol 24:10505–10514
38. Smith D, Query C, Konarska M (2008) Nought may endure but mutability: spliceosome dynamics and the regulation of splicing. Mol Cell 30:657–666
39. Nagel R, Lancaster A, Zahler A (1998) Specific binding of an exonic splicing enhancer by the pre-mRNA splicing factor SRp55. RNA 4:11–23
40. Damgaard C, Tange T, Kjems J (2002) hnRNP A1 controls HIV-1 mRNA splicing through cooperative binding to intro and exon splicing silencers in the context of a conserved secondary structure. RNA 8:1401–1415
41. Blencowe B (2000) Exonic splicing enhancers: mechanisms of action, diversity and role in human genetic diseases. Trends Biochem Sci 25:106–110
42. Matsuo M, Nishio H, Kitoh Y, Francke U, Nakamura H (1992) Partial deletion of a dystrophin gene leads to exon skipping and to loss of an intra-exon hairpin structure from the predicted mRNA precursor. Biochem Biophys Res Commun 182:495–500
43. Licatalosi D, Darnell R (2010) RNA processing and its regulation: global insights into biological networks. Nat Rev Genet 11:75–87
44. Barash Y, Calarco J, Gao W, Pan Q, Wang X, Shai O, Blencowe B, Frey B (2010) Deciphering the splicing code. Nature 465:53–59
45. Kornblihtt A (2006) Chromatin, transcription elongation and alternative splicing. Nat Struct Mol Biol 13:5–7
46. Proudfoot N (2003) Dawdling polymerases allow introns time to splice. Nat Struct Mol Biol 10:876–878
47. Li B, Carey M, Workman J (2007) The role of chromatin during transcription. Cell 128:707–719
48. Luco R, Pan Q, Tominaga K, Blencowe B, Pereira-Smith O, Misteli T (2010) Regulation of alternative splicing by histone modifications. Science 327:996–1000
49. Schwartz S, Meshorer E, Ast G (2009) Chromatin organization marks exon-intron structure. Nat Struct Mol Biol 16:990–996
50. Laurent L, Wong E, Li G, Huynh T, Tsirigos A, Ong C, Low H, Sung K, Rigoutsos I, Loring J, Wei C (2010) Dynamic changes in the human methylome during differentiation. Genome Res 20:320–331
51. Barski A, Cuddapah S, Cui K, Roh T-Y, Schones D, Wang Z, Wei G, Chepelev I, Zhao K (2007) High-resolution profiling of histone methylations in the human genome. Cell 129:823–837
52. Nishikura K (2010) Functions and regulation of RNA editing by ADAR deaminases. Annu Rev Biochem 79:2.1–2.29
53. Barak M, Levanon E, Eisenberg E, Paz N, Rechavi G, Church G, Mehr R (2009) Evidence for large diversity in the human transcriptome created by Alu RNA editing. Nucleic Acids Res 37:6905–6915
54. Laurencikiene J, Källman A, Fong N, Bentley D, Öhman M (2006) RNA editing and alternative splicing: the importance of co-transcriptional coordination. EMBO Rep 7:303–307

55. Gommans W, Mullen S, Maas S (2009) RNA editing: a driving force for adaptive evolution? BioEssays 31:1137–1145
56. Maas S, Kawahara Y, Tamburro K, Nishikura K (2006) A-to-I RNA editing and human disease. RNA Biol 3:1–9
57. Niswender C, Herrick-Davis K, Dilley G, Meltzer H, Overholser J, Stockmeier C, Emeson R, Sanders-Bush E (2001) RNA editing of the Human Serotonin 5-HT$_{2C}$ receptor: alterations in suicide and implications for serotonergic pharmacotherapy. Neuropsychopharmacology 24:478–491
58. Sebastiani P, Montano M, Puca A, Solovieff N, Kojima T, Wang M, Melista E, Meltzer M, Fischer S, Andersen S, Hartley S, Sedgewick A, Yasumichi A, Bergman A, Barzilai N, Terry D, Riva A, Anselmi C, Malovini A, Kitamoto A, Sawabe M, Arai T, Gondo Y, Steinberg M, Hirose N, Atzmon G, Ruvkun G, Bladwin C, Perls T (2009) RNA editing genes associated with extreme old age in humans and with lifespan in C. elegans. PLoS One 4:e8210
59. Li J, Levanon E, Yoon J-K, Aach J, Xie B, LeProust E, Zhang K, Gao Y, Church G (2009) Genome-wide identification of human RNA editing sites by parallel DNA capturing and sequencing. Science 324:1210–1213
60. Ebhardt H, Tsang H, Dai D, Liu Y, Bostan B, Fahlman R (2009) Meta-analysis of small RNA-sequencing errors reveals ubiquitous post-transcriptional RNA modifications. Nucleic Acids Res 37:2461–2470
61. Millevoi S, Vagner S (2010) Molecular mechanisms of eukaryotic pre-mRNA 3′ end processing regulation. Nucleic Acids Res 38:2757–2774
62. Spies N, Nielsen C, Padgett R, Burge C (2009) Biased chromatin signatures around polyadenylation sites and exons. Mol Cell 36:245–254
63. Mignone F, Gissi C, Liuni S, Pesole G (2002) Untranslated regions of mRNAs. Genome Biol 3:reviews0004.1–0004.10
64. Chepelev I, Wei G, Tang Q, Zhao K (2009) Detection of single nucleotide variations in expressed exons of the human genome using RNA-Seq. Nucleic Acids Res 37:e106
65. Jiang C, Pugh B (2009) Nucleosome positioning and gene regulation: advances through genomics. Nat Rev Genet 10:161–172
66. Campos E, Reinberg D (2009) Histones: annotating chromatin. Annu Rev Genet 43:559–599
67. Hake S, Allis C (2006) Histone H3 variants and their potential role in indexing mammalian genomes: the H3 barcode hypothesis. Proc Natl Acad Sci U S A 103:6428–6435
68. Godde J, Ura K (2008) Cracking the enigmatic linker histone code. J Biochem 143:287–293
69. Segal E, Fondufe-Mittendorf Y, Chen L, Thåström A, Field Y, Moore I, Wang J-P, Widom J (2006) A genomic code for nucleosome positioning. Nature 442:772–778
70. Segal E, Widom J (2009) What controls nucleosome positions? Trends Genet 25:335–343
71. Kaplan N, Moore I, Fondufe-Mittendorf Y, Gossett A, Tillo D, Field Y, LeProust E, Hughes T, Lieb J, Widom J, Segal E (2009) The DNA-encoded nucleosome organization of a eukaryotic genome. Nature 458:362–366
72. Field Y, Kaplan N, Fondufe-Mittendorf Y, Moore I, Sharon E, Lubling Y, Widom J, Segal E (2008) Distinct modes of regulation by chromatin encoded through nucleosome positioning signals. PLoS Comput Biol 4:e1000216
73. Cohanim A, Haran T (2009) The coexistence of the nucleosome positioning code with the genetic code on eukaryotic genomes. Nucleic Acids Res 37:6466–6476
74. Warnecke T, Batada N, Hurst L (2008) The impact of nucleosome code on protein-coding sequence evolution in yeast. PLoS Genet 4:e1000250
75. Washietl S, Machné R, Goldman N (2008) Evolutionary footprints of nucleosome positions in yeast. Trends Genet 24:583–587
76. Sasaki S, Mello C, Shimada A, Nakatani Y, Hashimoto S, Ogawa M, Matsushima K, Gu S, Kashara M, Ahsan B, Sasaki A, Saito T, Suzuki Y, Sugano S, Kohara Y, Takeda H, Fire A, Morishita S (2009) Chromatin-associated periodicity in genetic variation downstream of transcriptional start sites. Science 323:401–404
77. Tirosh I, Barkai N (2008) Two strategies for gene regulation by promoter nucleosomes. Genome Res 18:1084–1091

78. Cheung V, Chua G, Batada N, Landry C, Michnick S, Hughes T, Winston F (2008) Chromatin- and transcription related factors repress transcription from within coding regions throughout the Saccharomyces cerevisiae genome. PLoS Biol 6:2550–2562
79. Buratowski S (2008) Gene expression – where to start? Science 322:1804–1805
80. Cairns B (2009) The logic of chromatin architecture and remodelling at promoters. Nature 461:193–198
81. Albert I, Mavrich T, Tomsho L, Qi J, Zanton S, Schuster S, Pugh B (2007) Translational and rotational settings of H2A.Z nucleosomes across the Saccharomyces cerevisiae genome. Nature 446:572–576
82. Choi J, Kim Y-J (2009) Intrinsic variability of gene expression encoded in nucleosome positioning sequences. Nat Genet 41:498–503
83. Sandelin A, Carninci P, Lenhard B, Ponjavic J, Hayashizaki Y, Hume D (2007) Mammalian RNA polymerase II core promoters: insights from genome-wide studies. Nat Rev Genet 8:424–436
84. Irizarry R, Ladd-Acosta C, Wen B, Wu Z, Montano C, Onyango P, Cui H, Gabo K, Rongione M, Webster M, Ji H, Potash J, Sabunciyan S, Feinberg A (2009) The human colon cancer methylome shows similar hypo- and hypermethylation at conserved tissue-specific CpG island shores. Nat Genet 41:178–186
85. Carninci P, Sandelin A, Lenhard B, Katayama S, Shimokawa K, Ponjavic J, Semple C, Taylor M, Engström P, Frith M, Forrest A, Alkema W, Tan S, Plessy C, Kodzius R, Ravasi T, Kasukawa T, Fukuda S, Kanamori-Katayama M, Kitazume Y, Kawaji H, Kai C, Nakamura M, Konno H, Nakano K, Mottagui-Tabar S, Arner P, Chesi A, Gustincich S, Persichetti F, Suzuki H, Grimmond S, Wells C, Orlando V, Wahlestedt C, Liu E, Harbers M, Kawai J, Bajic V, Hume D, Hayashizaki Y (2006) Genome-wide analysis of mammalian promoter architecture and evolution. Nat Genet 38:626–635
86. Taylor M, Kai C, Kawai J, Carninci P, Hayashizaki Y, Semple C (2006) Heterotachy in mammalian promoter evolution. PLoS Genet 2:627–639
87. Tillo D, Kaplan N, Moore I, Fondufe-Mittendorf Y, Gossett A, Field Y, Lieb J, Widom J, Segal E, Hughes T (2010) High nucleosome occupancy is encoded at human regulatory sequences. PLoS One 5:e9129
88. Tilgner H, Nikolaou C, Althammer S, Sammeth M, Beato M, Valcárcel J, Guigó R (2009) Nucleosome positioning as a determinant of exon recognition. Nat Struct Mol Biol 16:996–1001
89. Choi J, Bae J-B, Lyu J, Kim T-K, Kim Y-J (2009) Nucleosome depletion and DNA methylation at coding region boundaries. Genome Biol 10:R89
90. Mellor J (2006) Dynamic nucleosomes and gene transcription. Trends Genet 22:320–329
91. Segal E, Widom J (2009) From DNA sequence to transcriptional behaviour: a quantitative approach. Nat Rev Genet 10:443–456
92. Ho L, Crabtree G (2010) Chromatin remodelling during development. Nature 463:474–484
93. Whitehouse I, Rando O, Delrow J, Tsukiyama T (2007) Chromatin remodelling at promoters suppresses antisense transcription. Nature 450:1031–1035
94. Clapier C, Cairns B (2009) The biology of chromatin remodelling complexes. Annu Rev Biochem 78:273–304
95. Berger S (2007) The complex language of chromatin regulation during transcription. Nature 447:407–412
96. Pennings S, Allan J, Davey C (2005) DNA methylation, nucleosome formation and positioning. Brief Funct Genomics Proteomics 3:351–361
97. Cedar H, Bergman Y (2009) Linking DNA methylation and histone modification: patterns and paradigms. Nat Rev Genet 10:295–304
98. Hemberger M, Dean W, Reik W (2009) Epigenetic dynamics of stem cells and cell lineage commitment: digging Waddington's canal. Nat Rev Mol Cell Biol 10:526–537
99. Suzuki M, Bird A (2008) DNA methylation landscapes: provocative insights from epigenomics. Nat Rev Genet 9:465–476

100. Lister R, Pelizzola M, Dowen R, Hawkins R, Hon G, Tonti-Filippini J, Nery J, Lee L, Ye Z, Ngo Q-M, Edsall L, Antosiewicz-Bourget J, Stewart R, Ruotti V, Millar A, Thomson J, Ren B, Ecker J (2009) Human DNA methylomes at base resolution show widespread epigenomic differences. Nature 462:315–322
101. Law JA, Jacobsen S (2010) Establishing, maintaining and modifying DNA methylation patterns in plants and animals. Nat Rev Genet 11:204–220
102. Takizawa T, Meaburn K, Misteli T (2008) The meaning of gene positioning. Cell 135:9–13
103. Misteli T (2007) Beyond the sequence: cellular organization of genome function. Cell 128:787–800
104. Kaiser T, Intine R, Dundr M (2008) De Novo formation of a subnuclear body. Science 322:1713–1717
105. Soutoglou E, Misteli T (2008) Activation of the cellular DNA damage response in the absence of DNA lesions. Science 320:1507–1510
106. Rajapakse I, Perlman M, Scalzo D, Kooperberg C, Groudine M, Kosak S (2009) The emergence of lineage-specific chromosomal topologies from coordinate gene regulation. Proc Natl Acad Sci U S A 106:6679–6684
107. Junier I, Martin O, Képès F (2010) Spatial and topological organization of DNA chains induced by gene co-localization. PLoS Comput Biol 6:e1000678
108. Cook P, Marenduzzo D (2009) Entropic organization of interphase chromosomes. J Cell Biol 186:825–834
109. Lanctôt C, Cheutin T, Cremer M, Cavalli G, Cremer T (2007) Dynamic genome architecture in the nuclear space: regulation of gene expression in three dimensions. Nat Rev Genet 8:104–115
110. Osborne C, Chakalova L, Brown K, Carter D, Horton A, Debrand E, Goyenechea B, Mitchell J, Lopes S, Reik W, Fraser P (2004) Active genes dynamically colocalize to shared sites of ongoing transcription. Nat Genet 36:1065–1071
111. Osborne C, Chakalova L, Mitchell J, Horton A, Wood A, Bolland D, Corcoran A, Fraser P (2007) Myc dynamically and preferentially relocates to a transcription factory occupied by Igh. PLoS Biol 5:e192
112. Schoenfelder S, Sexon T, Chakalova L, Cope N, Horton A, Andrews S, Kurukuti S, Mitchell J, Umlauf D, Dimitrova D, Eskiw C, Luo Y, Wei C-L, Ruan Y, Bieker J, Fraser P (2010) Preferential associations between co-regulated genes reveal a transcriptional interactome in erythroid cells. Nat Genet 42:53–62
113. Mitchell J, Fraser P (2010) Transcription factories are nuclear subcompartments that remain in the absence of transcription. Genes Dev 22:20–25
114. Sutherland H, Bickmore W (2009) Transcription factories: gene expression in unions? Nat Rev Genet 10:457–466
115. Cook P (2010) A model for all genomes: the role of transcription factories. J Mol Biol 395:1–10
116. Raj A, Peskin C, Tranchina D, Vargas D, Tyagi S (2006) Stochastic mRNA synthesis in mammalian cells. PLoS Biol 4:e309
117. Chubb J, Trcek T, Shenoy S, Singer R (2006) Transcriptional pulsing of a developmental gene. Curr Biol 16:1018–1025
118. Batada N, Hurst L (2007) Evolution of chromosome organization driven by selection for reduced gene expression noise. Nat Genet 39:945–949
119. Hurst L, Pál C, Lercher M (2008) The evolutionary dynamics of eukaryotic gene order. Nat Rev Genet 5:299–310
120. Janga S-C, Collado-Vides J, Babu M (2008) Transcriptional regulation constrains the organization of genes on eukaryotic chromosomes. Proc Natl Acad Sci U S A 105:15761–15766
121. Kosak S, Scalzo D, Alworth S, Li F, Palmer S, Enver T, Lee J, Groudine M (2007) Coordinate gene regulation during hematopoiesis is related to genomic organization. PLoS Biol 5:e309
122. Babu M, Janga S, de Santiago I, Pombo A (2008) Eukaryotic gene regulation in three dimensions and its impact on genome evolution. Curr Opin Genet Dev 18:1–12

123. Parada L, McQueen P, Misteli T (2004) Tissue-specific spatial organization of genomes. Genome Biol 5:R44
124. Meaburn K, Mistelli T (2007) Chromosome territories. Nature 445:379–381
125. Dekker J (2008) Gene regulation in the third dimension. Science 319:1793–1794
126. Gheldof N, Smith E, Tabuchi T, Koch C, Dunham I, Stamatoyannopoulos J, Dekker J (2010) Cell-type-specific long-range looping interactions identify distant regulatory elements of the CFTR gene. Nucleic Acids Res 38:4325–4336
127. Ott C, Blackledge N, Kerschner J, Leir S-H, Crawford G, Cotton C, Harris A (2009) Intronic enhancers coordinate epithelial-specific looping of the active CFTR locus. Proc Natl Acad Sci U S A 106:19934–19939
128. Fraser J, Rousseau M, Shenker S, Ferraiuolo M, Hayashizaki Y, Blanchette M, Dostie J (2009) Chromatin conformation signatures of cellular differentiation. Genome Biol 10:R37
129. Lanzuolo C, Roure V, Dekker J, Bantignies F, Orlando V (2007) Polycomb response elements mediate the formation of chromosome higher-order structures in the bithorax complex. Nat Cell Biol 9:1167–1174
130. Mishiro T, Ishihara K, Hino S, Tsutsumi S, Aburatani H, Shirahige K, Kinoshita Y, Nakao M (2009) Architectural roles of multiple chromatin insulators at the human apolipoprotein gene cluster. EMBO J 28:1234–1245
131. Ong C-T, Corces V (2009) Insulators as mediators of intra- and inter-chromosomal interactions: a common evolutionary theme. J Biol 8:73
132. Nikolaev L, Akopov S, Didych D, Sverdlov E (2009) Vertebrate protein CTCF and its multiple roles in a large-scale regulation of genome activity. Curr Genomics 10:294–302
133. Essien K, Vigneau S, Apreleva S, Singh L, Bartolomei M, Hannenhalli S (2009) CTCF binding site classes exhibit distinct evolutionary, genomic, epigenomic and transcriptomic features. Genome Biol 10:R131
134. Hou C, Dale R, Dean A (2010) Cell type specificity of chromatin organization mediated by CTCF and cohesion. Proc Natl Acad Sci U S A 107:3651–3656
135. Ohlsson R, Lobanenkov V, Klenova E (2010) Does CTCF mediate between nuclear organization and gene expression? BioEssays 32:37–50
136. Cuddapah S, Jothi R, Schones D, Roh T-Y, Cui K, Zhao K (2009) Global analysis of the insulator binding protein CTCF in chromatin barrier regions reveals demarcation of active and repressive domains. Genome Res 19:24–32
137. Fu Y, Sinha M, Peterson C, Weng Z (2008) The insulator binding protein CTCF positions 20 nucleosomes around its binding sites across the human genome. PLoS Genet 4:e1000138
138. Guelen L, Pagie L, Brasset E, Meuleman W, Faza M, Talhout W, Eussen B, de Klein A, Wessels L, de Laat W, van Steensel B (2008) Domain organization of human chromosomes revealed by mapping of nuclear lamina interactions. Nature 453:948–952
139. Hore T, Deakin J, Marshall-Graves J (2008) The evolution of epigenetic regulators CTCF and BORIS/CTCFL in amniotes. PLoS Genet 4:e1000169
140. Bose T, Gerton J (2010) Cohesinopathies, gene expression and chromatin organization. J Cell Biol 189:201–210
141. Feeney K, Wasson C, Parish J (2010) Cohesin: a regulator of genome integrity and gene expression. Biochem J 428:147–161
142. McNairn A, Gerton J (2008) The chromosome glue gets a little stickier. Trends Genet 24:382–389
143. Parelho V, Hadjur S, Spivakov M, Leleu M, Sauer S, Gregson H, Jarmuz A, Canzonetta C, Webster Z, Nesterova T, Cobb B, Yokomori K, Dillon N, Aragon L, Fisher A, Merkenschlager M (2008) Cohesins functionally associate with CTCF on mammalian chromosome arms. Cell 132:422–433
144. Williams A, Flavell R (2008) The role of CTCF in regulating nuclear organization. J Exp Med 205:747–750
145. Kapranov P, Cheung J, Dike S, Nix D, Duttagupta R, Willingham A, Stadler P, Hertel J, Hackermüller J, Hofacker I, Bell I, Cheung E, Drenkow J, Dumais E, Patel S, Helt G,

Ganesh M, Ghosh S, Piccolboni A, Sementchenko V, Tammana H, Gingeras T (2007) RNA maps reveal new RNA classes and a possible function for pervasive transcription. Science 316:1484–1488
146. Seila A, Calbrese J, Levine S, Yeo G, Rahl P, Flynn R, Young R, Sharp P (2008) Divergent transcription from active promoters. Science 322:1849–1851
147. Core L, Waterfall J, Lis J (2008) Nascent RNA sequencing reveals widespread pausing and divergent initiation at human promoters. Science 322:1845–1848
148. Preker P, Nielsen J, Kammler S, Lykke-Andersen S, Christensen M, Mapendano C, Schierup M, Jensen T (2008) RNA exosome depletion reveals transcription upstream of active promoters. Science 322:1851–1854
149. De Santa F, Barozzi I, Mietton F, Ghisletti S, Polletti S, Tusi B, Muller H, Ragoussis J, Wei C-L, Natoli G (2010) A large fraction of extragenic RNA PolII transcription sites overlap enhancers. PLoS Biol 8:e1000384
150. Kim T-K, Hemberg M, Gray J, Costa A, Bear D, Wu J, Harmin D, Laptewicz M, Barbara-Haley K, Kuersten S, Markenscoff-Papadimitriou E, Kuhl D, Bito H, Worley P, Kreiman G, Greenberg M (2010) Widespread transcription at neuronal activity-regulated enhancers. Nature 465:182–187
151. Wilusz J, Sunwoo H, Spector D (2009) Long noncoding RNAs: functional surprises from the RNA world. Genes Dev 23:1494–1504
152. Mercer T, Dinger M, Mattick J (2009) Long non-coding RNAs: insights into functions. Nat Rev Genet 10:155–159
153. Feng J, Bi C, Clark B, Mady R, Shah P, Kohtz J (2006) The Evf-2 noncoding RNA is transcribed from the Dlx-5/6 ultraconserved region and functions as a Dlx-2 transcriptional coactivator. Genes Dev 20:1470–1484
154. Wang X, Arai S, Song X, Reichart D, Du K, Pascual G, Tempst P, Rosenfeld M, Glass C, Kurokawa R (2008) Induced ncRNAs allosterically modify RNA-binding proteins in cis to inhibit transcription. Nature 454:126–130
155. Mariner P, Walters R, Espinoza C, Drullinger L, Wagner S, Kugel J, Goodrich J (2008) Human Alu RNA is a modular transacting repressor of mRNA transcription during heat shock. Mol Cell 29:499–509
156. Kondo T, Plaza S, Zanet J, Benrabah E, Valenti P, Hashimoto Y, Kobayashi S, Payre F, Kageyama Y (2010) Small peptides switch the transcriptional activity of Shavenbaby during Drosophila embryogenesis. Science 329:336–339
157. Bernstein E, Allis C (2005) RNA meets chromatin. Genes Dev 19:1635–1655
158. Dinger M, Amaral P, Mercer T, Pang K, Bruce S, Gardiner B, Askarian-Amiri M, Ru K, Soldà G, Simons C, Sunkin S, Crowe M, Grimmond S, Perkins A, Mattick J (2008) Long noncoding RNAs in mouse embryonic stem cell pluripotency and differentiation. Genome Res 18:1433–1445
159. Morris K, Santoso S, Turner A-M, Pastori C, Hawkins P (2008) Bidirectional transcription directs both transcriptional gene activation and suppression in human cells. PLoS Genet 4:e1000258
160. Nagano T, Mitchell J, Sanz L, Pauler F, Ferguson-Smith A, Feil R, Fraser P (2008) The Air noncoding RNA epigenetically silences transcription by targeting G9a to chromatin. Science 322:1717–1720
161. Pandey R, Mondal T, Mohammad F, Enroth S, Redrup L, Komorowski J, Nagano T, Mancini-DiNardo D, Kanduri C (2008) Kcnq1ot1 antisense noncoding RNA mediates lineage-specific transcriptional silencing through chromatin-level regulation. Mol Cell 32:232–246
162. Redrup L, Branco M, Perdeaux E, Krueger C, Lewis A, Santos F, Nagano T, Cobb B, Fraser P, Reik W (2009) The long noncoding RNA Kcnq1ot1 organises a lineage-specific nuclear domain for epigenetic gene silencing. Development 136:525–530
163. Tufarelli C, Sloane-Stanley J, Garrick D, Sharpe D, Ayyub H, Wood W, Higgs D (2003) Transcription of antisense RNA leading to gene silencing and methylation as a novel cause of human genetic disease. Nat Genet 34:157–165

164. Faghihi M, Wahlestedt C (2009) Regulatory roles of natural antisense transcripts. Nat Rev Mol Cell Biol 10:637–643
165. Yu W, Gius D, Onyango P, Muldoon-Jacobs K, Karp J, Feinberg A, Cui H (2007) Epigenetic silencing of tumour suppressor gene p15 by its antisense RNA. Nature 451:202–206
166. Kanduri C (2008) Functional insights into long antisense noncoding RNA Kncq1ot1 mediated bidirectional silencing. RNA Biol 5:208–211
167. Ohhata T, Hoki Y, Sasaki H, Sado T (2008) Crucial role of antisense transcription across the Xist promoter in Tsix-mediated Xist chromatin modification. Development 135:227–235
168. Rinn J, Kertesz M, Wang J, Squazzo S, Xu X, Brugmann S, Goodnough L, Helms J, Farnham P, Segal E, Chang H (2007) Functional demarcation of active and silent chromatin domains in human Hox loci by noncoding RNAs. Cell 129:1311–1323
169. Matsui K, Nishizawa M, Ozaki T, Kimura T, Hashimoto I, Yamada M, Kaibori M, Kamiyama Y, Ito S, Okumura T (2008) Natural antisense transcript stabilizes inducible nitric oxide synthase messenger RNA in rat hepatocytes. Hepatology 47:686–697
170. Rossignol F, Vaché C, Clottes E (2002) Natural antisense transcripts of hypoxia-inducible factor 1alpha are detected in different normal and tumour human tissues. Gene 299:135–140
171. Faghihi M, Modarresi F, Khalil A, Wood D, Sahagan B, Morgan T, Finch C, St. Laurent G III, Kenny P, Wahlestedt C (2008) Expression of a noncoding RNA is elevated in Alzheimer's disease and drives rapid feed-forward regulation of β-secretase. Nat Med 14:723–730
172. He Y, Vogelstein B, Velculescu V, Papadopoulos N, Kinzler K (2008) The antisense transcriptomes of human cells. Science 322:1855–1857
173. Hastings M, Milcarek C, Martincic K, Peterson M, Munroe S (1997) Expression of the thyroid hormone receptor gene, erbAα, in B lymphocytes: alternative mRNA processing is independent of differentiation but correlates with antisense RNA levels. Nucleic Acids Res 25:4296–4300
174. Watanabe T, Totoki Y, Toyoda A, Kaneda M, Kuramochi-Miyagawa S, Obata Y, Chiba H, Kohara Y, Kono T, Nakano T, Surani M, Sakaki Y, Sasaki H (2008) Endogenous siRNAs from naturally formed dsRNAs regulate transcripts in mouse oocytes. Nature 453:539–544
175. Faghihi M, Wahlestedt C (2006) RNA interference is not involved in natural antisense mediated regulation of gene expression in mammals. Genome Biol 7:R38
176. Peters N, Rohrbach J, Zalewski B, Byrkett C, Vaughn J (2003) RNA editing and regulation of Drosophila 4f-rnp expression by sas-10 antisense readthrough mRNA transcripts. RNA 9:698–710
177. Osato N, Suzuki Y, Ikeo K, Gojobori T (2007) Transcriptional interferences in cis natural antisense transcripts of human and mice. Genetics 176:1299–1306
178. Schmitt S, Prestel M, Paro R (2005) Intergenic transcription through a polycomb group response element counteracts silencing. Genes Dev 19:697–708
179. Sanchez-Elsner T, Gou D, Kremmer E, Sauer F (2006) Noncoding RNAs of trithorax response elements recruit Drosophila Ash1 to Ultrabithorax. Science 311:1118–1123
180. Chen K, Rajewsky N (2007) The evolution of gene regulation by transcription factors and microRNAs. Nat Rev Genet 8:93–103
181. Yu Z, Jian Z, Shen S, Purisima E, Wang E (2007) Global analysis of microRNA target gene expression reveals that miRNA targets are lower expressed in mature mouse and Drosophila tissues than in the embryos. Nucleic Acids Res 35:152–164
182. Kim V, Han J, Siomi M (2008) Biogenesis of small RNAs in animals. Nat Rev Mol Cell Biol 10:126–139
183. Kosik K (2009) MicroRNAs tell an evo-devo story. Nat Rev Neurosci 10:754–759
184. Christodoulou F, Raible F, Tomer R, Simakov O, Trachana K, Klaus S, Snyman H, Hannon G, Bork P, Arendt D (2010) Ancient animal microRNAs and the evolution of tissue identity. Nature 463:1084–1088
185. Heimberg A, Sempere L, Moy V, Donoghue P, Peterson K (2008) MicroRNAs and the advent of vertebrate morphological complexity. Proc Natl Acad Sci U S A 105:2946–2950
186. Liu N, Okamura K, Tyler D, Phillips M, Chung W-J, Lai E (2008) The evolution and functional diversification of animal microRNA genes. Cell Res 18:985–996

187. Wheeler B, Heimberg A, Moy V, Sperling E, Holstein T, Heber S, Peterson K (2009) The deep evolution of metazoan microRNAs. Evol Dev 11:50–68
188. Yu X, Lin J, Zack D, Mendell J, Qian J (2008) Analysis of regulatory network topology reveals functionally distinct classes of microRNAs. Nucleic Acids Res 36:6494–6503
189. Couzin J (2008) MicroRNAs make big impression in disease after disease. Science 319:1782–1784
190. Hébert S, de Strooper B (2009) Alterations of the microRNA network cause neurodegenerative disease. Trends Neurosci 32:199–206
191. Passetti F, Ferreira C, Costa F (2009) The impact of microRNAs and alternative splicing in pharmacogenomics. Pharmacogenomics J 9:1–13
192. Milo R, Shen-Orr S, Itzkovitz S, Kashtan N, Chklovski D, Alon U (2002) Network motifs: simple building blocks of complex networks. Science 298:824–827
193. Johnston R, Chang S, Etchberger J, Ortiz C, Hobert O (2005) MicroRNAs acting in a double-negative feedback loop to control a neuronal cell fate decision. Proc Natl Acad Sci U S A 102:12449–12454
194. Martinez N, Ow M, Barrasa M, Hammell M, Sequerra R, Doucette-Stamm L, Roth F, Ambros V, Walhout A (2008) A C. elegans genome-scale microRNA network contains composite feedback motifs with high flux capacity. Genes Dev 22:2535–2549
195. Hobert O (2008) Gene regulation by transcription factors and microRNAs. Science 319:1785–1786
196. Cohen S, Brennecke J, Stark A (2006) Denoising feedback loops by thresholding – a new role for microRNAs. Genes Dev 20:2769–2772
197. To T, Maheshri N (2010) Noise can induce bimodality in positive transcriptional feedback loops without bistability. Science 327:1142–1145
198. Hornstein E, Shomron N (2006) Canalization of development by microRNAs. Nat Genet Suppl 38:S20–S24
199. Melton C, Judson R, Blelloch R (2010) Opposing microRNA families regulate self-renewal in mouse embryonic stem cells. Nature 463:621–626
200. Muddashetty R, Bassell G (2009) A boost in microRNAs shapes up the neuron. EMBO J 28:617–618
201. Schratt G, Tuebing F, Nigh E, Kane C, Sabatini M, Kiebler M, Greenberg M (2006) A brain-specific microRNA regulates dendritic spine development. Nature 439:283–289
202. Schratt G (2009) microRNAs at the synapse. Nat Rev Neurosci 10:842–849
203. Brion P, Westhof E (1997) Hierarchy and dynamics of RNA folding. Annu Rev Biophys Biomol Struct 26:113–137
204. Lopez de Silanes I, Zhan M, Lal A, Yang X, Gorospe M (2004) Identification of a target RNA motif for RNA-binding protein HuR. Proc Natl Acad Sci U S A 101:2987–2992
205. Long D, Lee R, Williams P, Chan C, Ambros V, Ding Y (2007) Potent effect of target structure on microRNA function. Nat Struct Mol Biol 14:287–294
206. Gorodkin J, Hofacker IL, Torarinsson E, Yao Z, Havgaard JH, Ruzzo WL (2010) De novo prediction of structured RNAs from genomic sequences. Trends Biotechnol 28(1):9–19
207. Zuker M (2003) Mfold web server for nucleic acid folding and hybridization prediction. Nucleic Acids Res 31:3406–3415
208. Knudsen B, Hein J (1999) RNA secondary structure prediction using stochastic context-free grammars and evolutionary history. Bioinformatics 15:446–454
209. Lombark G, Bensi D, Fernandez-Zapico M, Urrutia R (2006) Evidence for the existence of an HP1-mediated subcode within the histone code. Nat Cell Biol 8:407–415
210. Tay Y, Zhang J, Thomson A, Lim B, Rigoutsos I (2008) MicroRNAs to Nanog, Oct4 and Sox2 coding regions modulate embryonic stem cell differentiation. Nature 455:1124–1128
211. Tan S, Guo J, Huang Q, Chen X, Li-Ling J, Li Q, Ma F (2007) Retained introns increase putative microRNA targets within $3'$ UTRs of human mRNA. FEBS Lett 581:1081–1086
212. Shomron N, Levy C (2009) MicroRNA-biogenesis and pre-mRNA splicing crosstalk. J Biomed Biotechnol 2009:594678

213. Kedde M, Strasser M, Boldajipour B, Oude Vrielink J, Slanchev K, le Sage C, Nagel R, Voorhoeve P, van Duijse J, Ørom U, Lund A, Perrakis A, Raz E, Agami R (2007) RNA-binding protein Dnd1 inhibits microRNA access to target mRNA. Cell 131:1273–1286
214. Nishikura K (2006) Editor meets silencer: crosstalk between RNA editing and RNA interference. Nat Rev Mol Cell Biol 7:919–931
215. Faulkner G, Kimura Y, Daub C, Wani S, Plessy C, Irvine K, Schroder K, Cloonan N, Steptoe A, Lassmann T, Waki K, Hornig N, Arakawa T, Takahashi H, Kawai J, Forrest A, Suzuki H, Hayashizaki Y, Hume D, Orlando V, Grimmond S, Carninci P (2009) The regulated retrotransposon transcriptome of mammalian cells. Nat Genet 41:563–571
216. Bell G, Hey T, Szalay A (2009) Beyond the data deluge. Science 323:1297–1298
217. Cochrane G, Akhtar R, Bonfield J, Bower L, Demiralp F, Faruque N, Gibson R, Hoad G, Hubbard T, Hunter C, Jang M, Juhos S, Leinonen R, Leonard S, Lin Q, Lopez R, Lorenc D, McWilliam H, Mukherjee G, Plaister S, Radhakroshnan R, Robinson S, Sonhany S, Hoopen P, Vaughan R, Zalunin V, Birney E (2009) Petabyte-scale innovations at the European Nucleotide Archive. Nucleic Acids Res 37(Database Issue):D19–D25
218. Shumway M, Cochrane G, Sugawara H (2010) Archiving next generation sequencing data. Nucleic Acids Res 38(Database Issue):D870–D871
219. Sangket U, Phongdarra A, Chotigeat W, Nathan D, Kim WY, Bhak J, Ngamphiw C, Tongsima S, Khan A, Lin H, Tan T (2008) Automatic synchronization and distribution of biological databases and software over low-bandwidth networks among developing countries. Bioinformatics 24:299–301
220. Richter B, Sexton P (2009) Managing and analyzing next-generation sequence data. PLoS Comput Biol 5:e1000369
221. Bateman A, Wood M (2009) Cloud computing. Bioinformatics 25:1475
222. Brooksbank C, Cameron G, Thornton J (2010) The European Bioinformatics Institute's data resources. Nucleic Acids Res 38(Database Issue):D17–D25
223. Luger K, Mäder AW, Richmond RK, Sargent DF, Richmond TJ (1997) Crystal structure of the nucleosome core particle at 2.8 A resolution. Nature 389(6648):251–260

Part II
Information Fusion and Retrieval

Chapter 3
Information Retrieval in Life Sciences: A Programmatic Survey

Matthias Lange, Ron Henkel, Wolfgang Müller, Dagmar Waltemath, and Stephan Weise

Abstract Biomedical databases are a major resource of knowledge for research in the life sciences. The biomedical knowledge is stored in a network of thousands of databases, repositories and ontologies. These data repositories differ substantially in granularity of data, storage formats, database systems, supported data models and interfaces. In order to make full use of available data resources, the high number of heterogeneous query methods and frontends requires high bioinformatic skills. Consequently, the manual inspection of database entries and citations is a time-consuming task for which methods from computer science should be applied.

Concepts and algorithms from information retrieval (IR) play a central role in facing those challenges. While originally developed to manage and query less structured data, information retrieval techniques become increasingly important for the integration of life science data repositories and associated information. This chapter provides an overview of IR concepts and their current applications in life sciences. Enriched by a high number of selected references to pursuing literature, the following sections will successively build a practical guide for biologists and bioinformaticians.

M. Lange (✉) • S. Weise
Leibniz Institute of Plant Genetics and Crop Plant Research, Bioinformatics and Information Technology, OT Gatersleben, Corrensstraße 3, 06466 Stadt Seeland, Germany
e-mail: lange@ipk-gatersleben.de; weise@ipk-gatersleben.de

R. Henkel • D. Waltemath
Department of Systems Biology and Bioinformatics, University of Rostock, Ulmenstraße 69, 18057 Rostock, Germany
e-mail: ron.henkel@uni-rostock.de; dagmar.waltemath@uni-rostock.de

W. Müller
HITS gGmbH, Schloss-Wolfsbrunnenweg 35, 69118 Heidelberg, Germany
e-mail: wolfgang.mueller@h-its.org

Keywords Information retrieval • Data management • Search engines • Relevance ranking • Recommender systems • Semantic data networks • Data integration

3.1 Motivation: Information Systems in Life Sciences

The progress in molecular biology, ranging from experimental data acquisition on individual genes and proteins, over postgenomic technologies, such as RNA-seq, phenotyping, proteomics, systems biology and integrative bioinformatics aims to capture the big picture of entire biological systems [55]. As a consequence of this revolution, the amount of data in the life sciences has exploded. The wave of new technologies, for example, in genomics, is enabling data to be generated at unprecedented scales [85]. As of February 2013, NCBI GenBank provides access to 162,886,727 sequences, and PubMed comprises over 22 million citations for biomedical literature from MEDLINE, life science journals and online books. The number of public available databases passed recently the high water mark of 1,512 [32]. This data deluge must now be harnessed and exploited.

Another aspect is the continuous developments in information procurement, preparation and processing as shown in Fig. 3.1. Over the past years, information processing techniques evolved from library research and individual data archives to web-based systems using intercontinental high-speed network links for an ad hoc data exchange, cloud computing and distributed databases. This continuous and

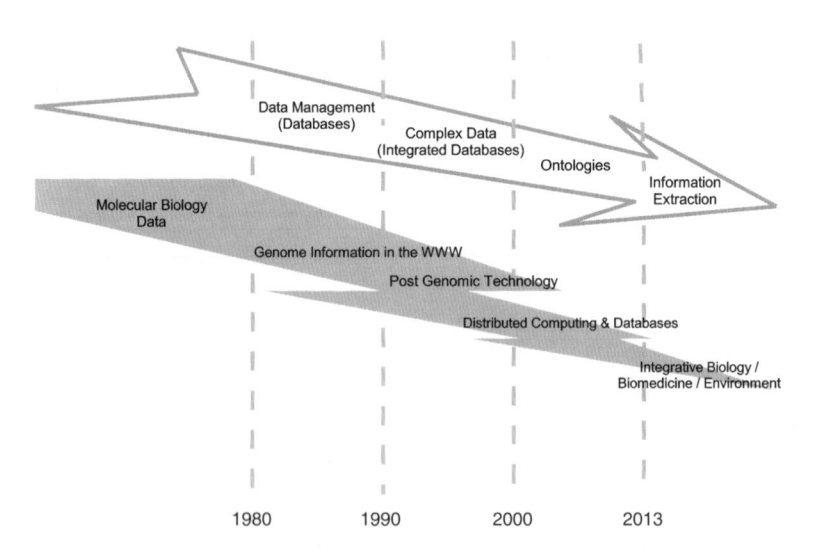

Fig. 3.1 The development of information processing in life sciences adapted from [101] (Reprinted by permission from Macmillan Publishers Ltd, copyright 2002) – Classic database management systems and the domain-specific modelling of project databases are replaced by integrative technologies, i.e. data warehouses, data networks and information retrieval

ongoing shift is attended by the use of *database management systems (DBMS)* which are applied to the management of increasingly complex data structures and voluminous content [98]. The key concepts in bioinformatics with regard to data handling are a consistent classification and unambiguous definition of the modelled biological objects in the databases, the raising use of ontologies, connected with methods of knowledge processing, information extraction and data mining [82, 97].

The consequences of this development are new requirements for information retrieval methods. Typically, life scientists and bioinformaticians formulate their queries rather vaguely. This does not necessarily happen due to inexperience or ignorance but because their search is often explorative with no clear idea of the expected answer. Vague queries though pose a problem on current databases and information systems as these queries cannot be semantically interpreted, without comprehensive semantic document tagging or the use of controlled vocabulary. Much more specific problems such as data distribution and isolation, structural heterogeneity, less metadata, interfaces query languages and deep (invisible) web are further examples of the underlying challenges.

In this context, *information retrieval (IR)* is getting increased importance as technology to face heterogeneities in data representation, storage and organisation towards an efficient information access. The methods for representation and organisation of information items should be designed in accordance to provide users an easy access to the information of their interest [8]. The first step towards this formulated aims is a raising need to find, extract, merge and synthesise information from multiple, disparate sources [56]. In particular, the convergence of biology, computer science and information technology will accelerate this multidisciplinary endeavour. The basic needs for IR are summarised in [58]:

1. On-demand access and retrieval of the most up-to-date biological data and the ability to perform complex queries across multiple heterogeneous databases to find the most relevant information
2. Access to the best-of-breed analytical tools and algorithms for extraction of useful information from the massive volume and diversity of biological data
3. A robust information integration infrastructure that connects various computational steps involving database queries, computational algorithms and application software

Information retrieval in life science databases exhibits some fundamental differences from the way people search in the web or in a general-purpose digital library. First of all, links play a central role for data integration. Not only a single article to a specific entity is of relevance, but all linked articles may be relevant. However, articles just mentioning the entity of relevance may be irrelevant. Second, life science databases are organised in a domain-centric manner, usually concentrating around specific entity types (e.g. metabolomics). It is easy to extract all domain information related to one entity. In contrast, it is very difficult to collect comprehensive, cross-domain information on an entity if the knowledge is spread across entities of different domains, e.g. genome structure-focused databases versus metabolite or pathway-centric ones. A similar picture of heterogeneity can be observed in data access and querying. Methods spread among Boolean queries; predefined queries in web information systems, also known as canned queries; semantic web; keyword-based retrieval in text documents; relevance ranking; and recommender systems are commonly used in life science dry labs.

In this chapter, we will subsequently introduce relevant concepts for information retrieval in the life sciences. It is organised as follows: The Sect. 3.2 provides an overview of basic concepts for data storage, metadata formats and query interfaces, as well as data integration. The Sect. 3.3 then introduces the theoretical foundations, the core concepts of information retrieval and the specific implementation in life sciences. Here, the focus is on characteristics of information retrieval in the life sciences, exploratory information retrieval, recommender systems, human–computer interfaces and semantic aspects with an emphasis on model databases and data networks. The life science search engine LAILAPS is presented as example for an exploratory IR system. The last section contains a comprehensive summary of this chapter.

3.2 Information Systems and Databases

In general, the term *information system (IS)* describes a somehow connected compound of information [89]. In computer science, an information system aims, manages and provides information to support all necessary processes and workflows, especially in companies. Usually, an information system consists of different applications, which are interacting with a *database management system (DBMS)*. Information systems are a main focus in business information technology.

In computer science, a *database (DB)* is a well-structured and functionally associated set of data [29]. A database is managed by a special software – the so-called database management system (DBMS). Together, DB and DBMS form a *database system (DBS)*. The majority of database systems are using the *relational database model* [18].

In life sciences, the term database is often used as a synonym for the term information system. Since the data volume in life sciences is growing rapidly [82], e.g. due to high-throughput technologies (see also Sect. 3.1), the importance of information systems in this area of research is increasing continuously. Often information systems in life sciences use a data basis that is not organised in database management systems [17], but flat files, markup files, HTML or XML files instead. Moreover, the systems are specific to only one data domain. A third characteristic of information systems in life sciences is that they provide different means of access, e.g. web interfaces, web services or static HTML pages, and provide different data exchange formats. The resulting challenges will be described in the following sections.

3.2.1 Data Domains

A *data domain* comprises all data of a specific area, e.g. the domain of the sequence data or the domain of the phenotypic data. Even though data domains can be analysed separately, a combined analysis of multiple data domains, e.g. genotype–phenotype correlations, provides a much higher chance for success. Subsequently, some examples of data domain are listed. Without the intention of providing a comprehensive classification of life science data domains, this list will give an impression about their wide range and diversity.

- *Sequence data:* In biology, this term refers to sequences of nucleotides (DNA sequence) or sequences of amino acids (amino acid sequence/protein sequence), which are the result of a sequencing. Here, sequencing means the determination of all sub-elements. Several sequencing technologies have been developed. Examples are "classical" techniques, such as Sanger sequencing (chain-termination method) [84], Maxam–Gilbert sequencing [73] or EST-based sequencing [2], and next-generation sequencing (NGS) techniques, such as 454 pyrosequencing [72] or Illumina (Solexa) sequencing [11].
- *Variation and marker data:* In genetics, a marker is a piece of DNA with a known location in the genome, which has different expressions in different organisms. Examples are restriction fragment length polymorphism (RFLP) markers [13] or single nucleotide polymorphism (SNP) markers [103]. Today, large amounts of marker data can be obtained by high-throughput technologies.
- *Expression data:* Gene expression means the transformation of DNA information into structures or functions of cells, e.g. the synthesis of enzymes. Depending on different criteria, such as special tissues or compartments, developmental stages or environmental effects, varying amounts of gene products are produced (expressed). With array technologies [86] or by help of RNA-seq, a multitude of product concentrations can be analysed simultaneously (expression profiling).
- *Metabolic network data:* Metabolic networks (pathways) are sequences of biochemical reactions. They can be different depending on the organism, developmental stages, subcellular loci, etc. Data about these networks is an important basis for the understanding of biological subjects at a systems level [104].
- *Phenotypic data:* The phenotype of an organism comprises all characteristics (traits) which can be observed directly and indirectly. It covers a large variety of traits. Besides traits that are mostly determined genetically (e.g. the hair colour), there are also many traits which depend on environmental effects, such as biotic or abiotic stresses.
- *Passport data:* Not often used in the "classical" bioinformatics, but for the management of *plant genetic resources (PGR)* in the so-called gene banks, passport data is indispensable. Passport data contains information, which is used to uniquely identify genotypes.

- *Literature data:* In science, the structured management of literature references is of high importance. Central databases, such as NCBI PubMed[1] or DBLP,[2] collect millions of references from thousands of journals, proceedings, etc. and provide this data to the scientific community. Such information is often used for text mining approaches.

3.2.2 Data Interfaces and Query Methods

Data is only useful if it can be found on request. Consequently, appropriate *query mechanisms* are a prerequisite to reusing existing knowledge in databases. In this respect, queries should be independent from the physical data format, and it should be possible to extract data by specific criteria or to perform database operations, respectively. For performing database operations, query languages can be used, which are based on a data model. Here, it can be distinguished between *procedural* and *declarative* query languages. The former case can be implemented using sequential programming or nesting of database operators, whereas in the latter case only the structure of the results needs to be defined. In other words, only the "what" will be specified, but not the "how".

Data interfaces are necessary for linking applications and data management. These interfaces can be implemented as so-called *application programming interfaces (APIs)*. Common communication interfaces for linking applications and databases are:

- *(Local) File-based access:* A simple method to access data is the use of files from a local file system. This also includes network file systems, e.g. NFS, and file access via data transfer protocols, e.g. FTP. For the data access, the whole file must be parsed. Since the data format is known, data elements can be extracted and then be transferred into data structures. Several parsers have been implemented and are available via APIs (see Sect. 3.2.3).
- *Remote procedure call (RPC):* Another possibility for accessing data is the call of distant (or remote) methods. These comprise protocols such as REST,[3] SOAP,[4] DCOM [16], .NET or CORBA [93]. These methods provide extended functionality, ranging from simple method calls to distributed object networks, web services or persistence frameworks. An essential feature of these standards is the independence of programming languages.
- *DBMS query APIs:* A combination of data query languages and APIs enables remote data access, similarly to DBMS functionality. The technology behind

---

[1] http://www.ncbi.nlm.nih.gov/pubmed/
[2] http://dblp.uni-trier.de/
[3] http://www.ics.uci.edu/~fielding/pubs/dissertation/rest_arch_style.htm
[4] http://www.w3.org/TR/soap/

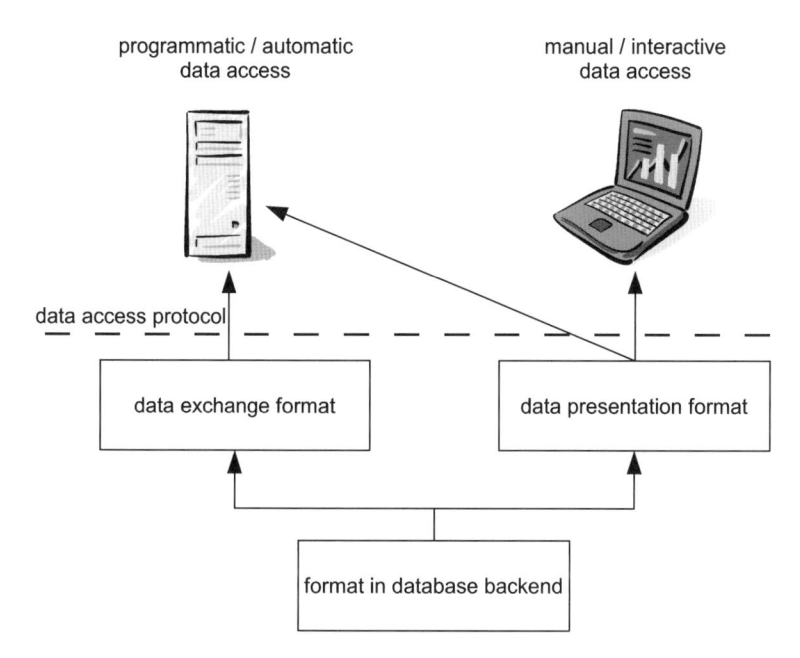

Fig. 3.2 Abstract schema to data storage and format layer in life sciences

either embeds special database access commands into the programming language or integrates the data query language with function calls using APIs. Here already existing programming language-specific APIs and DBMS-specific APIs can be reused. Moreover, DBMS abstracting architectures, such as JDBC [94] or ODBC [33], are available.

3.2.3 Data Formats

A *data format* is a well-defined structure to persistently store data in one or more files. File-based data formats are widely used for the exchange and presentation of data in life sciences [1]. The actual data format is dependent on the storage level and the required access patterns. As shown in Fig. 3.2, it is useful to distinguish different storage layers, which are backend, data exchange and data presentation. The backend layer has a particular emphasis on effective persistence and efficient access structures. In contrast, the data exchange layer is focused on supporting a platform-independent format enriched with structural and semantic metadata. The presentation layer is optimised for an optimal layout and should be flexible to support different HCI technologies and devices.

Whether the *data backend* is a DBMS or it is based on flat file techniques, data independence can be assumed. Thus, data formats used here shall not be dealt with in detail. For *data presentation*, HTML is widely used as a data format. While the content of HTML pages can also be extracted using parsers, however, HTML only plays a minor role for *data exchange*. This is because HTML is mainly used to present and structure elements and the focus is more on the visual layout of data. This hampers the machine-based processing. A more suitable format for data exchange is the Extensible Markup Language (XML).

In addition, the use of domain-specific, not necessarily formal, defined text *flat files* plays an important role. Popular databases use such formats, e.g. EMBL [52]. Another example is the FASTA format [79] which was originally developed for a bioinformatics tool for sequence comparisons. Today, it is a de facto standard for sequence data exchange. A third example is the so-called two-letter code for databases from the European Bioinformatics Institute (EBI) which uses attribute–value pairs.

In the case of flat files, only an indispensable *format description* enables the development of parsers. Such a description should contain the following elements:

- *Allowed constructs:* All allowed words are specified as combinations of valid characters.
- *Syntax description:* The syntax specifies rules for constructing valid combinations, sequences and structures of the constructs described above.
- *Data schema semantics:* Here, rules for mapping the data format structures into elements and relationships of data schemata are specified.

For molecular biological databases, *formal and informal descriptions* of the format are common practice for both, allowed constructs and syntax description. In contrast, data schema semantics are only rarely described. An example is the UniProt database [9] which provides an XML schema for the mapping of UniProt's XML format onto hierarchical structures of XML databases.

Informal descriptions allow to develop parsers manually by interpreting the given rules, but they are not suitable to generate parsers automatically. For automatic parser generation, however, a formal format description is indispensable. Formal descriptions enable machine processing. Examples for appropriate notations from computer science and bioinformatics are the Document Type Definition (DTD) for XML or the Abstract Syntax Notation One (ASN.1). ASN.1 is, for example, used at the National Center for Biotechnology Information (NCBI) for the specification of data types. The UniProt consortium uses XML/DTD to format flat files, e.g. the data exchange format of the UniProt database.

Especially for molecular biological databases, XML plays an important role in data formatting. The following list contains several XML-based data formats [1]:

- *Biopolymer Markup Language (BioML) [31]:* BioML was developed for modelling the hierarchical structures of organisms.
- *Chemical Markup Language (CML) [75]:* CML aims at managing different chemical information in connection with additional information, e.g. publications.

- *KEGG Markup Language (KGML)*[5]*:* KGML contains a DTD for the representation of metabolic pathways including metabolites and enzymes.
- *Systems Biology Markup Language (SBML) [47]:* SBML is a markup language for the representation of computational models in biology. It contains structures for describing subcellular loci (compartments), biochemical reactions and chemical entities involved. Parameters can be declared both globally (for all reactions) and locally (for a single reaction only). Furthermore, units and mathematical rules can be specified.
- *Taxonomic Markup Language [34]:* The Taxonomic Markup Language contains a DTD for the description of taxonomic relationships between organisms.

Apart from the above mentioned, many more XML-based data formats exist, e.g. CellML (Cell Markup Language) [20] or MAGE-ML (MicroArray and Gene Expression Markup Elements).[6] The ongoing development of standard formats for model representation is internationally being coordinated by the COMBINE initiative.[7]

3.2.4 Metadata

Not only business companies are losing hundreds of billions of US dollars per year due to bad *data quality* [27], this also holds true for other areas, including the research sector. For a meaningful use of data – not only in running projects, but also beyond – a high data quality is indispensable. Reaching this aim can be supported by the substantial use of *metadata*. Metadata is additional information provided together with the generated data. One major advantage of the availability of metadata is that they help to perform promising data analysis using data from different life science domains. Metadata is (structured) data describing a resource, an entity, an object or other data. It is used to retrieve, use and maintain a resource, an entity, etc. Unfortunately, often the acquisition of (primary) data and its subsequent processing are not well documented. For example, additional information, such as genotype, development and growth conditions, environmental conditions, tissue or treatment of biological objects, is missing at all or is described using different vocabularies. Further relevant information includes statistical methods or software tools and the parameters applied onto the data. Frequently, this lack of metadata leads to extra costs or additional personnel expenditures when aiming to reuse data or reproduce a result, e.g. when being forced to perform the same experiment multiple times.

---

[5]http://www.kegg.jp/kegg/xml/
[6]http://www.mged.org/Workgroups/MAGE/
[7]The computational modelling in biology network, COMBINE, http://co.mbine.org/.

The problems described above can be downsized by a complete and well-structured documentation of all steps starting with the acquisition of raw data and ending with the publication of results. Thus, the annotation of data with metadata is one important factor for its interpretation, reusability and structuring. This is reflected by manifold metadata schemata that are used in life sciences, mostly under the umbrella of the Minimum Reporting Guidelines for Biological and Biomedical Investigations (MIBBI) project [99]. Reporting guidelines define the minimum information necessary to be provided with a biological or biomedical experiment. The textual description of these information guidelines is often complemented by a data format encoding exactly that information in XML format (see Sect. 3.2.3) and providing mechanisms to link these XML elements with metadata in external resources, such as bio-ontologies, or technical information (e.g. file creators or modification dates for files). In general, it can be subdivided into *semantic* or *technical* metadata.

3.2.4.1 Semantic Metadata

Semantic metadata is closely connected to the scientific data domains and comprises an own universe of several hundreds of metadata schemata. For instance, in systems biology, a review summarises 30 different standards for metadata and data exchange formats [14]. *Ontologies* belong to semantic metadata. In computer science, an ontology is a definition of classes (concepts, objects) and their relationships (attributes, roles) [40]. It is well defined and contains the vocabulary of a data domain, thus improving the interoperability between systems or the communication between human beings.

Due to the growing amount of data in life sciences, it gets more and more important to put this data into relation. Therefore, ontologies are increasingly used [10]. Examples for life science ontologies are:

- *Gene Ontology (GO) [6]:* Molecular functions, biological processes and cellular components
- *Trait Ontology (TO) [50]:* Phenotypic traits of plants
- *Plant Ontology (PO) [7]:* Anatomy and developmental stages of plants
- *MGED Ontology (MO) [105]:* Annotation of microarray experiments

The BioPortal [106] maintains and integrates bio-ontologies that adhere to the requirements of the OBO foundry for open biological, high-quality ontologies [96]. Ontologies in the BioPortal can be browsed visually, and they contain cross-links to other OBO ontologies, enabling extensive exploration of biological knowledge, as well as thorough annotation of data. An *annotation* is a piece of meta-information accompanying a data set. It describes or explains the subject or content it refers to.

3.2.4.2 Technical and Administrative Metadata

Technical and administrative metadata cover aspects of management and processing of digital scientific resources. The collection and storage of structured technical metadata is an important prerequisite for the automatic management and processing of life science data sets. Technical metadata comprise aspects of how to access files, i.e. information about the system requirements for use in terms of hardware and software as well as the unique identification and documentation of the file format in which the resource exists. Each data set should have a unique, persistent identifier, which is identified regardless of its location.

For example, in life sciences, there is a deficiency of generally accepted conventions for referencing data records. Proprietary identifiers, such as so-called accession numbers, are designed as a unique combination of alphanumeric characters. For example, the proprietary identifier Q8W413 in the UniProt database [69] refers to the protein beta-fructofuranosidase.[8] The enzyme 3.2.1.26[9] points to the same entry but is interpreted as standard nomenclature for enzymes. In The Arabidopsis Information Resource (TAIR), the locus tag At2g36190[10] is an identifier for the coding gene of the same protein in the plant *Arabidopsis thaliana* (prefix *At*). Furthermore, the gene synonym AtFruct6 is an example for a semantically enriched acronym of a gene: *At* denotes *Arabidopsis thaliana* and *Fruct* beta-fructofuranosidase.

To overcome this problem, tools have been designed that resolve identifiers and approaches to standardise technical metadata. Known resolver systems are, for example, identifiers.org [51] and the UniProt database identifier mapping.[11] Popular schemata for technical metadata are the Dublin Core Metadata Element Set (DCMES),[12] accepted as ISO standard 15836, as well as the closely related DataCite Metadata Schema.[13] DCMES was developed by scientists and librarians to homogeneously describe digital objects using 15 mandatory elements. The DataCite schema is less strict and comprises only 5 mandatory and 12 optional elements. However, the most popular way of primary data annotation remains to be semantically enriched file names.

---

[8] http://www.uniprot.org/uniprot/Q8W413

[9] http://www.expasy.org/enzyme/3.2.1.26

[10] http://www.arabidopsis.org/servlets/TairObject?type=locus&name=AT2G36190

[11] http://www.uniprot.org/?tab=mapping

[12] http://dublincore.org/documents/dces

[13] http://schema.datacite.org/meta/kernel-2.2/index.html

3.2.5 Database Integration

In general, *data integration* is a service combining contents of multiple, often heterogeneous, data sources, thus enabling to gain new insights [107]. In contrast to the integration of information systems in business companies, data integration in life sciences mainly focuses on combining data of *heterogeneous sources*, e.g. from the World Wide Web. According to [87], heterogeneity can be classified as (i) heterogeneity on systems level (different system properties of the sources, e.g. optimiser strategies), (ii) heterogeneity on data model level (use of different database models, e.g. relational or object-oriented model), (iii) heterogeneity on schema level (e.g. different representation of similar data) and (iv) heterogeneity on data level (e.g. different data for similar database objects).

Research in life sciences typically distinguishes two integration approaches [21]:

1. *Virtual (or logical) data integration:*
   This type of integration is often used for web-based data sources. Here, an integration system sends a query to several data sources and combines the results into a report at runtime. Since no data is stored locally, the results are always up to date, but the query performance is usually lower than with the materialised integration.
2. *Materialised (or physical) data integration:*
   Following this approach, data sources are queried for new data at regular intervals, and this data is stored locally. The integration system then queries the local data only, which has a higher performance than querying distributed sources as with the virtual integration. However, the timeliness of the locally stored data depends on the update intervals.

In the recent past, typical approaches using the virtual integration were *multi-database systems (MDBS)* and *mediator-based systems*. Multi-database systems extract data from several separate database systems and present this data using a homogeneous view [83]. In contrast to these systems, which focus on data stored in database systems, mediator-based systems [108] aim at integrating data stored outside of databases, e.g. HTML or flat files. The latter approach is widely used in bioinformatics. Examples for virtual integration in life sciences are Entrez [90], the Sequence Retrieval System (SRS) [30] and the Distributed Annotation System [26].

The typical approach using materialised integration is the *data warehouse (DWH)* approach which gained popularity in the end of the 1980s [23]. In contrast to *OnLine Transactional Processing (OLTP)* systems, which are designed for management of operative data (no historical data), data warehouses aim at providing non-volatile, aggregated and time-dependent data for analyse purposes, e.g. decision support. For setting up a data warehouse, data from different sources is extracted into a so-called staging area, transformed and then integrated into the data warehouse. *Data marts* are department-specific or application-specific and complement DWH, aiming at answering particular questions. Here, the two contradictory approaches of Inmon [49] (top-down approach) and Kimball [54] (bottom-up approach) are distinguished. According to Inmon, all necessary data is stored in the data warehouse. Data marts are then derived from the data warehouse. In contrast, Kimball regards the creation of data marts as the beginning of the warehousing process. Thus, the data warehouse is a virtual collection of all data marts. Examples for materialised integration in life sciences are Atlas [92], BioMart [53, 95] or BioWarehouse [63].

The need for data integration in life sciences is increasing continuously [36]. So far, the aim of data integration was to provide a homogeneous view onto the integrated data. Recently, a paradigm change can be observed. As described in [19], it gets more and more accepted that different users need different kinds of data integration, because the semantics of data depends on its context. This change in thinking grounds in the fact that the number of scientific questions asked on the available data increased tremendously (e.g. due to high-throughput technologies). Consequently, extended possibilities of retrieving relevant information are necessary.

3.3 Information Retrieval

The increasing popularity of information retrieval as a method to handle semi-structured data and to formulate fuzzy queries correlates with the growth of data that is available online. This development is also reflected in milestones such as the triumphant throughout of PubMed as the world's most important biomedical literature search engine since 1996 [100].

Because of heterogeneity in both, the schema and the system, it is hardly possible to use structured query languages, i.e. SQL or OQL, to access the above-mentioned distributed data. In contrast, the tendency is to apply search engines or information systems to acquire speedily and precisely the information needed [24, 60, 68]. This promising technology is effective for knowledge and data published in journal articles or in its condensed form as hundreds of life science databases [32, 38].

Search engine technology provides efficient and intuitive IR methods to find relevant data in a collection of distributed, heterogeneously structured and modelled data repositories. *Desktop search engines*[14] like Windows Search or Strigi are popular at the scientists' desktops. Frameworks like Apache Solr[15] allow to embed full text search and relevance ranking into data repositories, as well as faceted search. The increasing availability and performance of this technology support the trend to replace classic query forms and Boolean query languages by keyword-based search and relevance filtering. This replacement gets increasingly important in life science information systems and is also implemented in primary data repositories, e.g. the DataCite Metadata Search.

---

[14] http://en.wikipedia.org/wiki/List_of_search_engines#Desktop_search_engines
[15] http://lucene.apache.org/solr/

Instead of referring to relevance ranking, in the following, the term *ranked retrieval* will be used, which expresses the necessity to provide an order for results from a data retrieval process. The interpretation of the term *order* is one central concept of ranked retrieval. Mathematically, it is a partially ordered set R, where R includes the result of a data retrieval query. Furthermore, for R a binary relation $<$ indicates that, for certain pairs of elements in the set, one of the elements precedes the other. In the context of ranked retrieval, the relation $r_1 < r_2 \mid r_1, r_2 \in R$ may have different definitions. The definition of this order relation is the focus of the ranking.

The order of query results becomes particularly important when a query comprises a high number of results. The user should have the possibility to structure and filter data, which are usually displayed as list of data records. If the data records comprise many fields with a high number of individual values, the result listing comprises data excerpts or even a list of access numbers, i.e. IDs. In that case, it is of particular importance to provide a useful order.

Empirically, the word "useful" could have very different meanings. This meaning is hardly dependent on the user's *pertinence*. There are cases when the order is defined by ordinal numbers, like publication date or serial numbers. Another order criteria is the lexicographic order. But numeric or lexicographic ordering is not necessarily a sufficient ranking criterion. Thus, defining relevance functions to determine the relevance of a data item and mapping it to an orderable p-value is one of the major challenges in IR.

In the following sections, two major categories of relevance ranking in life sciences will be discussed. The first category is the *explorative information retrieval* with the focus on an explorative and unbiased retrieval of data over a maximum set of databases, where the relevance ranking is mainly based on popularity and structure in the data itself. The second category, *semantic information retrieval*, is based on the presence of a model in a predefined network of data records that matches best to a very focused query. The model uses word associations and property lists.

3.3.1 Explorative Information Retrieval

Explorative information retrieval is a concept which bases on the idea of exploratory search [70] and represents the activities performed by researchers who are either:

- Unfamiliar with the domain of their goal
- Unsure about the ways to achieve their goals or
- Even unsure about their goals in the first place

In Fig. 3.3, the three major types of search are summarised as *lookup, learn, investigate* and classified into the activities *lookup search* and *exploratory search*.

Following this argumentation, explorative IR combines diverse methods of information retrieval, i.e. domain-specific text indexing, relevance feedback, relevance prediction or recommender systems, with human-computer interaction (HCI) in order to help users exploring data rather than performing lookup searches.

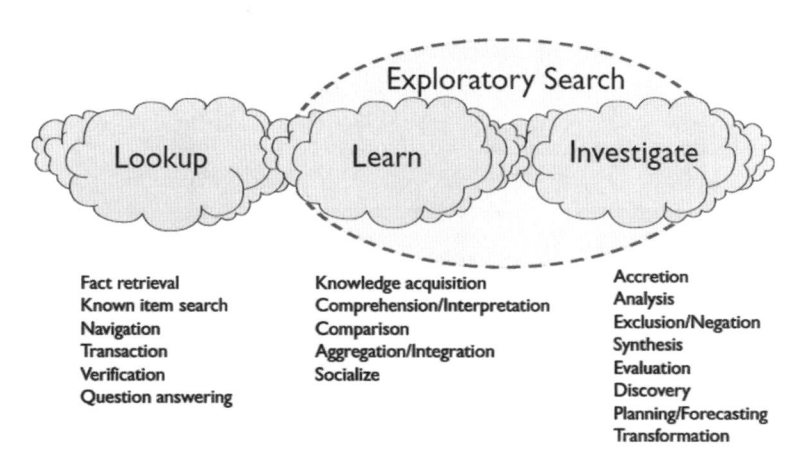

3 Information Retrieval in Life Sciences: A Programmatic Survey

Fig. 3.3 Common search activities in web search, which are labelled as lookup–learn–investigate in [70] (©2006 Association for Computing Machinery, Inc. Reprinted by permission)

An up-to-date overview about the research activities on explorative IR can be found at http://en.wikipedia.org/wiki/Exploratory_search. Studies as the one described in Marti Hearst's book on Search User Interfaces [43] show that search behaviour evolves over time and is strongly influenced by the presence and capabilities of search engines. The main search engine experience of users is still contact with relevance-ranked search. To our experience, current prevalent strategy in bio information retrieval is ranked or Boolean search, combined with metadata-driven browsing and recommendation for exploration of data sets. However, new types of interfaces that emphasise exploratory search are also up-and-coming.

3.3.1.1 Relevance Ranking

"Just head for Google or Entrez and get the related web page or database entry." This is being said among biologists who search information about a certain object [24]. However, issues like finding reliable information about the function of a protein, or identifying the protein that is involved in a certain activity of the cell cycle, are much more challenging tasks. One has to choose (or screen) more than 1,512 life science databases and billions of database records [32].

Intuitively, the first choice for information acquisition are web search engines. Web site ranking techniques order query hits by relevance. However, trying to apply ranking methods that were developed to rank natural language text or WWW sites to life science content and databases is questionable [81]. For example, the top-ranked Google hit for `arginase` is a Wikipedia page. This is because the page is referenced by a high number of web pages or Google assigned a manual defined priority rank. Here, the hypothesis is: *A high hyperlink in-degree of a page means high popularity and high popularity means high relevance* [61].

In order to find scientifically relevant database entries, scientists need strong scientific evidence in relation to the specific research field. A dentist has other relevance criteria than a plant biologist or a patent agent. The intuitive and commonly used way at the scientist's desktop is query refinement. Criteria like who published in which journal, for which organism, evidence scores and surrounding keywords are of major importance. Even complete search guides are published, e.g. for dentists [22].

Other ranking algorithms use *term frequency – inverse document frequency (TF–IDF)* as ranking criteria. Apache Lucene[16] is a popular implementation of this concept and is frequently used in bioinformatics, like LuceGene from the GMOD project [77], which is used for the EBI search frontend EB-eye. The TF-IDF approach works well but misses the semantic context between the database entries and the query.

Another approach is probabilistic relevance ranking [48], where probabilistic values for the relevance of database fields and word combinations have to be predefined. In combination with a user feedback system, the probabilistic approach shows promising ranking performance [4].

Semantic search engines use methods from natural language processing, semantic tagging and dictionaries to predict the semantically most similar database entries. Such conceptual search strategies, implemented in GoPubMed [25] or ProMiner [41], are frequently used algorithms in text mining projects.

After choosing a ranking algorithm for a search engine, the next task is to define possible ranking criteria. Conventional search engines use several ranking criteria. Andrade and Silva consider the similarity between the result entry and the search query itself as a top-ranking criterion [5]. The importance of linkage in ranking has been put forward by PageRank, its variations and ranking extensions [81], which now constitute a mature field.

Greifeneder [39] proposes several possible relevance criteria, including the absolute or relative frequencies of the keyword(s) of the search query, the scope or the actuality of the web page constituting the query result.

Schöch also mentioned the shortness of a URL and the order and the proximity of the search query terms as a criterion [88]. Both Greifeneder and Schöch suggest to check the entries for their popularity [39, 88]. This idea is based on centrality computation, which is an important research area in network analysis. One popular example for this usage is the PageRank algorithm of Google [15, 61].

3.3.1.2 Recommender Systems

In its most common formulation, the recommendation problem is reduced to the problem of estimating ratings for the items that have not been seen by a user and would be of interest. Intuitively, this estimation is usually based on the ratings given

---

[16]http://lucene.apache.org

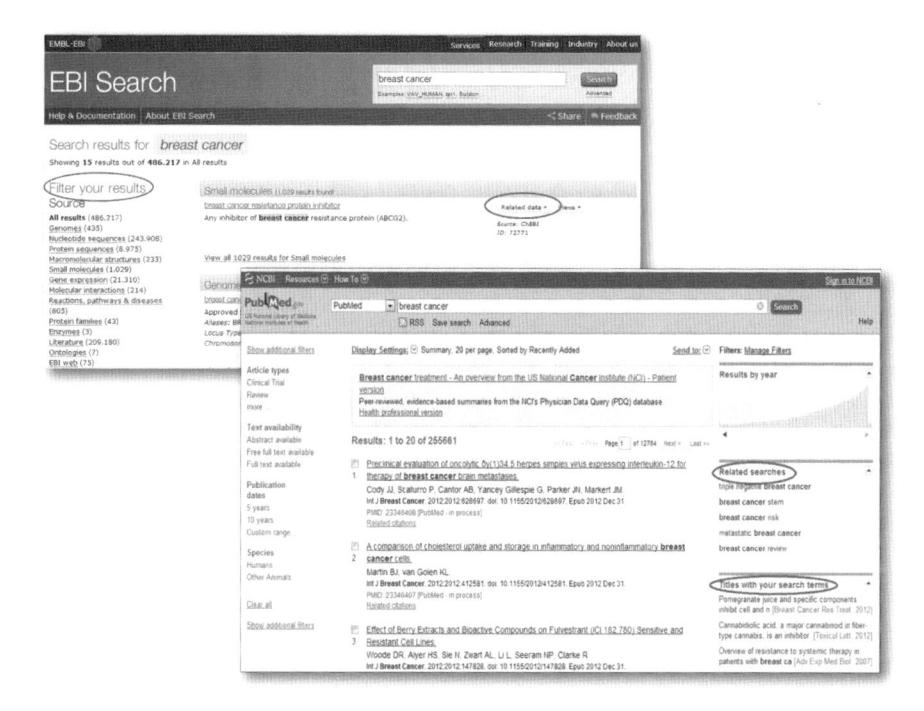

Fig. 3.4 Recommender systems used in the EBI's EB-eye IR system [37] (*left*) and NCBI PubMed literature search (*right*) – cross database search data or abstracts for the term "breast cancer" result in more than 486,000 hits in EBI databases and more than 255,000 in PubMed abstracts. The queries were executed at 2013/01/25. In PubMed, "Related searches" and "Titles with your search terms" suggest references using collaborative filtering. EB-eye makes intensive use of facets, which may be applied to incrementally refine the query and related documents using vector space model

by this user to other items and on some other information [3]. In *recommender systems*, the utility of a data record is usually represented by a rating, which indicates how a particular user liked a particular data set. An example of a user-item rating is PubMed's "Related searches" and "Titles with your search terms" (see Fig. 3.4).

Recommendation in life science IR can be divided into the phases *query expansion* and *related documents prediction*.

The first phase is *query expansion*. It describes the process of adding terms to or deleting terms from the original query. Here, a recommender system should anticipate from users strategies to find a pearl – the *citation pearl growing strategy* and the *building blocks strategy* [28]. In case of the building blocks strategy, the user divides the information retrieval problem into different concepts and assigns one or more reference terms to each concept. This is embedded into an incremental process of refinements until the most relevant document is selected by the user as local optimum. The *citation pearl growing strategy* uses intermediate query result, which is retrieved by a broad query, and interactively pick terms to expand the original query. The concepts can be implemented in automatic query expansion systems which make use of thesauri, ontologies and synonym lists and, in the case of pearl picking, use top-ranked query results, for example, by collaborative user rating, and pick the relatively most frequent terms in the top documents to expand the query. An add-on is the syntactic expansion of single terms. This is done by computing edit distances to words in a dictionary, phonetic or word stem expansions. A popular implementation of these concepts is the facets. EBI's EB-eye IR system [37] and the information retrieval portal GoPubMed [25], which use the Gene Ontology [6] as thesaurus, are examples of successful application of facets in bioinformatics. Section 3.3.1.3 include some more elaborations to HCI, in particular facets.

The second phase is *related documents prediction* (also known as "more like this" or "page like this"). Based on a query result with relevance-ordered database records, the task of the recommender system now is to extend the result set with related documents. These related documents are not necessarily part of the core result set. There are five major methods proposed to predict such neighbour documents:

1. *Shared terminology:* Significant number of shared words; distance scoring using vector space model.
2. *Part-of data cluster:* Data records are part of the same data partition, i.e. synthetic genes and same species.
3. *Cross references:* Identifiers or explicit hyperlinks build data networks; distance scoring is used to predict neighbours [74].
4. *Collaborative filtering:* Follow users, who already (successfully) refined queries; filter user by client clustering, i.e. origin domain, country and user profile.
5. *Content-based recommendation:* Suggest data records, which were selected in past in a close query session/time context.

The above methods are rarely implemented in life science IR systems. Some of them apply shared terminology, cross references and part-of clusters, e.g. PubMed or EB-eye.

3.3.1.3 Human–Computer Interfaces

Marti Hearst gives in her book a literature-based overview about challenges in information retrieval interface design [43]. One interesting observation that she makes and that is easily verified is that even after 15 years of HCI in web search, general-purpose web search interfaces are still based on a one-line entry of search terms coupled with some query suggestions.

However, in the past 10 years, a new search paradigm emerged, called *Hierarchical Faceted Search (HFS)* [42]. This search paradigm is especially convincing for small, hand-picked data sets, i.e. the classic Nobel Prize Winners example available.[17] However, it has shown viability also for huge data sets such as search results in online stores.

---

[17] http://flamenco.berkeley.edu/demos.html

The goal of HFS is to enable users to explore data sets. It does so by guiding the user, as well as efficiently communicating progress of the search and a position within the collection. HFS is an improvement on classical hierarchical search. How this works can simply be illustrated using a search for car by brand, size class, and engine type. Each car has a given brand, a size class and engine type. They are facets describing the car.

Classifying a given set of cars into one hierarchy, one would have to choose which facet to put first. For example, should be browsed by engine type or rather by size class first? Once the hierarchy is chosen, every user will have to go down the predefined path to browse the cars collection.

The base innovation in HFS is to avoid this decision; instead it is accepted that each item in the cars collection has multiple facets. Each facet corresponds to a hierarchy of subsets, and each car is member of one subset for each of its facets. The faceted search interface enables the user to choose the important facets and to choose to which subsets a query result has to belong at the same time. For example, users want a small car, they do not care about the engine type and it must be a Chevrolet. They thus picked one subset of the size facet and one for the brand facet.

To get a feeling of the amazingly simple and intuitive browsing that can be achieved this way, try the flamenco Nobel Prize Winners demo. Please note how details play a big role in faceted search, for example, the display of query result sizes before the query in order to give a preview of what can be expected when clicking on a given facet.

While this example shows the advantages of faceted search, there are some inconveniences that keep faceted search from wider use for large data collections:

- Too many facets and too large fan-out of facet hierarchies: In free data collections, there is a huge amount of potential facets. It is impossible to show all of them on a screen.
- Absence of high-quality facet hierarchies: Annotated by hand, one can design high-quality facets; however, automatic classification in high-quality facets is hard.

GoPubMed (see example at Fig. 3.5) exemplifies strengths and challenges of faceted search for biologists: On the one hand, the interface enables browsing via facets, using the well-developed taxonomies that biology has to offer; on the other hand, browsing uses a lot of its intuitivity with the huge fan-out of bio-ontologies. GoPubMed counters this via emphasised display of *top concepts* and the possibility for logged-in users to define *favourite terms*. Other possibilities of countering the fan-out problem are subject of ongoing research. However, some systems recently started to include elements of faceted search in addition to classic search, e.g. the "browse targets" functionality in ChEMBL,[18] or autocompletion with display of result size previews in SABIO-RK.[19]

---

[18] https://www.ebi.ac.uk/chembl/malaria/target/browser/classification

[19] http://sabio.h-its.org

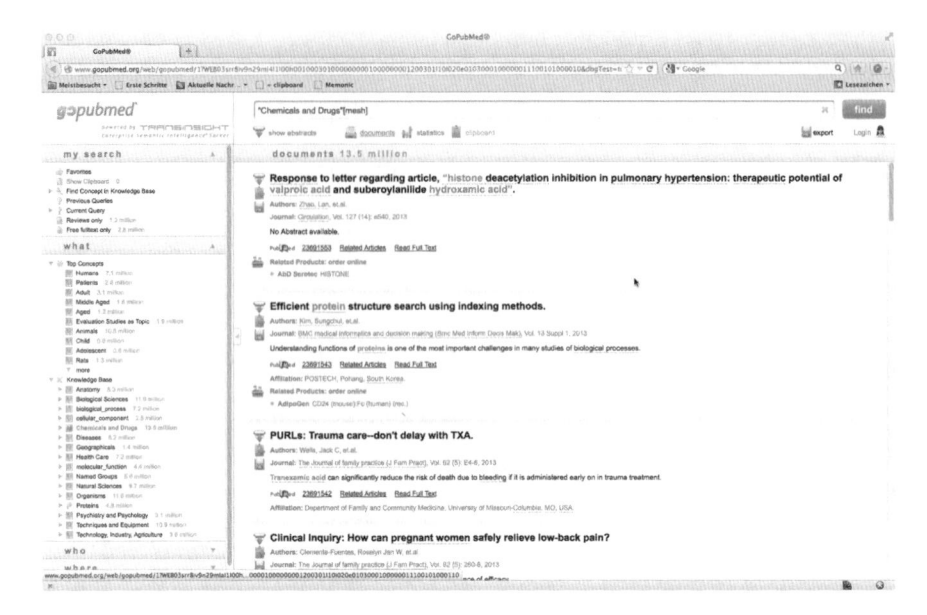

Fig. 3.5 GoPubMed example search. Notice how care is taken to limit the fan-out of trees, keeping it down to only 20 children of the "Knowledge Base" tree. However, already 20 entries have to be read one by one. Logged-in users could counter this by using bookmarked terms for future searches, thus creating search trails

3.3.1.4 The Explorative IR System LAILAPS

LAILAPS stands for "*L*ife Science *A*pplication for *I*nformation Retrieval and *L*ightweight *A*PI for *P*ortable *S*earch Engines" and as metaphor for the Greek mythological dog who never failed to catch the prey what he was hunting. In IR semantics, the aim is to provide a tool that supports the information discovery in the world's life science databases. This bold goal must meet continuously changing requirements. Some are gained from over 10 years experience in dozens of data management, database integration and analysis projects. The result is the development of the LAILAPS IR system. This project has been running for 6 years and combines state-of-the-art methods and concepts from the computer sciences, life sciences and bioinformatics. Empirically collected user requirements from bioinformaticians, IT-skilled biologists as well as less experienced students are used to design an intuitive user interface and feedback system. The first LAILAPS version was released in 2007 as an project that was coordinated by an European plant science company. Motivated by insufficient relevance ranking and the high number of unsorted query results from database query systems, the aim was to implement a search engine for protein databases with a user-specific relevance ranking model.

The approach was to import major public protein databases – i.e. UniProt, PIR and KEGG – into an in an EAV schema, decompose and tokenise the text,

Table 3.1 LAILAPS feature set to score database entries

Feature class	Description
Attribute	Attribute in which the query term was found
Database	Database origin of the entry
Frequency	Frequency of all query terms in the entry and attribute
Co-occurrence	Expresses how close and in which order the query term were found
Keyword	Rating of keyword semantics sorrounding the query hits
Organism	Organism to which the entry relates to
Raw data length	Length of the raw data, which is embedded in the database entry
Text position	Portion of the attribute covered by the query term
Synonym	Information if the hit was produced by an automatic synonym expansion

compute a reverse text index and compute scores for data entities. The concept of the LAILAPS query system is to support lists of search terms and phrases. A search result is a relevance-ranked list of database entries. Each entry is displayed in form of an rich snipped that summarised the content in one text line. The basis of the relevance ranking is a set of nine classes of features, which are shown in Table 3.1. The quantification of these features is computed for each result record as static entry properties or as from the properties of the text index search itself. The parameterisation of the relevance prediction algorithm is based on user feedback. The user may explicitly rate the page quality or the web browser tracks the user actions and estimates the page quality. This reference data is used to train user-specific neural networks, which predict from feature scores the page relevance. The initial training has been performed with a set of 1,089 manually relevance-rated protein entries that results from 19 queries [60]. A 80/20 cross validation shows a precision between 0.62 and 0.81, a recall of 1.00 and an f-score between 0.76 and 0.90.

The screenshots in Fig. 3.6 display the major components of the LAILAPS web application. A portlet version is available to embed LAILAPS into a custom web page.

Since 2011, the LAILAPS development is focused to support the explorative IR in a genomic context. Here, LAILAPS is used to bridge genomic metadata, like functional annotation to genes or other regions at a genome. The concept is:

1. Compile a domain specific list of data hubs, which acts as information retrieval core.
2. Text search and relevance ranking.
3. Reverse identifier lookup.

The implementation of this concept for the genomic data domain underlines the flexibility of LAILAPS concept. Here, the world's major resources of genomic data annotation are compiled in a list of eight major databases: Trait Ontology, Pfam, Gramene, Plant Ontology, SwissProt, TrEMBL, Gene Ontology and PDB. Those are indexed and linked back to the genomics data, i.e. the Genebank Informa-

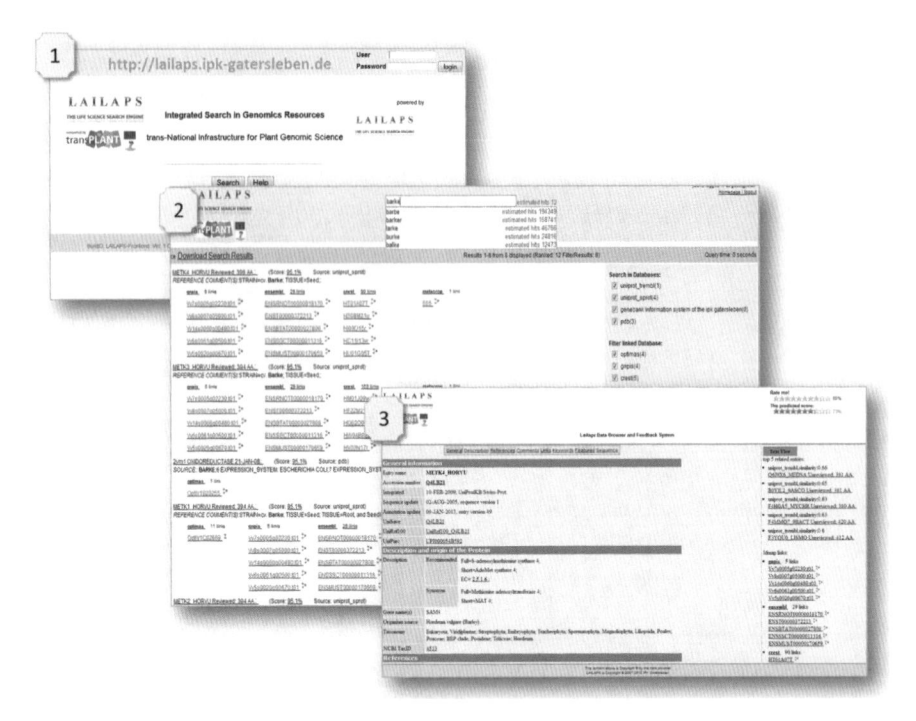

Fig. 3.6 The LAILAPS search engine for integrated search in transPLANT genomics data network. Part (*1*) shows the entry point of the search engine. In screenshot (*2*), a result of a keyword search for "barke", a genotype of barley, is shown. The result contains relevance-ranked hits in indexed genome annotation data hubs (UniProt, Gene Ontology, PFAM, etc.) and related linked genomic resources, i.e. Ensembl, GnpIS, CR-EST. In screenshot (*3*), the integrated data browser and feedback system, which act as input for the incremental training of the relevance predicting neural network

tion System (GBIS) of the German ex-situ Genbank,[20] EBI integrated genomics information system Ensembl,[21] and the INRA integrated genomics information system GNpIS[22] by the French INRA institute. The results of search queries are relevance-ordered links to genomic data. LAILAPS is part of the transPlant consortium to build a transnational plant genomic infrastructure and supported by the European Commission within its 7th Framework Programme, under the thematic area "Infrastructures". The implementation of this IR infrastructure is available at http://lailaps.ipk-gatersleben.de.

---

[20]http://gbis.ipk-gatersleben.de/gbis_i/home.jsf

[21]http://www.ensembl.org

[22]http://urgi.versailles.inra.fr/gnpis

3.3.2 Semantic Information Retrieval

The focus of this book chapter has so far been on the integration and retrieval of large-scale bioinformatics data. Another type of data that needs to be integrated are computational simulation models. During the past decades, modelling and simulation techniques have been used to answer biological questions. A consequence is the development of computational models, often in the area of systems biology. Systems biology is the study of complex biological systems by means of computational approaches and methods. A computational model of a biological system then represents aspects of that system, using, for example, mathematical equations. The number of available models has grown steadily over the last decade, and so has the models' complexity [44]. Models are being shared and reused in standard formats [102], so-called model representation formats (see Sect. 3.3.2.1). The increasing number of models is stored and managed in model repositories such as BioModels Database or PMR2 (see Sect. 3.3.2.2). To handle the models' increasing complexity, semantic annotation has been established as a tool to describe a model's nature. The novel research field of semantic systems biology investigates how to use these annotations to improve model management tasks such as model retrieval, model combination or version control. Section 3.3.2.3 focuses on annotation-based model retrieval and ranking.

3.3.2.1 Model Representation Formats and Standards

To reuse existing model code, the code itself must, first, be made available in model databases. Second, it must be encoded in exchangeable standard formats, which can then be interpreted by software tools. BioModels Database [66] is one example of an open model repository that freely distributes models in standard formats. Frequently used model representation formats are all XML based; examples are the aforementioned Systems Biology Markup Language (SBML [47]), CellML [20] or NeuroML [35] for models of neuroscientific investigations. These standard formats encode the necessary information to rebuild the model structure and underlying mechanisms in a software environment, e.g. for simulation studies.

Together with the model, a whole plethora of meta-information is provided, including the reference publication, the model authors, the semantics of the encoded entities, the model curation state, the underlying mathematics or the graphical representation of the model. Often, meta-information is encoded in bio-ontologies [12] (e.g. Gene Ontology, GO [6], the Systems Biology Ontology (SBO) [65] or the NCBI Taxonomy[23]) and linked to model entities through semantic annotations.

Model annotations mostly refer to technical and administrative information (see Sect. 3.2.4.2), while annotations of model components point to background

---

[23]http://www.ncbi.nlm.nih.gov/Taxonomy/

knowledge from biology or chemistry. The annotation information may either be contained in the model or it may be stored in an external file (see Sect. 3.2.4.1). As well as the model encoding itself, the annotation would best be provided in a standardised form, e.g. using the Resource Description Framework (RDF) [62]. RDF can be interpreted by a computer, and therefore RDF-encoded meta-information can automise tasks such as mode search, comparison, merging or clustering [44, 57, 91]. The ontology terms are in addition highly linked and therefore allow to infer further knowledge about the model.

Semantic annotations in RDF should follow the recommendation for model annotations, called MIRIAM guidelines [64]. The MIRIAM guidelines describe which additional information should be provided together with the model code and how it should be encoded. The SBML standard follows these recommendations and stores annotations as triplets of model entities, qualifiers and URIs pointing to an ontology entry (a so-called identifier [59]). For example, the XML element species represents an entity taking part in a biochemical reaction. The relation between the annotated XML element, e.g. the species, and the ontology reference, e.g. a GO identifier, is expressed also using standardised *qualifiers*.[24] The strongest relation is build up by the IS qualifier, i.e. the XML element *IS* exactly what is described in the ontology entry pointed to by the URI. Several weaker qualifiers exist, e.g. isVersionOf.

The meta-information encoded in model annotations is a major resource for information retrieval tasks. One prominent example is improved model search. For example, a user searching for models dealing with caffeine may express this search by typing caffeine or $C_8H_{10}N_4O_2$, or 1,3,7-trimethylpurine-2,6-dione. A retrieval system is capable of finding the URIs pointing to ontology entries dealing with caffeine and relating them back to models that contain these URIs in their annotations. The basis is the creation of an index of terms from available ontology information. Researchers may use these terms, which best describe the nature of a particular molecule, to perform keyword-based searches. Keywords are more intuitive than cryptic model URIs or computer-generated entity names. If a model is properly annotated with ontology information about caffeine, then the IR-based search will also cover synonyms and external descriptions. Consequently, it is possible to retrieve models based on keywords that do not necessarily occur in the model code itself.

3.3.2.2 Exemplary Model Databases and Repositories

Models in exchangeable standard formats need also be stored and made publicly available to the modelling community to foster reuse. A number of databases and repositories have been established over the past years. The following is a brief review of selected model repositories [102].

---

[24]http://www.ebi.ac.uk/miriam/main/mdb?section=qualifiers

One distributor of freely available SBML models is BioModels Database [66]. To date it contains 436 curated and 497 non-curated models[25] and several thousands of automatically generated pathway models.[26] The majority of models in BioModels Database are concerned with signal transduction and metabolic processes. All models of the curated branch are guaranteed to be valid SBML and to reproduce the results described in the accompanying paper. Internally, metadata is extracted and stored in a MySQL database. Metadata includes information about the submission and modification dates of a model file, authors' information, references and annotations encoded as the aforementioned MIRIAM identifiers. Additionally, Apache Lucene is used to index a subset of model elements and metadata. BioModels Database supports browsing and searching for models. One way to browse is the list of available models (sorted by BioModels Database ID (BMID), model name, publication ID or date of last modification). Another way is to use a tree-structured browser that is based on GO terms. When searching for a model, a so-called multistep search is performed [66]. The system works in three sequential steps. Given a search term, first, the metadata, publications and the annotations stored in the MySQL database are queried. The result of this search is a set of BMIDs. Secondly, the stored SBML XML files are queried, using the previously generated indexes and parsing information such as the SBML `notes` tag. The returned BMIDs are added to the result set. If the search included query terms from external resources, then, thirdly, supplementary information is searched, using either information available in the local MySQL database or web services. For the specific case of searching for a term in a taxonomy, the taxonomy tree is also traversed for neighbour terms, and model IDs associated with that term are added to the result set. The output is generated by using the BMIDs to query the MySQL database for the formerly extracted metadata that is necessary for display on the web site. Search results are returned in an unordered result set.

The Physiome Model Repository (PMR2, [109]) is an online repository for CellML models at different stages of curation. The Plone-based Content Management System contains models of a wide range of different biological processes, including signal transduction pathways, metabolic pathways, electrophysiology, immunology, cell cycle, muscle contraction and mechanical models [67]. PMR2 intends to foster the processes of model curation and annotation so that ideally all models replicate the results in the published paper and the search for models and elements within models is facilitated. Models in the CellML Model Repository are browsed by different (physiological) categories, including cell cycle, signal transduction or metabolism. A CMS-wide full-text search allows for simple free text search. A search by particular model features (e.g. specifically by author or publication year) is not possible. Search results are returned in an unordered result set.

---

[25]Twenty-fourth release of BioModels Database, December 2012.

[26]http://code.google.com/p/path2models/

ModelDB [45] is a format-independent database for curated models related to computational neuroscience. It provides authors a repository for the storage of models, in particular in preparation for submission in neuroscience journals. ModelDB accepts models in any language for any environment [45]. It keeps the originally submitted model files, that is, the complete code specifying the attributes of the original biological system represented in the model, including interface and control code to run the model in the associated simulation environment, and a non-standardised readme text file explaining briefly how to use the provided computer code. Additionally, ModelDB stores model meta-information, including a concise statement of the model purpose, how to use it and a complete citation of the reference publication [45]. The underlying database management system is Oracle 10. as an instance of the Entity–Attribute–Value/Classes–Relationship framework (EAV/CR, [71]) for data representation. The search functionality in ModelDB relies on the meta-information entered by the model submitter. Search by author name or accession number (ModelDB ID) is supported. The complete list of models can be returned sorted by the model name or by the author. Additionally, some predefined queries regarding different criteria such as cell type or simulators are available. However, the queries do not incorporate the model files themselves; as such a search on the model code is not possible. The meta-information is not standardised, but consists of partially predefined strings and partially manually entered data. Third-party knowledge is not incorporated in the search process; the submitted models are not annotated.

JWS Online Model Database is part of the JWS Online Simulator [78], a web-based simulator for biochemical kinetic models. The model repository serves as the maintainer for a number of kinetic models that can be interactively run online. It supports the search for SBML models by a limited number of characteristics, including the author, publication title and journal, organism or model type. A web-based tool offers a searchable categorisation of models in the repository, distinguishing, for example, between cell cycle models and metabolism. A full-text search is not supported. Search results are returned ordered by author name. As there does not exist a publication on the technical background of the model repository, further information about the backend of the provided interface cannot be provided.

3.3.2.3 Model Retrieval and Ranking

A common shortcoming of all above mentioned model repositories is their limited ability to retrieve and rank models. A query containing domain-specific keywords retrieves an unordered set of models. Thus, it is up to the user to browse the results and inspect the models manually. The keywords searched for are not necessarily present in a model itself; however, they might be related to a model by an annotation. Progress in model search has been made with recently developed IR methods for ranked model retrieval [44]. We elucidate here how a keyword-based model search retrieves ranked results using the aforementioned model from

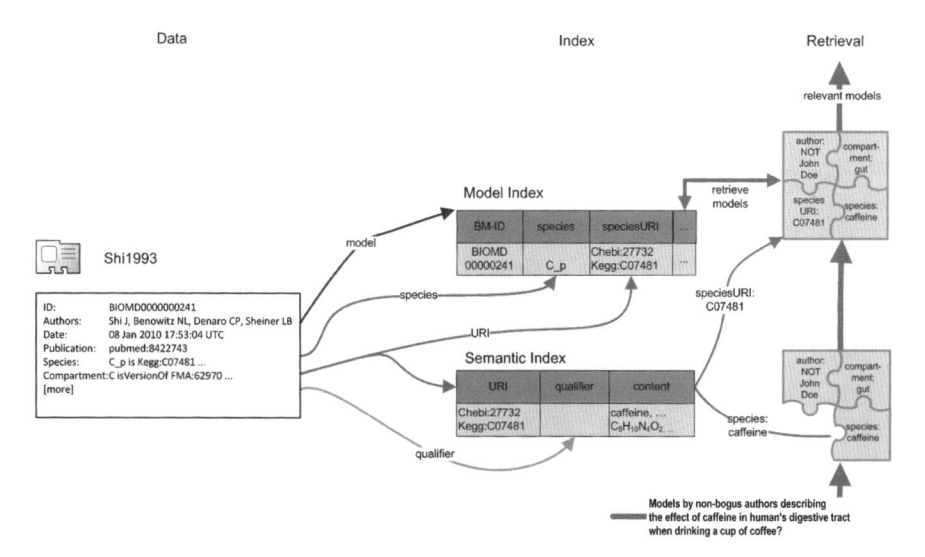

3 Information Retrieval in Life Sciences: A Programmatic Survey   99

Fig. 3.7 This figure shows what model information is stored into the model and semantic index. Additionally, the search is expanded to retrieve models according to their biological content

BioModels Database.[27] This SBML-encoded model contains five compartments, five species, five rate rules and one assignment rule. Even though the model is all about `caffeine` (see example from Sect. 3.3.2.1), related keywords like $C_8H_{10}N_4O_2$, `1,3,7-trimethyl-3,7-dihydro-1H-purine-2,6-dione` or `guaranine` will not retrieve the model at all. This problem is solved by incorporating a model annotation. Figure 3.7 shows an excerpt of the example model. The model index holds information directly encoded in the model, i.e. the model's name, species or compartment names and also URIs used to annotate model entities. The semantic index in addition stores all URIs and links back to models. Here the textual content behind each URI is resolved and indexed.

The model retrieval is then performed using multiple steps. First, the specific query is sent to the model index. If no models or only models matching poorly on the query are retrieved, the search can be refined using the semantic index. Here, the keywords are used to identify matching URIs used to annotate models. As URIs link back to their corresponding models, it is possible to retrieve models using keywords not encoded in the model itself. Such a query expansion is shown in Fig. 3.7 where the term caffeine is used to add URIs to the original query. After all matching models are retrieved, a score is computed for each match. The score mostly relies on the concept of term frequency and inverse document frequency (see Sect. 3.3 for explanation). However, also the importance of certain model components is taken into account, e.g. a species is more important than a parameter value. In case of

---

[27]http://www.ebi.ac.uk/biomodels-main/BIOMD0000000241

URIs, also the relation between URI and annotated entity denoted by the qualifier is taken into account. A deeper explanation is given in [44]. The described approach can be tested on BioModels Database.

Additional possibilities for model search emerge if the networks spanned by several ontologies are integrated. Here, the so-called cross-links can be established and evaluated. One approach is the Bio2RDF[28] project which makes use of the vast information encoded in life science databases. The basic idea is to convert and to link the database contents with semantic web technologies [76]. After converting and linking, each database provides a SPARQL point [80]. The SPARQL point allows to create sophisticated queries on multiple data providers who also offer a SPARQL point. As a result, a number of RDF-triples matching the query are retrieved. Bio2RDF heavily uses semantic web technologies, allowing for automatic traversal through the network. An integrated network of ontologies can be used with OWL-based reasoning methods to identify model similarities (e.g. [46]).

In the ranked retrieval approach, which is closely related to a hierarchical faceted search from Sect. 3.3.1.3, the starting point when querying such a network of ontologies is one particular ontology entry, e.g. xanthine. If a user is interested in models revealing information about xanthine and its derivatives, a URI pointing to the xanthine entry is fed into the system. Thus, the descendants are retrieved and added up, along with inter-ontology links for the specific entry, to form a query. Finally, the query is sent to the model index, and a ranked list of models is retrieved.

3.4 Summary

Due to the increasing demands for data management in the life sciences, information retrieval is no longer just a buzzword. It has instead become a core concept in bioinformatics and related research fields. However, while project proposals still continue to ask for more storage in their budget plans, the aim should be to develop methods for more efficient use of storage. The mere drop of files to the largest possible secondary storage devices, i.e. hard drives or cloud storage solutions, could mean a dead end. Current practice is the storage of working files using a sophisticated naming system for files in combination with Microsoft Excel sheets to link some metadata. This is particularly true for many wet lab desks, and it may be suitable for personal- or even-group level data maintenance. The drawbacks of this system, however, become obvious in its publication process. Highly personalised data representation makes the data only discoverable by insiders, computer scientists or skilled bioinformaticians. The data of interest first needs to be transferred into well-modelled, granular structured and well-interfaced database systems before being reused. A main argument for data reuse is that the distribution of knowledge and later processing by computational analysis is essential to all scientific work.

---

[28]http://bio2rdf.org/

In order to meet the demands expressed above, this chapter gave an overview of core methods and technologies for modern information management in life sciences. The first focus was on databases and information systems. In this context, the change from flat file data exchange to relational database modelling over static database integration approaches to flexible data networks using semantic technologies has been described. Particularly exciting is the vision of a holistic view of a universe of thousands of single yet integrated, well-structured databases. This is, in fact, the real value of the data collected so far. It is not in the form of daily reinvented project-related scripts. The development of such scripts demands time and expert knowledge, and sometimes magic parameters and access paths are used. In contrast, reusable frameworks such as open templates for a workflow-driven data analysis should be preferred. The objective here is a sufficient standardisation and semantic enrichment of the data.

Obviously, the creation of reusable frameworks is a laborious and costly process. However, the overall gain for science will be even bigger. Therefore, lab staff needs to be motivated to use lab information systems and to maintain their protocols, observations and files in database systems. It continues at the scientist's level, where the data streams should be consolidated and properly semantically tagged, long-term citable stored and linked in a scientific publication as supplemental material, preferable in the already established domain databases. Finally, bioinformaticians should place emphasis on the code and interface quality. Besides coding, scripting and data analysis under time pressure, the potential lies in well-documented, object-oriented developed and well-tested software as well as in the use of standard data access protocols and interfaces. This enables the global scientific community to extract all possible knowledge from the existing data.

In addition to the granular and integrated access to globally distributed data, the selective access to information and their extraction is very important. Not the mere of data volume matters. The high number of, on the first view separated, but from a different perspective overlapping, data domains is often the most important cost factor for information retrieval. It could be argued that the actual core of the information retrieval is to find data and ultimately obtain information. This concern is mainly reflected in the section information retrieval. The section has been written with a focus on techniques and actual systems. Here, two most interesting aspects were described in summary – the exploratory and the semantic retrieval.

The focus of the first is on relevance ranking in a set of data query results and recommender systems to improve the query sensitivity and to filter the most important data items in respect to the user's needs. The second focus is on semantic information retrieval, such as the use of metadata or semantic networks and, finally, semantically interpreted data queries.

In this chapter, no evaluations of or recommendations for specific methods or systems were made. This is due to the fact that such evaluations strongly depend on actual applications, which are existing in a wide variety in life sciences. Instead, an extensive list of references of relevant sources in primary literature as well as of web sources was added, which should be seen as a starting point of own detailed studies of the readers.

Acknowledgements This work was supported by the European Commission within its 7th Framework Programme, under the thematic area "Infrastructures", contract number 283496, by the BMBF e:bio programme (University of Rostock) and the Leibniz Institute of Plant Genetics and Crop Plant Research (IPK).

WWW Link List

Resource	Brief description	WWW link
PubMed	PubMed comprises citations for biomedical literature	http://www.ncbi.nlm.nih.gov/pubmed
DBLP	The Computer Science Bibliography provides bibliographic information on major computer science journals and proceedings	http://dblp.uni-trier.de
SOAP	The Simple Object Access Protocol is a protocol specification for exchanging structured information in computer networks	http://www.w3.org/TR/soap
REST	Representational State Transfer is a style of software architecture for distributed systems such as the World Wide Web	http://www.ics.uci.edu/~fielding/pubs/dissertation/rest_arch_style.htm
KGML	KEGG Markup Language (KGML) is an exchange format of the KEGG pathway maps	http://www.kegg.jp/kegg/xml
MAGE	MicroArray and Gene Expression MAGE aims to provide a standard for the representation of microarray expression data	http://www.mged.org/Workgroups/MAGE
COMBINE	COMBINE (Computational Modeling in Biology Network) is an initiative to coordinate the development of the various community standards and formats for computational models	http://co.mbine.org
UniProt	UniProt provides a comprehensive, high-quality and freely accessible resource of protein sequence and functional information	http://www.uniprot.org/uniprot
ENZYME	The Enzyme nomenclature database (ENZYME) is a repository of information relative to the nomenclature of enzymes	http://www.expasy.org/enzyme
TAIR	The Arabidopsis Information Resource (TAIR) maintains a database of genetic and molecular biology data for the model plant Arabidopsis thaliana	http://www.arabidopsis.org
DCES	The Dublin Core Metadata Element Set (DCES) is a vocabulary of 15 properties for use in resource description	http://dublincore.org/documents/dces

(continued)

3 Information Retrieval in Life Sciences: A Programmatic Survey

(continued)

Resource	Brief description	WWW link
MDS	The DataCite Metadata Store (MDS) is a service for data publishers to mint DOIs and register associated metadata	http://mds.datacite.org/
Wikipedia List of Search Engines	List of search engines, including web search engines, selection-based search engines, metasearch engines, desktop search tools and web portals and vertical market web sites that have a search facility for online databases	http://en.wikipedia.org/wiki/List_of_search_engines
Apache Solr	Solr™ is the popular, blazing fast open-source enterprise search platform from the Apache Lucene™ project	http://lucene.apache.org/solr
Explorative IR	Wikipedia overview about the research activities on explorative information retrieval	http://en.wikipedia.org/wiki/Exploratory_search
Apache Lucene	The Apache Lucene™ project develops open-source search software	http://lucene.apache.org
Flamenco	Flamenco search interface framework has the primary design goal of allowing users to move through large information spaces in a flexible manner	http://flamenco.berkeley.edu
Malaria Data Target Classification Hierarchy	Example of faceted search in Malaria Data in addition to classic search	https://www.ebi.ac.uk/chembl/malaria/target/browser/classification
SABIO-RK	SABIO-RK is a curated database that contains information about biochemical reactions and their kinetic rate equations with parameters and experimental conditions	http://sabio.h-its.org
LAILAPS	LAILAPS (Life Science Application for Information Retrieval and Lightweight API for Portable Search Engines) aims to support the information discovery in the world's life science databases	http://lailaps.ipk-gatersleben.de
Ensembl	The Ensembl project produces genome databases for vertebrates and other eukaryotic species and makes this information freely available online	http://www.ensembl.org
GBIS/I	Query portal to retrieve information from the German federal ex situ seed collection	http://gbis.ipk-gatersleben.de/gbis_i/home.jsf
GnPIS	Genetic and Genomic Information System is a tool aiming to provide simple and fast access to the data located in all URGI (plant and fungi data integration) databases	http://urgi.versailles.inra.fr/gnpis
NCBI Taxonomy	The Taxonomy Database is a curated classification and nomenclature for all of the organisms in the public sequence databases	http://www.ncbi.nlm.nih.gov/Taxonomy

(continued)

(continued)

Resource	Brief description	WWW link
BioModels.net qualifiers	The qualifier of an annotation should reflect the relationships between the biological objects represented by the model element and the annotation	http://biomodels.net/qualifiers
path2models	The purpose of the project is to systematically generate mathematical models corresponding to the entire KEGG pathways and submit them to BioModels Database	http://code.google.com/p/path2models
BioModels Database	BioModels Database is a repository hosting computational models of biological systems	http://www.ebi.ac.uk/biomodels-main
Bio2RDF	Integration of ontology networks into biomodel search	http://bio2rdf.org/
Identifiers.org	Identifiers.org is a system providing resolvable persistent URIs used to identify data	http://identifiers.org
SPARQL Query Language	SPARQL can be used to express queries across diverse data sources, whether the data is stored natively as RDF or viewed as RDF via middleware	http://www.w3.org/TR/rdf-sparql-query

References

1. Achard F, Vaysseix G, Barillot E (2001) XML, bioinformatics and data integration. Bioinformatics 17(2):115–125
2. Adams M, Kelley J, Gocayne J, Dubnick M, Polymeropoulos M, Xiao H, Merril C, Wu A, Olde B, Moreno R, Kerlavage A, McCombie W, Venter J (1991) Complementary DNA sequencing: expressed sequence tags and human genome project. Science 252(5013):1651–1656
3. Adomavicius G, Tuzhilin A (2005) Toward the next generation of recommender systems: a survey of the state-of-the-art and possible extensions. IEEE Trans Knowl Data Eng 17(6):734–749
4. Agichtein E, Brill E, Dumais S (2006) Improving web search ranking by incorporating user behavior information. In: SIGIR'06: proceedings of the 29th annual international ACM SIGIR conference on research and development in information retrieval, Seattle. ACM, New York, pp 19–26
5. Andrade L, Silva MJ (2006) Relevance ranking for geographic IR. In: Workshop on geographic information retrieval, SIGIR'06, Seattle
6. Ashburner M, Ball CA, Blake JA, Botstein D, Butler H, Cherry JM, Davis AP, Dolinski K, Dwight SS, Eppig JT, Harris MA, Hill DP, Issel-Tarver L, Kasarskis A, Lewis S, Matese JC, Richardson JE, Ringwald M, Rubin GM, Sherlock G (2000) Gene ontology: tool for the unification of biology. Nat Genet 25(1):25–29
7. Avraham S, Tung CW, Ilic K, Jaiswal P, Kellogg EA, McCouch S, Pujar A, Reiser L, Rhee SY, Sachs MM, Schaeffer M, Stein L, Stevens P, Vincent L, Zapata F, Ware D (2008) The plant ontology database: a community resource for plant structure and developmental stages controlled vocabulary and annotations. Nucl Acids Res 36(suppl_1):D449–D454
8. Baeza Yates RA, Neto BR (1999) Modern information retrieval. Addison-Wesley Longman Publishing Co., Inc., Boston
9. Bairoch A, Apweiler R, Wu CH, Barker WC, Boeckmann B, Ferro S, Gasteiger E, Huang H, Lopez R, Magrane M, Martin MJ, Natale DA, O'Donovan C, Redaschi N, Yeh LL (2005) The universal protein resource (UniProt). Nucl Acids Res 33(suppl_1):D154–D159

10. Bard JBL, Rhee SY (2004) Ontologies in biology: design, applications and future challenges. Nat Rev Genet 5(3):213–222
11. Bentley DR, Balasubramanian S, Swerdlow HP, Smith GP, Milton J, Brown CG, Hall KP, Evers DJ, Barnes CL, Bignell HR et al (2008) Accurate whole human genome sequencing using reversible terminator chemistry. Nature 456(7218):53–59
12. Bodenreider O, Stevens R (2006) Bio-ontologies: current trends and future directions. Brief Bioinform 7(3):256–274
13. Botstein D, White R, Skolnick M, Davis R (1980) Construction of a genetic linkage map in man using restriction fragment length polymorphisms. Am J Hum Genet 32(3):314–331
14. Brazma A, Krestyaninova M, Sarkans U (2006) Standards for systems biology. Nat Rev Genet 7:593–605
15. Brin S, Page L (1998) The anatomy of a large-scale hypertextual Web search engine. In: Proceedings of the seventh international conference on world wide web 7, Brisbane, vol 30. Elsevier, Amsterdam, pp 107–117
16. Brockschmidt K (1995) Inside OLE, 2nd edn. Microsoft Press, Redmond
17. Bry F, Kröger P (2003) A computational biology database digest: data, data analysis, and data management. Distrib Parallel Databases 13(1):7–42
18. Codd EF (1970) A relational model of data for large shared data banks. Commun ACM 13(6):377–387
19. Cohen-Boulakia S, Leser U (2011) Next generation data integration for life sciences. In: Proceedings of the 2011 IEEE 27th international conference on data engineering (ICDE'11), Hannover. IEEE Computer Society, Los Alamitos, pp 1366–1369
20. Cuellar A, Lloyd C, Nielsen P, Bullivant D, Nickerson D, Hunter P (2003) An overview of cellmL 1.1, a biological model description language. Simulation 79(12):740–747
21. Davidson S, Overton C, Buneman P (1995) Challenges in integrating biological data sources. J Comput Biol 2(4):557–572
22. Day J (2001) The quest for information: a guide to searching the internet. J Contemp Dent Pract 2(4):033–043
23. Devlin B, Murphy P (1988) An architecture for a business and information system. IBM Syst J 27(1):60–80
24. Divoli A, Hearst M, Wooldridge MA (2008) Evidence for showing gene/protein name suggestions in bioscience literature search interfaces. In: Pacific symposium on biocomputing, Kohala Coast, vol 13, pp 568–579
25. Doms A, Schroeder M (2005) GoPubMed: exploring PubMed with the Gene ontology. Nucl Acids Res 33(suppl_2):W783–W786
26. Dowell R, Jokerst R, Day A, Eddy S, Stein L (2001) The distributed annotation system. BMC Bioinform 2(1):7
27. Eckerson WW (2002) Data quality and the bottom line: achieving business success through a commitment to high quality data. TDWI report series, The Data Warehousing Institute, Seattle
28. Efthimiadis EN (2000) Interactive query expansion: a user-based evaluation in a relevance feedback environment. J Am Soc Inf Sci 51(11):989–1003
29. Elmasri R, Navathe SB (2000) Fundamentals of database systems, 3rd edn. Addison-Wesley, Reading
30. Etzold T, Harris H, Beaulah S (2003) SRS: an integration platform for databanks and analysis tools in bioinformatics. In: Lacroix Z, Critchlow T (eds) Bioinformatics: managing scientific data. Morgan Kaufmann, San Francisco, pp 109–145
31. Fenyö D (1999) The Biopolymer markup language. Bioinformatics 15(4):339–340
32. Fernández-Suárez XM, Galperin MY (2013) The 2013 nucleic acids research database issue and the online molecular biology database collection. Nucl Acids Res 41(D1):D1–D7
33. Geiger K (1995) Inside ODBC: [Der Entwicklerleitfaden zum Industriestandard für Datenbank-Schnittstellen]. Microsoft Press, Unterschleissheim
34. Gilmour R (2000) Taxonomic markup language: applying XML to systematic data. Bioinformatics 16(4):406–407

35. Gleeson P, Crook S, Cannon R, Hines M, Billings G, Farinella M, Morse T, Davison A, Ray S, Bhalla U et al (2010) Neuroml: a language for describing data driven models of neurons and networks with a high degree of biological detail. PLoS Comput Biol 6(6):e1000815
36. Goble C, Stevens R (2008) State of the nation in data integration for bioinformatics. J Biomed Inform 41(5):687–693
37. Goujon M, Valentin F, Miyar T, McWilliam H, Lopez R (2007) The EB-eye. EMBnetnews 13(4):18–21
38. Gray J (2007) Jim gray on eScience: a transformed scientific method. Retrieved from http://research.microsoft.com/en-us/collaboration/fourthparadigm/4th_paradigm_book_jim_gray_transcript.pdf
39. Greifeneder H (2010) Erfolgreiches SuchmaschinenMarketing: Wie Sie bei Google, Yahoo, MSN & Co. ganz nach oben kommen, 2nd edn. Gabler Verlag
40. Gruber TR (1993) A translation approach to portable ontology specifications. Knowl Acquis 5(2):199–220
41. Hanisch D, Fundel K, Mevissen HT, Zimmer R, Fluck J (2005) Prominer: rule-based protein and gene entity recognition. BMC Bioinform 6(Suppl_1):S14
42. Hearst M (2006) Design recommendations for hierarchical faceted search interfaces. In: ACM SIGIR workshop on faceted search, Seattle
43. Hearst M (2009) Search user interfaces. Cambridge University Press, Cambridge/New York
44. Henkel R, Endler L, Peters A, Le Novère N, Waltemath D (2010) Ranked retrieval of computational biology models. BMC Bioinform 11(1):423
45. Hines M, Morse T, Migliore M, Carnevale N, Shepherd G (2004) Modeldb: a database to support computational neuroscience. J Comput Neurosci 17(1):7–11
46. Hoehndorf R, Dumontier M, Gennari JH, Wimalaratne S, de Bono B, Cook DL, Gkoutos GV (2011) Integrating systems biology models and biomedical ontologies. BMC Syst Biol 5(1):124
47. Hucka M, Bergmann F, Keating S, Schaff J, Smith L (2010) The systems biology markup language (SBML): language specification for level 3 version. http://sbml.org/Documents/Specifications/SBML_Level_3/Version_1/Core
48. Ide NC, Loane RF, Demner-Fushman D (2007) Essie: a concept-based search engine for structured biomedical text. J Am Med Inform Assoc 14(3):253–263
49. Inmon W (2005) Building the data warehouse, 4th edn. Wiley, Indianapolis
50. Jaiswall P, Ware D, Ni J, Chang K, Zhao W, Schmidt S, Pan X, Clark K, Teytelman L, Cartinhour S, Stein L, McCouch S (2002) Gramene: development and integration of trait and gene ontologies for rice. Comparative and Functional Genomics 3(2):132–136
51. Juty N, Le Novère N, Laibe C (2012) Identifiers.org and miriam registry: community resources to provide persistent identification. Nucl Acids Res 40(D1):D580–D586
52. Kanz C, Aldebert P, Althorpe N, Baker W, Baldwin A, Bates K, Browne P, van den Broek A, Castro M, Cochrane G, Duggan K, Eberhardt R, Faruque N, Gamble J, Diez FG, Harte N, Kulikova T, Lin Q, Lombard V, Lopez R, Mancuso R, McHale M, Nardone F, Silventoinen V, Sobhany S, Stoehr P, Tuli MA, Tzouvara K, Vaughan R, Wu D, Zhu W, Apweiler R (2005) The EMBL nucleotide sequence database. Nucl Acids Res 33(suppl_1):D29–D33
53. Kasprzyk A (2011) Biomart: driving a paradigm change in biological data management. Database 2011:bar049
54. Kimball R (1998) Bringing up supermarts – a step-by-step approach to building a data warehouse from granular data. DBMS and Internet Syst 11(1):47–53
55. Kitano H (2002) Systems biology: a brief overview. Science 295:1662–1664
56. Krallinger M, Valencia A, Hirschman L (2008) Linking genes to literature: text mining, information extraction, and retrieval applications for biology. Genome Biol 9(Suppl 2):S8
57. Krause F, Uhlendorf J, Lubitz T, Schulz M, Klipp E, Liebermeister W (2010) Annotation and merging of SBML models with semanticsbml. Bioinformatics 26(3):421–422
58. Lacroix Z, Critchlow T (2003) Bioinformatics: managing scientific data. Morgan Kaufmann, San Francisco

59. Laibe C (2011) Identifiers. org and miriam registry: perennial identifiers for crossreferencing purposes. Available from Nature Precedings. http://dx.doi.org/10.1038/npre.2011.6479.1
60. Lange M, Spies K, Bargsten J, Haberhauer G, Klapperstück M, Leps M, Weinel C, Wünschiers R, Weißbach M, Stein J, Scholz U (2010) The LAILAPS search engine: relevance ranking in life science databases. J Integr Bioinform 7(2):e110
61. Langville AN, Meyer CD (2006) Google's PageRank and beyond: the science of search engine rankings. Princeton University Press, Princeton
62. Lassila O, Swick RR, Consortium WWW (1998) resource description framework (RDF) model and syntax specification. http://www.w3.org/1998/10/WD-rdf-syntax-19981008
63. Lee T, Pouliot Y, Wagner V, Gupta P, Stringer-Calvert D, Tenenbaum J, Karp P (2006) BioWarehouse: a bioinformatics database warehouse toolkit. BMC Bioinform 7(1):170
64. Le Novère N, Finney A, Hucka M, Bhalla U, Campagne F, Collado-Vides J, Crampin E, Halstead M, Klipp E, Mendes P et al (2005) Minimum information requested in the annotation of biochemical models (MIRIAM). Nat Biotechnol 23(12):1509–1515
65. Le Novère N, Courtot M, Laibe C (2006) Adding semantics in kinetics models of biochemical pathways. In: Proceedings of the 2nd international symposium on experimental standard conditions of enzyme characterizations, Ruedesheim
66. Li C, Donizelli M, Rodriguez N, Dharuri H, Endler L, Chelliah V, Li L, He E, Henry A, Stefan M et al (2010) Biomodels database: an enhanced, curated and annotated resource for published quantitative kinetic models. BMC Syst Biol 4(1):92
67. Lloyd C, Lawson J, Hunter P, Nielsen P (2008) The cellmL model repository. Bioinformatics 24(18):2122–2123
68. Lu Z (2011) PubMed and beyond: a survey of web tools for searching biomedical literature. Database 2011:baq036
69. Magrane M, UniProt Consortium (2011) UniProt Knowledgebase: a hub of integrated protein data. Database 2011:bar009
70. Marchionini G (2006) Exploratory search: from finding to understanding. Commun ACM 49(4):41–46
71. Marenco L, Tosches N, Crasto C, Shepherd G, Miller P, Nadkarni P (2003) Achieving evolvable web-database bioscience applications using the EAV/CR framework: recent advances. J Am Med Inform Assoc 10(5):444–453
72. Margulies M, Egholm M, Altman WE, Attiya S, Bader JS, Bemben LA, Berka J, Braverman MS, Chen YJ, Chen Z et al (2005) Genome sequencing in microfabricated high-density picolitre reactors. Nature 437(7057):376–380
73. Maxam A, Gilbert W (1977) A new method for sequencing DNA. Proc Natl Acad Sci 74(2):560–564
74. Mehlhorn H, Lange M, Scholz U, Schreiber F (2012) IDPredictor: predict database links in biomedical database. J Integr Bioinform 9(2):e190
75. Murray-Rust P, Rzepa H (1999) Chemical markup, XML, and the World Wide Web. 1. Basic principles. J Chem Inf Comput Sci 39(6):928–946. http://www.xml-cml.org
76. Nolin MA, Ansell P, Belleau F, Idehen K, Rigault P, Tourigny N, Roe P, Hogan JM, Dumontier M (2008) Bio2RDF network of linked data. In: Semantic web challenge; international semantic web conference (ISWC 2008), Karlsruhe
77. O'Connor B, Day A, Cain S, Arnaiz O, Sperling L, Stein L (2008) Gmodweb: a web framework for the generic model organism database. Genome Biol 9(6):R102
78. Olivier B, Snoep J (2004) Web-based kinetic modelling using JWS online. Bioinformatics 20(13):2143–2144
79. Pearson W, Lipman D (1988) Improved tools for biological sequence comparison. Proc Natl Acad Sci USA 85:2444–2448
80. Prud'hommeaux E, Seaborne A (2008) SPARQL query language for RDF. http://www.w3.org/TR/rdf-sparql-query/
81. Richardson M, Prakash A, Brill E (2006) Beyond pagerank: machine learning for static ranking. In: WWW'06: proceedings of the 15th international conference on World Wide Web, Edinburgh. ACM, New York, pp 707–715

82. Roos DS (2001) Bioinformatics-trying to swim in a sea of data. Science 291(5507):1260–1261
83. Saake G, Heuer A (1999) Datenbanken: Implementierungstechniken, 1st edn. MITP, Bonn
84. Sanger F, Nicklen S, Coulson AR (1977) DNA sequencing with chain-terminating inhibitors. Proc Natl Acad Sci 74(12):5463–5467
85. Schadt E, Linderman M, Sorenson J, Lee L, Nolan G (2010) Computational solutions to large-scale data management and analysis. Nat Rev Genet 11(9):647–657
86. Schena M, Shalon D, Davis RW, Brown PO (1995) Quantitative monitoring of gene expression patterns with a complementary DNA microarray. Science 270(5235):467–470
87. Schmitt I (1998) Schemaintegration für den Entwurf Föderierter Datenbanken. infix, Sankt Augustin
88. Schöch V (2001) Die Suchmaschine Google. Seminararbeit, Institut für Informatik, Freie Universität zu Berlin
89. Schönsleben P (2001) Integrales Informationsmanagement: Informationssysteme für Geschäftsprozesse – Management, Modellierung, Lebenszyklus und Technologie, 2nd edn. Springer, Berlin/Heidelberg
90. Schuler GD, Epstein JA, Ohkawa H, Kans JA (1996) Entrez: molecular biology database and retrieval system. In: Doolittle RF (ed) Computer methods for macromolecular sequence analysis. Methods in enzymology, vol 266. Academic, San Diego, pp 141–162
91. Schulz M, Krause F, Le Novère N, Klipp E, Liebermeister W (2011) Retrieval, alignment, and clustering of computational models based on semantic annotations. Mol Syst Biol 7(1):512
92. Shah S, Huang Y, Xu T, Yuen M, Ling J, Ouellette BFF (2005) Atlas – a data warehouse for integrative bioinformatics. BMC Bioinform 6(1):34
93. Siegel J (1996) CORBA fundamentals and programming. Wiley, New York
94. Siple MD (1998) The complete guide to Java database programming with JDBC. McGraw-Hill, New York/London
95. Smedley D, Haider S, Ballester B, Holland R, London D, Thorisson G, Kasprzyk A (2009) BioMart – biological queries made easy. BMC Genomics 10(1):22
96. Smith B, Ashburner M, Rosse C, Bard J, Bug W, Ceusters W, Goldberg L, Eilbeck K, Ireland A, Mungall C et al (2007) The OBO foundry: coordinated evolution of ontologies to support biomedical data integration. Nat Biotechnol 25(11):1251–1255
97. Stein L (2010) The case for cloud computing in genome informatics. Genome Biol 11(5):207
98. Stephens SM, Chen JY, Davidson MG, Thomas S, Trute BM (2005) Oracle database 10 g: a platform for BLAST search and regular expression pattern matching in life sciences. Nucl Acids Res 33(suppl_1):D675–D679
99. Taylor C, Field D, Sansone S, Aerts J, Apweiler R, Ashburner M, Ball C, Binz P, Bogue M, Booth T et al (2008) Promoting coherent minimum reporting guidelines for biological and biomedical investigations: the MIBBI project. Nat Biotechnol 26(8):889–896
100. United States National Library of Medicine (2011) Pubmed celebrates its 10th anniversary. http://www.nlm.nih.gov/pubs/techbull/so06/so06_pm_10.html
101. Valencia A (2002) Search and retrieve: large-scale data generation is becoming increasingly important in biological research. But how good are the tools to make sense of the data? EMBO Rep 3(5):396–400
102. Waltemath D, Henkel R, Winter F, Wolkenhauer O (2013) Reproducibility of model-based results in systems biology. In: Prokop A, Csukás B (eds) Systems biology: integrative biology and simulation tools. Springer, Dordrecht
103. Weiner M, Hudson T (2002) Introduction to SNPs: discovery of markers for disease. Biotechniques 32(Supplement):S4–S13
104. Weise S, Grosse I, Klukas C, Koschützki D, Scholz U, Schreiber F, Junker B (2006) Meta-all: a system for managing metabolic pathway information. BMC Bioinform 7(1):e465
105. Whetzel PL, Parkinson H, Causton HC, Fan L, Fostel J, Fragoso G, Game L, Heiskanen M, Morrison N, Rocca-Serra P, Sansone SA, Taylor C, White J, Stoeckert CJ (2006) The MGED ontology: a resource for semantics-based description of microarray experiments. Bioinformatics 22(7):866–873

106. Whetzel P, Noy N, Shah N, Alexander P, Nyulas C, Tudorache T, Musen M (2011) BioPortal: enhanced functionality via new web services from the national center for biomedical ontology to access and use ontologies in software applications. Nucl Acids Res 39(suppl_2):W541–W545
107. Wiederhold G (1996) Intelligent integration of information – foreword. J Intell Inf Syst 6(2/3):93–98
108. Wiederhold G (1997) Mediators in the architecture of future information systems. In: Huhns MN, Singh MP (eds) Readings in agents. Morgan Kaufmann, San Francisco, pp 185–196
109. Yu T, Lloyd C, Nickerson D, Cooling M, Miller A, Garny A, Terkildsen J, Lawson J, Britten R, Hunter P et al (2011) The physiome model repository 2. Bioinformatics 27(5):743–744

Chapter 4
Data Warehouses in Bioinformatics

Benjamin Kormeier

Abstract The progress in the area of biological research in recent years has led to a multiplicity of different databases and information systems. Molecular biology deals with complex problems and an enormous amount of versatile data will be produced by high-throughput techniques. Hence, the total number of databases, as well as the data itself, is continuously increasing, and with it the distribution and heterogeneity of the data rises. The importance of database integration has been recognized for many years. Therefore, this chapter presents the problems in database integration as well as a small selection of well-known existing integration systems which have been developed. Finally, this chapter presents an in-house data warehouse approach for biological data. Integrated data is the basis for network analysis, reconstruction, and visualization.

4.1 Introduction

One of the main challenges in bioinformatics which began with research for the Human Geome Project, is the integration of molecular data. Currently, high throughput analysis delivers data of complete genomes, for instance short sequences of all genes in an organism or thousands of expression patterns of a cell in shortest time. Analysis of this high-throughput data by manual investigation using publications or relevant databases is no longer possible. Consequently, the biologist is supported by tools and methods that can accumulate experimental data with complementary data sources, estimate the data and compare or classify the data.

B. Kormeier (✉)
Bioinformatics/Medical Informatics Department, Bielefeld University, Universitätsstraße 25,
D-33501 Bielefeld, Germany
e-mail: bkormeie@techfak.uni-bielefeld.de

4.2 Problems in Database Integration

Molecular biological data has a high *semantic heterogeneity* that is usually caused by experimental data extracted from a series of experiments. Molecular biology deals with complex problems. Therefore, enormous amounts of versatile data will be produced by experiments. The total number of databases, and of course the data itself, is continuously increasing. Hence, the distribution and heterogeneity of the data rises as well. Particularly, data heterogeneity is one of the main problems in molecular biological data integration. Furthermore, *technical heterogeneity* is caused by a high number of different formats and interfaces of the different data sources. The data is mostly not available in a standard format which causes *structural heterogeneity*. Missing standards and consensus for basic biological terms produces semantic heterogeneity. Beside this problem, there are some more problems in data integration. In the next sections, basic problems of data integration in the field of *distribution*, *autonomy*, and *heterogeneity* will be discussed. Leser and Naumann define those fields as the so-called orthogonal dimensions of data integration [16]. For this reason, in each dimension problems can occur independently.

4.2.1 Distribution

Usually, data sources of an integrated system are distributed. That means each and every source is located on a separate system and/or different locations. It will be distinguished between *physical* and *logical distributions*. Physical distribution is data that is physically and geographically organized on different distributed systems. The following problems can be caused by physical distribution in data integration: localization of data, data which is represented in multiple schemata, and the optimization of distributed queries. The concept of data warehousing can solve these problems of physical distribution. Data warehouses will be discussed later in Sect. 4.3.4.

Homogeneous data of a system that is located at different logical places leads to logical distribution. This means the system is redundant and several problems can occur. The localization of this data is very difficult and ambiguous. For instance, if a user has problems tracking the origin of the data. A possible solution could be to provide metadata, for example, a global schema. Additionally, duplicates and conflicts can occur with logical distribution. The system has to be identified in order to fix these problems to guarantee consistent data.

4.2.2 Autonomy

The distribution of several data sources leads automatically to the problem of autonomy. Autonomy in data integration means the independence of the data source that refers to access, configuration, development, and administration. Overall, autonomy can be divided into the following four types [16]: design autonomy, interface autonomy, access autonomy, and legal autonomy. A data source has *design autonomy* when it has the freedom to decide how its data can be provided and represented. This autonomy is also related to the data model, schema, and transaction management or if a data source has the freedom to define the method of access. For instance, defining a protocol for the query language of the system is called *interface autonomy*. Interface autonomy is strongly related to design autonomy, because the storage of data typically determines the data access. If the system is able to decide who can access which data, then the data source is access autonomous. Legal autonomy is achieved if the integration of a resource is prohibited. Additional kinds of autonomy can be found in [7].

4.2.3 Heterogeneity

The major problem of data integration is heterogeneity that is caused by autonomy. Distribution can also cause heterogeneity, but not in general. For instance, two information systems that have identical methods, but do not provide identical models and structures for data access, are called heterogeneous. Leser and Naumann enumerated different kinds of heterogeneities: technical heterogeneity, syntactic heterogeneity, data model heterogeneity, structural heterogeneity, schematic heterogeneity, and semantic heterogeneity [16].

Technical heterogeneity is the implementation of different access methods to the data source. This kind of problem is solved if the integrated system is able to query the data source and the request returns a correct result set. Different representations of the same issue are called *syntactic heterogeneity*. Different character encodings in a data set are good examples of syntactic heterogeneity. This problem can easily be solved by converting the data into a common format. *Data model heterogeneity* exits, if data sets of a data source can be managed by different data models. One data model is, for instance, object oriented and another one is relational. Hence, if both data models are equivalent, then a data model heterogeneity is nonexistent. Design autonomy often causes structural, schematic, and semantic heterogeneity in data integration. Structural differences in the representation of data are called *structural heterogeneity*. A special case of structural heterogeneity is *schematic heterogeneity*, where different concepts of a data model describe the same issue or data. Semantic heterogeneity characterizes the differences in sense, interpretation, types of terms, and concepts. In particular, synonyms and homonyms play a major role in these conflicts. These problems can be solved if schema elements have the same meaning and an identical name.

4.3 Approaches of Database Integration

The development of an integrated database system is a complex task, if a large number of heterogeneous databases have to be integrated. A blueprint of the architecture of the integrated system is essential for success. In general, two architectures for data integration exist. They are divided into *materialized integration* and *virtual integration*. Materialized integration stores the whole data set of source persistent in a database management system (DBMS). Periodic update strategies transfer updated data and extensions to the global system. Then, the integration system has to normalize the data and duplicates and failures have to be removed. Finally, the central database will be updated to provide an up-to-date data set. Materialized integration has the advantage of high velocity, because there is no communication between different data sources, as well as no restriction of queries, which could be the case in virtual integration systems.

Virtual integration does not store data in a persistent way. Usually, the data is located on different local systems and queried by a global schema. A complex normalization and transformation process is not necessary, as compared to materialized integration. Queries are managed by a global schema, while the underlying data is "virtually" available. The main task is to generate complex queries to get, transform, and aggregate adequate data from different data sources. If data sources provide only restricted interfaces it is a problem of virtual integration and queries of the global schema cannot be answered or executed.

Major approaches of database integration in bioinformatics have been discussed and reviewed in the last years.

- *Hypertext navigation systems.* HTML frontends linked to molecular biological databases.
- *Federated database systems and mediator-based systems* are virtual integration systems. They do not store any data in a global schema. Federated systems integrate multiple autonomous database systems into a virtual single federated database. Typically, each database is interconnected via a computer network or in some cases the World Wide Web. Hence, the databases can be geographically decentralized.
- *Multi-database systems* do not have a global schema. These systems interactively generate queries for several databases simultaneously.
- *Data warehouses* are materialized integration approaches. They store data persistent in a global data repository, which is typically a relational DBMS.

All these approaches have the same aim: providing techniques to overcome several kinds of heterogeneous data and to provide a retrieval system for scientists to support their research activities and experiments.

4.3.1 Hypertext Navigation Systems

Nowadays, most databases are connected to the World Wide Web and can be accessed with a common browser. Usually, many of these databases provide links to other databases. Accession numbers (AC), unique identifiers, or other database identifiers are used for linking database entries. Actually, many databases use different identifiers or terms for the same entries; hence, interlinking databases is a difficult task. Therefore, pair-wise or binary mappings between database entries have to be generated to provide links between different databases. Mostly, databases only provide cross-links with the most relevant databases.

Many other database attributes can be used for linking databases to each other, for instance, EC numbers, CAS (Chemical Abstracts Service) registry numbers, GO (Gene Ontology) terms, or other controlled vocabularies. Often, databases are not linked with each other, even if they use the same controlled vocabulary. However, it is not possible to link all the databases with each other. One reason is that providers are not aware of all other relevant databases. Nevertheless, interlinked web sites are a common way of database "integration".

4.3.2 Multi-database Systems

Multi-database systems are usually a network of database systems [7]. The management of the whole data set is not controlled by the overall system. Independent partitions control the data. Therefore, the user has access to the different data sources using a common query language. Examples of conflicts of integration offered by the provided query language are multiple redundant data, structural differences between data sources or semantic heterogeneities.

Systems are called federated database systems if data sources maintain a certain level of autonomy. In contrast, a central system takes control of data sets when the system is no longer federated. It is not exactly defined by which level of autonomy the border between federated database systems and multi-database systems is arranged.

In a multi-database system, the schemata are divided into the internal layer, the conceptual layer, and the external layer. The user has to define a view for the required data. A query spanning multiple databases is specified by the multi-database query language. In a central unit, the query is fractionalized and sent to the different databases. The result sets are sent to the processing unit and returned as a merged result to the user.

4.3.3 Federated Database Systems

A *federated database system* (FDBS) consists of multiple cooperating component systems that are autonomous. Moreover, it consists of a federated database management system that controls the component systems. Federated architectures differ based on levels of integration with the component database systems. Furthermore, they differ in the services offered by the federation, as well as in the extension of the systems. A detailed definition was given by Sheth and Larson [19].

Federated database systems can be categorized as loosely or tightly coupled systems depending on the level of coupling. In a loosely coupled FDBS, each user is the administrator of his own federated schema. Each user is responsible for understanding the semantics of objects in the export schemata. Users are also responsible for the elimination of heterogeneities from the DBMS. Finally, users are able to store their schema under their own accounts. A schema can be deleted at any time by the user [19].

In tightly coupled FDBS, export schemata are created between the component database administrator and federation database administrator. Usually, the component database administrator has control over the export schemata, while the federation database administrator has the authority to read the database to help determine what data is available and where it is located. The federation database creates and controls the federated schemata [19].

4.3.4 Data Warehouse Systems

Data warehouses (DWH) are the widely used architectures of materialized integration in informatics and especially in bioinformatics. Basically, data warehouses are used in the field of information management. Data analysis, data mining, and the long-term storage of business intelligence in companies are the major advantages of data warehouse systems. In bioinformatics DWHs are usually used for data integration. DWHs are often preferred in contrast with virtual integration approaches, which have some disadvantages: no write access, poor speed of request handling, problems in availability of data sources, and complexity of queries.

A general definition of a DWH was defined by Bauer and Günzel: "A data warehouse is a physical database that provides an integrated view of arbitrary data for analysis." [1]. A DWH cannot be assigned to classical OLTP (online transaction processing) systems, which are optimized for fast and reliable transaction handling. Typically, most of the OLTP interactions are involved in a relatively small number of rows a larger group of tables, by comparison with data warehouse systems. DWHs are assigned to OLAP (online analytical processing) systems, which are able to quickly answer multidimensional analytical queries. OLAP systems can be classified into the category business intelligence, which also includes relational reporting and data mining. Usually, in DWH, new data will be added; already stored data will not be manipulated or overwritten.

Typically, a data warehouse process is divided into four phases:

1. Data from different resources will be obtained. This means the data will be extracted and transformed. This phase is called *ETL process (extraction-transform-load process)*.
2. The data will be saved persistently in the DWH.
3. The separated data will be divided into several data marts, if necessary.
4. The data of the DWH or data marts will be analyzed. Finally, the data will be provided to external applications.

4.4 Data Warehouses

In this section, relevant and widely used data integration approaches and data sources in the field of bioinformatics will be introduced. Usually, the data is distributed in multiple data sources. Those sources differ in the biological context, internal representation, used underlying systems, access possibility, and complexity.

Relevant integration approaches in the context of biological data warehouses will be introduced. The focus lies on the data warehouse approaches Atlas, Columba, BioWarehouse, and CoryneRegNet, because the systems are equipped with important requirements in biological data integration that are relevant for our data warehouse approach that will be discussed in the next section. Moreover they are well-known examples in the literature of data integration approaches in bioinformatics. Additionally, in Table 4.1, all data integration approaches will be compared. Generally, in bioinformatics, integration approaches can be divided into four classes [16]:

- *Indexing systems:* e.g., SRS (Sequence Retrieval System) [8], Entrez, and BioRS
- *Multi-databases:* e.g., OPM (Object Protocol Model) [4] and DiscoveryLink [9]

Table 4.1 Comparison of different data warehouse approaches. Advantages are marked in bold letters

Property	Atlas	BioWarehouse	Columba	CoryneRegNet
Integration	Tightly coupled	Tightly coupled	Loosely coupled	Tightly coupled
DBMS	MySQL	MySQL, Oracle	PostgreSQL	PostgreSQL
Programming language	**Java, C++, Perl**	Java, C	Python, Perl	PHP, Java
Architecture	Application	Application	**Web interface**	**Web interface**
Platform independence	No (Unix systems)	No (Linux systems)	Yes	Yes
Updates	Manually	Manually	Old	Unknown
Maintenance and development	Unknown	Periodical	Project ended	Periodical
License	**GNU**	MPL	Freely available on request	AFL
Open source	**Yes**	**Yes**	Unknown	**Yes**

- *Ontology-based integration:* e.g., TAMBIS (Transparent Access to Multiple Bioinformatics Information Sources) [20] and ONDEX [12]
- *Data warehouse:* e.g., Atlas [18], BioWarehouse [15], Columba [22], CoryneRegNet [2], ONDEX [12], and SYSTOMONAS [6]

To integrate medical and molecular biological data, the first step is to structure and evaluate the amount of available data sources. A comparison and choice of data sources is only possible on the basis of an adequate set of criteria. Currently, the most important sources for medical and biological data, from our point of view, are KEGG, BRENDA, EMBL-Bank, ENZYME, GO, HPRD, OMIM, UniProt, and Transfac/Transpath. Based on relevant biological database and information systems, data integration is an essential step in constructing biological networks.

4.5 Related Data Integration Approaches

The data warehouse infrastructures Atlas and BioWarehouse will be introduced in this section. Both systems provide a software infrastructure that can be installed and configured locally. They give a good insight into biological data integration. Additionally, the Columba data warehouse with its web application (Sect. 4.5.3) and the ontology-based data warehouse approach CoryneRegNet (Sect. 4.5.4) will be discussed. Both systems give an insight in building web-based data warehouses that become more and more popular. Another widely used ontology-based data warehouse approach is ONDEX, which has already been introduced in Chap. 5. For that purpose, ONDEX will not be described in this section.

4.5.1 Atlas

The Atlas system was developed at UBC Bioinformatics Centre (University of British Columbia) in Canada. Atlas is freely available and is protected under the terms of the GNU General Public License. A Unix operation system is required for running the Atlas system.

The goal of Atlas is to provide data as well as a software infrastructure for bioinformatics research and development. The biological data warehouse locally stores and integrates the data from biological sequences, molecular interactions, homology information, as well as functional annotations of genes and biological ontologies.

The system architecture of Atlas consists of the *data sources*, an *ontology system*, the *relational data models*, different *APIs* (application programming interfaces), and *applications*. Figure 4.1 illustrates the system architecture of Atlas. The data sources of Atlas are categorized into four classes: *sequence, molecular interactions, gene-related resources*, and *ontology*. A complete list of the Atlas data sources could be found in [18].

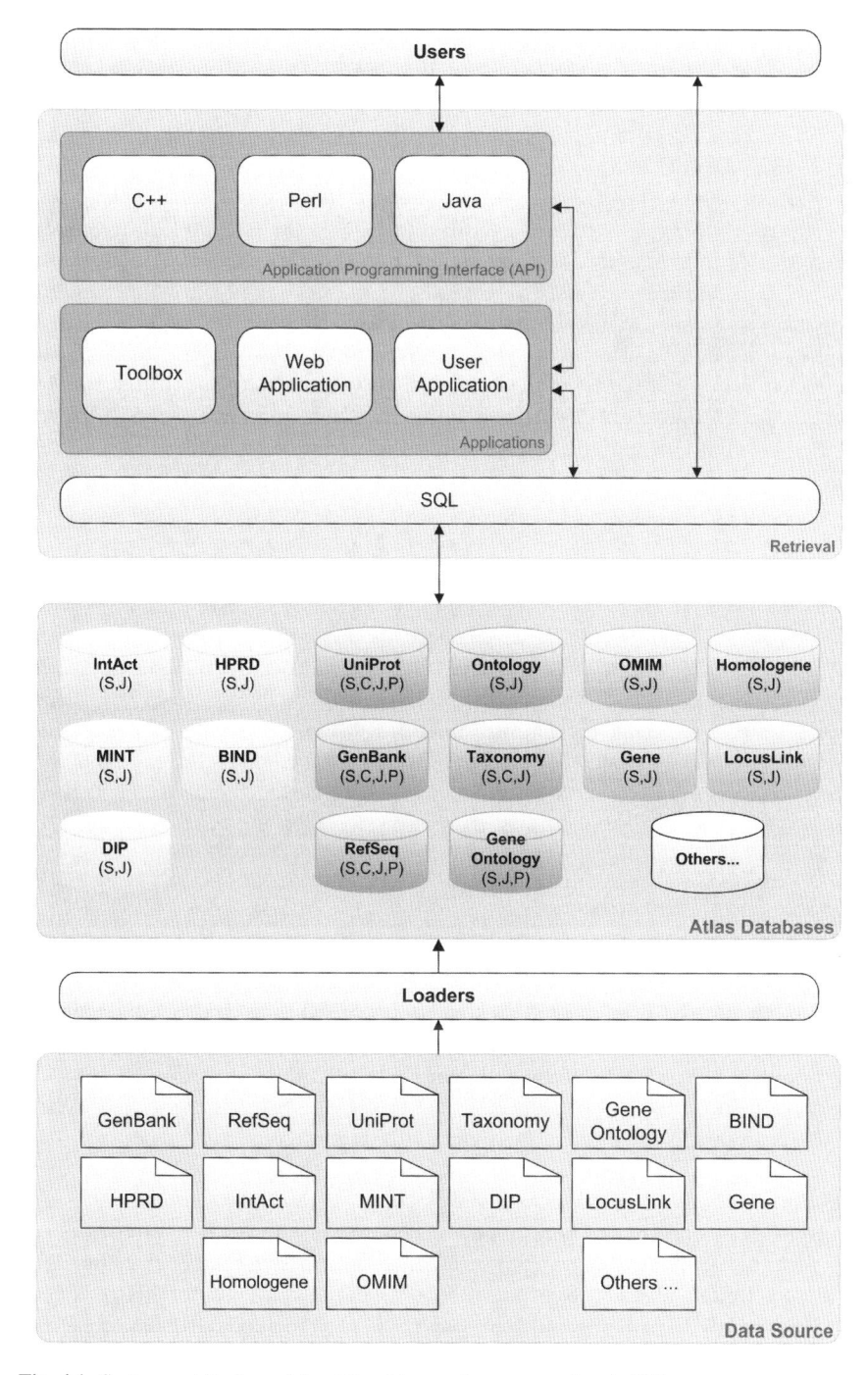

Fig. 4.1 System architecture of the Atlas data warehouse according to [18]

Each mentioned category has its own database schema in the Atlas relational database model. A complete entity relationship schema of the data warehouse approach is presented in [18]. Depending on the level of coupling, this approach can be categorized as a tightly coupled system. As relational backend, Atlas uses the MySQL DBMS.

The different APIs of Atlas are developed in three programming languages: C++, Java, and Perl. But not every API is available in the respective programming language. Furthermore, the APIs are divided into two classes: *loader* and *retrieval*. The first is the loader APIs that consists of parsers to populate the relational schemata and store the data in the Atlas databases. And the second class of APIs is the retrieval APIs. Hence, APIs allow retrieval of the data and store it in the data warehouse. Furthermore, they are necessary for developing custom retrieval applications. Moreover, Atlas provides numerous Unix command line tools, such as *ac2seq* which is able to find a sequence in FASTA format[1] on the basis of accession numbers. A user is also able to send direct SQL queries via MySQL client to the data warehouse.

Atlas is designed to run as a service on a local computer system or server. According to [18], it is also possible to access the system via the World Wide Web, although this web site is currently unavailable. In summary, Atlas has a couple of advantages: many tools, integrative database schema, short response time (because of local installation), and complete access to the database. Some disadvantages of the Atlas system have also been identified: extensive maintenance of the system, high system requirement, not platform independent, the actuality of the data depends on the user or administrator, tools are only available for command line, missing web interface, and only MySQL is supported as database management system.

4.5.2 BioWarehouse

The BioWarehouse system was developed by the Bioinformatics Research Group (SRI International), Computer Science Laboratory (SRI International), and Stanford Medical Informatics (Stanford University). It is also part of the Bio-SPICE (Biological Simulation Program for Intra- and Inter-Cellular Evaluation) project. Bio-SPICE is an open-source framework and software toolset for systems biology. BioWarehouse is an open-source toolkit that integrates different biological databases, such as ENZYME, KEGG, GO, and UniProt. The software is protected under terms of the MPL (Mozilla Public License) and is currently available in version 4.6. The software runs only under Linux operating systems, because the software is not platform independent. BioWarehouse facilitates the creation of user-defined and user-specific data warehouse instances. All available data sources for this toolkit can be found in [15]. Similar to the Atlas system, different relational database schemata

---

[1] http://www.ncbi.nlm.nih.gov/blast/fasta.shtml

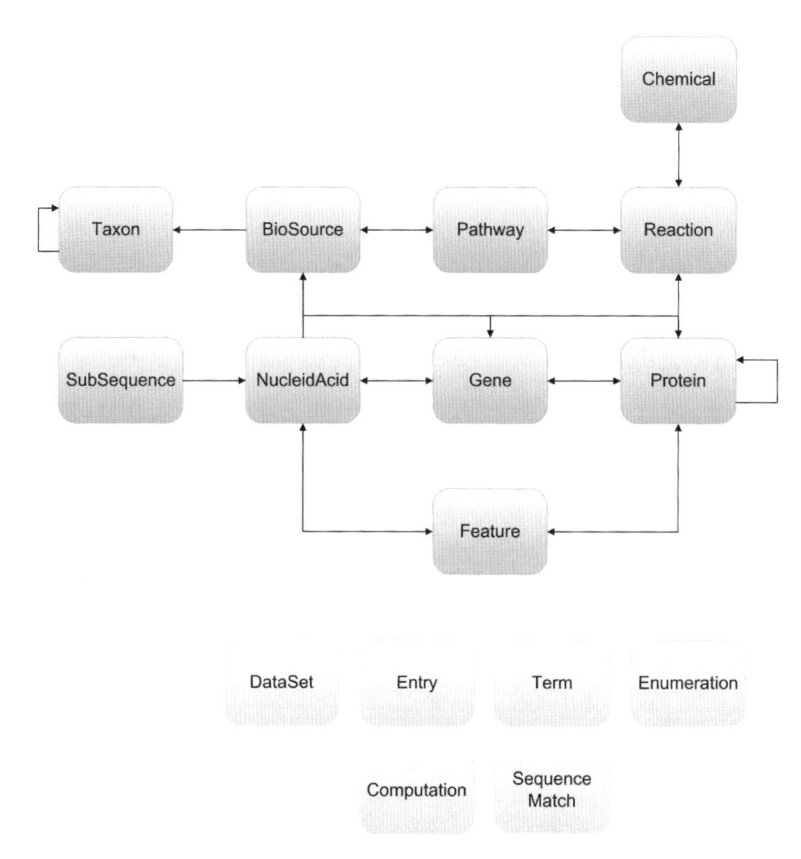

Fig. 4.2 The main data types in the BioWarehouse schema according to [15], and the relationships between them. An arc represents a connection between two data types. For instance, the data type *Gene* contains a column that references data type *Protein*

exist according to the different data types. BioWarehouse supports the MySQL and Oracle database management systems. Figure 4.2 illustrates the BioWarehouse database schema, whereas the entities symbolize the particular data type. Integration of the different data sources is realized by a specific loader. Each loader is adapted for a particular data source. The data is transformed into a consistent format, because of heterogeneities between different data sources. Afterwards the data is transferred into the database schema. Loaders have been implemented in the programming languages C and Java. A unique feature of the BioWarehouse loaders is the error tolerance during integration. In case an error occurs, the integration process will be finished and incorrect data sets will be marked. Moreover, BioWarehouse provides a set of Java utility classes that are useful for developers who want to construct their own loaders or applications.

The BioWarehouse system can be used in two different ways: First, publicly available versions of BioWarehouse called PublicHouse and EcoliHouse which are available via the Internet. In addition, a user has to register to access public servers of BioWarehouse. Access to the system occurs via a MySQL client, such as phpMyAdmin. In this way the user is able to manage the data sources and have not only read access. The advantages of the system are an integrative database schema, short response time (because of local installation), full access to the database, and MySQL and Oracle supported. On the other hand, the disadvantages of this data warehouse approach are, extensive maintenance of the system, high system requirements, it is the administrators are responsible for keeping the data up-to-date, a missing web interface and SQL knowledge is required.

4.5.3 Columba

The Columba data warehouse system was developed by the Department of Computer Science, Humboldt-Universität zu Berlin, Department of Biochemistry, Charité Universitätsmedizin Berlin, Zuse Institute Berlin, and University of Applied Sciences Berlin. Columba integrates data from 12 heterogeneous biological data sources in the field of protein structures and protein annotations. Proteins are the most important aspect in Columba. The protein structure from the database PDB plays a major role in the system. Furthermore, the PDB data is extracted and enriched with additional information such as protein sequence, protein function and involvement in biological networks, as well as membership of protein families. Typically, several data sources are provided in different exchange formats, as described in Sect. 4.2.3. If parsers from other projects, such as BioSQL and BioPython, are not available for those formats, new parsers has to be developed. Therefore, the programming languages Python and Perl were used. Columba uses PostgreSQL as relational backend to manage the data. The Columba database schema, as shown in Fig. 4.3, is comparable with a star schema. Hence, it is clear that the core of Columba is the protein structures of PDB. Information related to the protein structures are included into specific sub-schemata that comes from different sources that are arranged around the main table. Each data source is modeled as a different dimension and has its own sub-schema within the overall schema of Columba. The Columba data warehouse approach has the advantage of simple system maintenance, intuitive query model, and high recognition value of the information.

Columba is accessible via a web interface which uses a common browser. The web interface allows full-text search as well as attribute specific searches. The full-text search engine is implemented with the extension Tsearch2 in PostgreSQL. Furthermore, the search engine supports Boolean operators to link keywords with each other.

In summary, Columba integrates different data sources for proteins which simplify research in this area. On the other hand, Columba does not provide comprehensive knowledge about molecular biology. This is a crucial disadvantage. Redundancies could not be excluded, because every data source is an independent dimension in the model.

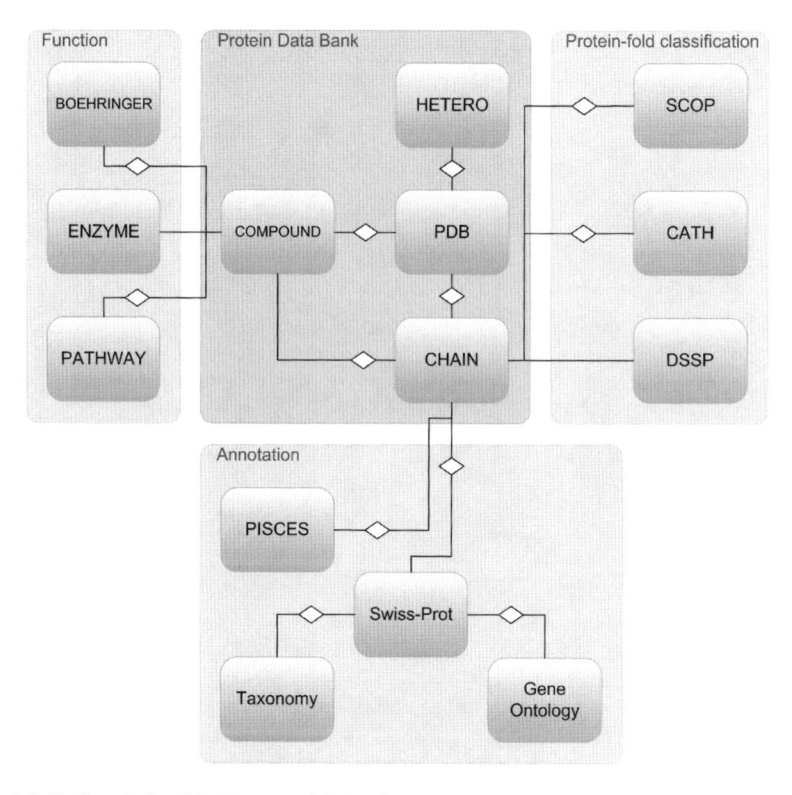

Fig. 4.3 Entity relationship diagram of Columba according to [22]

4.5.4 CoryneRegNet

The CoryneRegNet system was developed at the Center for Biotechnology (CeBiTec), Bielefeld University. CoryneRegNet (*Coryne*bacterial Transcription Factors and *Reg*ulatory *Net*works) is an ontology-based data warehouse approach which provides data about transcription factors and gene regulatory networks. The system provides data about all recently sequenced corynebacteria and model organism *Escherichia coli*, whereas the tool focuses on corynebacterium [17]. Figure 4.4 illustrates the system architecture of CoryneRegNet. Different data sources are transformed by parsers into a consistent object-oriented ontology-based data structure. The data structures of CoryneRegNet are closely related to ONDEX, as described in Chap. 5. Finally, the ontology-based data structures are transformed into a relational database model, because CoryneRegNet uses a MySQL database as relational backend. Figure 4.5 shows the entity relationship diagram of CoryneRegNet that consists of *generalized data structure* and *ontological data structure*. All essential data such as genes and proteins are stored into the ontological data structure.

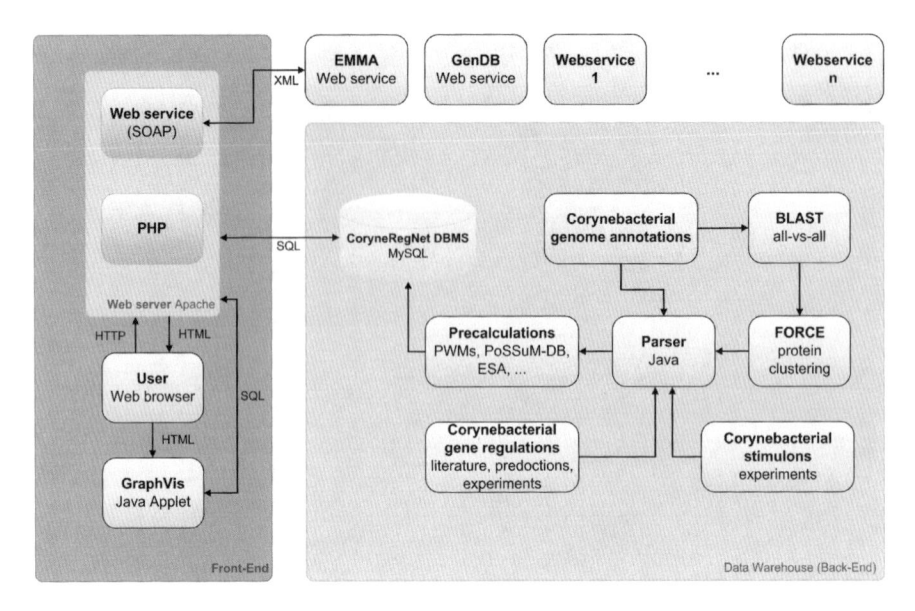

Fig. 4.4 System architecture of CoryneRegNet according to [2]

The CoryneRegNet application is accessible via a web service that is illustrated in Fig. 4.4. Using this web service, it is possible to access additional and up-to-date data dynamically. Overall, a web service can also have disadvantages in terms of performance and security. Performance delays can occur because of high traffic.

CoryneRegNet is developed using PHP and Java programming language and has a MySQL database management system as relational backend. The web interface provides several search and analysis possibilities. Furthermore, transcription factors and regulatory networks will be visualized in a Java applet.

In summary, CoryneRegNet has a user-friendly and intuitive web application, and other applications can access CoryneRegNet via web services and provide additional information. However, a Java installation is required to visualize networks in a Java applet. A disadvantage of the Java applets are longer loading times as well as the web service. Similar to the Columba, CoryneRegNet is limited to bacteria as it does not have general molecular knowledge.

4.5.5 Summary

In this section, the data integration approaches Atlas, BioWarehouse, Columba, and CoryneRegNet were discussed. All projects use the data warehouse technique for data integration. Atlas and BioWarehouse provide a software infrastructure for data integration. In contrast, Columba and CoryneRegNet provide a web interface, and therefore they are directly useable. Only CoryneRegNet additionally provides a web

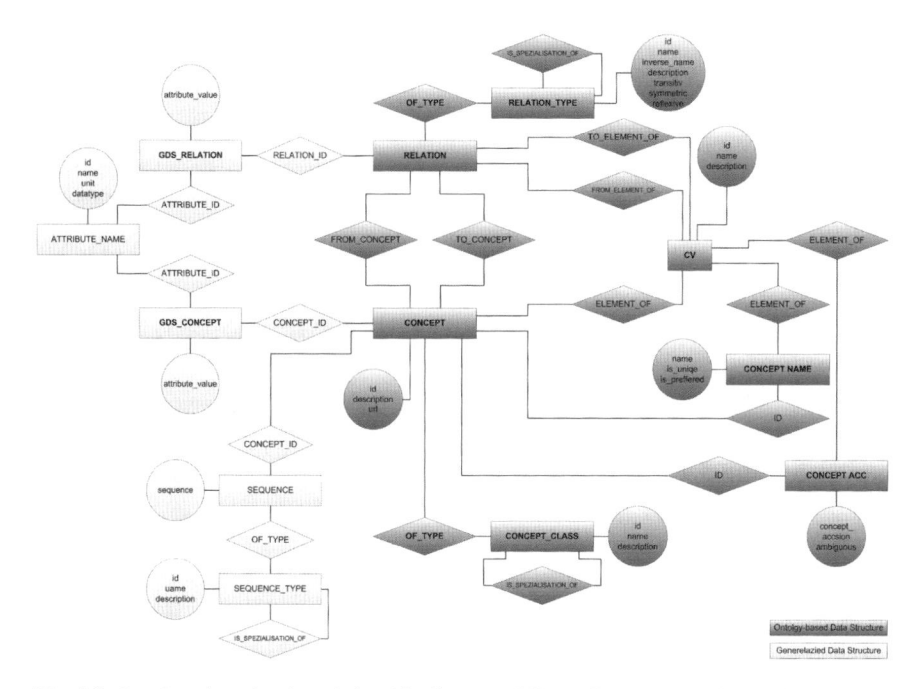

Fig. 4.5 Ontology-based entity relationship diagram of CoryneRegNet according to [3]

service. In Table 4.1 a comparison of the data warehouse systems on the basis of different criteria is given. The advantages and disadvantages of the particular software solution are marked. Today the Atlas and Columba projects are no longer funded and will therefore cease to exist.

Platform independence and a user-friendly system are important criteria for integration approaches. Typically, web applications are preferred as they are very flexible and interactive. The term *Web 2.0* characterizes flexible and interactive web applications. Atlas, Columba, and CoryneRegNet support only one specific DBMS. In comparison, BioWarehouse provides a choice between the MySQL and Oracle database management systems. Columba and CoryneRegNet only provide a restricted access to the database, but provide suitable query forms. In contrast, Atlas and BioWarehouse support a complete database access.

4.6 BioDWH and DAWIS-M.D.

On the basis of the previously discussed problems in data integration and the advantages/disadvantages of data warehouses approaches in bioinformatics, we want to introduce the data warehouse infrastructure BioDWH [21] as well as our web-based data warehouse information system DAWIS-M.D. [10]. The BioDWH infrastructure is the basis for the DAWIS-M.D. information system. However,

BioDWH is a Java implemented DWH infrastructure that uses a relational database management systems, such as MySQL or Oracle, as backend. A unique feature of BioDWH is a graphical user interface for parsing, loading, and updating the source data into the data warehouse. Our infrastructure provides parsers for important and widely used molecular databases, such as BRENDA, EMBL-Bank, ENZYME, EPD, GO, HPRD, IntAct, iProClass, JASPAR, KEGG, MINT, OMIM, Reactome, SCOP, TRANSFAC, TRANSPATH, and UniProt. Another key feature of the system is a persistence layer that maintains the independence from the relational database management system (RDBMS) and the application logic (i.e., the parser). The Java application, object-relational mapping as a persistence method is a powerful paradigm to represent objects in a relational database system. That means a mapping between objects and metadata of the database is described. Basically, ORM works with reversible transformation of data from one representation into another. Moreover, the BioDWH infrastructure provides a plug-in architecture to include a new parser into the system. The extraction, transformation, and loading (ETL) process is implemented by the parser for the data source. Additionally, a monitor component recognizes changes within the data sources and is able to start the ETL process again. To keep the data warehouse up-to-date, updates of the data sources have to be incrementally propagated. Therefore, our system uses a timestamp-based and file-size-based monitor strategy.

Based on the BioDWH data warehouse infrastructure, the web-based data warehouse information system DAWIS-M.D. (Data Warehouse Information System for Metabolic Data) was implemented. For the information system, we developed an n-tier architecture that realizes a platform independent web application (see Fig. 4.6). A unique feature of DAWIS-M.D., in comparison with other systems as already discussed in the previous section, is the interdependence between the application logic and the RDBMS. Therefore, again the technique of object-relational mapping is used. Hibernate was used as the framework to realize the ORM within the persistence layer. As relational backend, we use a MySQL database. The following data sources are integrated into DAWIS-M.D.: BRENDA, EMBL-Bank, ENZYME, EPD, GO, HPRD, JASPAR, KEGG, OMIM, SCOP, TRANSFAC, TRANSPATH, and UniProt. BioDWH maintains the up-to-dateness of the information/data in DAWIS-M.D. The data from the databases is divided into 13 different domains: Compound, Disease, Drug, Enzyme, Gene, Gene Ontology, Genome, Glycan, Pathway, Protein, Reaction, Reactant Pair, and Transcription Factor. For each domain a specific search form with auto-complete function is available via the web application. The large number of biomedical databases and their various information contents make the acquisition of information very time consuming, inefficient, and difficult. Therefore, the web application provides an integrated view of comprehensive biomedical knowledge from different data sources. Hence, it is possible for the scientist to navigate quickly and efficiently in a large result set. Moreover, DAWIS-M.D. provides a quick and efficient search in a large data set. In most of the systems, relationships between other biological elements are not identified or clearly represented. But it is essential for scientists to understand the complex biological mechanisms and their interactions. Therefore, DAWIS-M.D. is

4 Data Warehouses in Bioinformatics

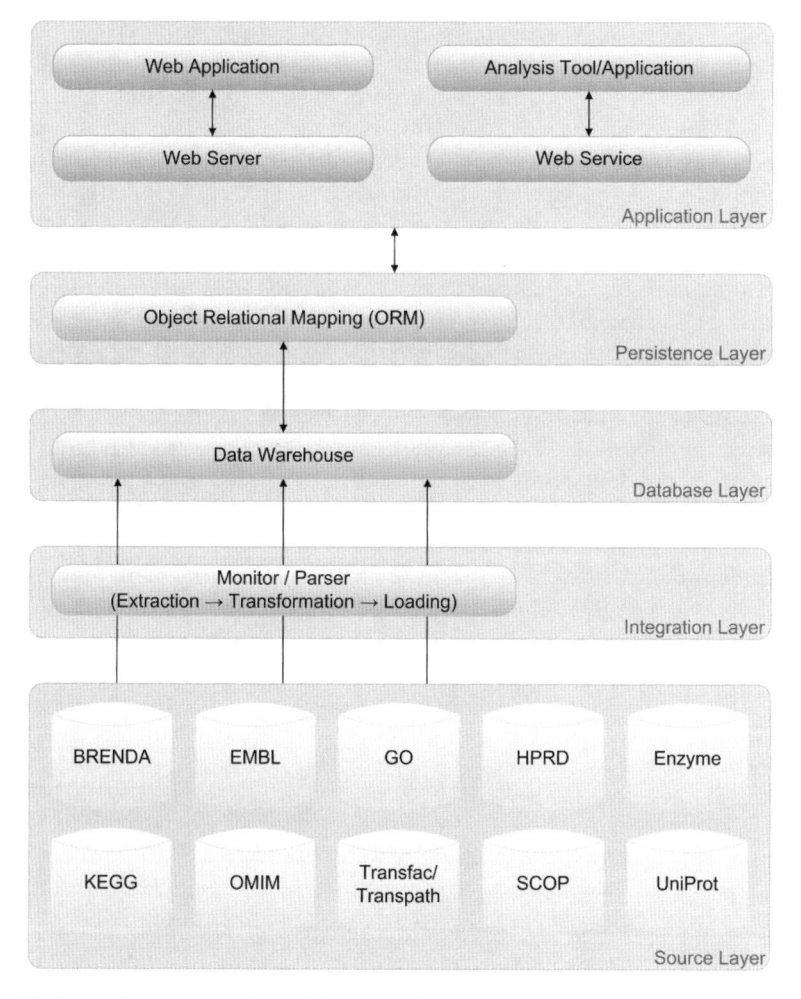

Fig. 4.6 Schematic representation of the DAWIS-M.D. n-layer system architecture from the original heterogeneous data sources to the web application layer

the identification and representation of relationships and interactions between other biological elements or mechanisms.

4.7 Reconstruction and Visualization of Biological Networks Based on Biological Data Warehousing

Using our data warehouse approach, described in the previous section, we have already shown that data integration is useful for several biological questions. We were able to build project-specific data warehouses for cardiovascular diseases [13], as well as general data warehouse systems (e.g., for metabolic data) [10]. Furthermore, it is possible using the data from BioDWH to generate and predict biological networks and enrich them with additional information [5, 14]. Using the network editor VANESA, different fields of studies are combined such as life science, database consulting, modeling, and simulation for a semiautomatic reconstruction of complex biological networks [11]. The idea of the reconstruction, visualization, and analysis of molecular networks using the network editor VANESA is described in Chap. 8.

CELLmicrocosmos 4.2 PathwayIntegration (CmPI) is an approach to visualize and analyze intercellular and intra-compartmental relationships by correlating pathways with an abstract cell environment in 3D space. By using data coming from our data warehouse, metabolic pathways from KEGG can be parsed. The pathway structure, consisting of enzymes, their substrates, and products with the connecting reactions, can be shown directly in a 3D layout within the cell. For the enzymatic localization, terms from the databases BRENDA and UniProt are used. Usually information exists on the subcellular level but also mapping information about the intra-compartmental mapping may be derived. Sometimes the localization information from the database contains comments specifying more precisely the whereabouts of a protein then the regular cell component information. In this case, CmPI uses the comment for mapping. More detailed information about visualization and analysis of intercellular and intra-compartmental relationships by correlating pathways with an abstract cell environment in 3D space can be found in Chap. 10.

4.8 Summary

One of the major challenges in bioinformatics is the integration of molecular biological information from heterogeneous, autonomous, and distributed data. Data integration has been an important research field for the past decades and will be for years to come, since the number of molecular databases is continuously increasing. It is important that scientists can analyze information from different data sources to meet their objectives. Consequently, existing data warehouse systems Atlas, Columba, BioWarehouse, and CoryneRegNet were presented in detail. On the basis of the discussed problems in data integration and the advantages/disadvantages of the presented data warehouses approaches, we introduced our own data warehouse approach DAWIS-M.D. that was constructed with the BioDWH data warehouse infrastructure. Furthermore, this system is the basis for the network modeling application VANESA and the interactive 3D cell visualization tool CELLmicrocosmos. Overall, with this chapter, we gave an overview of a set of systems that can solve the discussed problems in the field of biological data integration.

References

1. Bauer A, Günzel H (2004) Data-Warehouse-Systeme, XVI, 592 S. : graph. Darst. dpunkt-Verl., Heidelberg. 3-89864-251-8
2. Baumbach J (2007) CoryneRegNet 4.0 – a reference database for corynebacterial gene regulatory networks. BMC Bioinform 8:429
3. Baumbach J, Brinkrolf K, Czaja LF, Rahmann S, Tauch A (2006) CoryneRegNet: an ontology-based data warehouse of corynebacterial transcription factors and regulatory networks. BMC Genomics 7:24
4. Chen I-MA, Markowitz VM (1995) An overview of the object protocol model (OPM) and the OPM data management tools. Inf Syst 20(5):393–418. issn:0306-4379, doi:10.1016/0306-4379(95)00021-U, Elsevier Science, Oxford
5. Chen M, Hariharaputran S, Hofestädt R, Kormeier B, Spangardt S (2009) Petri net models for the semi-automatic construction of large scale biological networks. Nat Comput. Springer, Netherlands, 1567-7818 (Print) 1572-9796 (Online), doi:10.1007/s11047-009-9151-y
6. Choi C, Munch R, Leupold S, Klein J, Siegel I, Thielen B, Benkert B, Kucklick M, Schobert M, Barthelmes J, Ebeling C, Haddad I, Scheer M, Grote A, Hiller K, Bunk B, Schreiber K, Retter I, Schomburg D, Jahn D (2007) SYSTOMONAS–an integrated database for systems biology analysis of Pseudomonas. Nucl Acids Res 35:D533–D537
7. Conrad S (1997) Föderierte Datenbanksysteme – Konzepte der Datenintegration, XI, 331 S. : graph. Darst., Springer, Berlin [u.a.]. 3-540-63176-3
8. Etzold T, Ulyanov A, Argos P (1996) SRS: information retrieval system for molecular biology data banks. Meth Enzymol 266:114–128
9. Haas LM, Schwarz PM, Kodali P, Kotlar E, Rice JE, Swope WC (2001) DiscoveryLink: a system for integrated access to life sciences data sources. IBM Syst J 40(2):489–511. 0018-8670, IBM Corp., Riverton
10. Hippe K, Kormeier B, Janowski S, Töpel T, Hofestädt R (2010) DAWIS-M.D. – a data warehouse system for metabolic data. GI Jahrestag 2:720–725
11. Janowski S, Kormeier B, Töpel T, Hippe K, Hofestädt R, Willassen N, Friesen R, Rubert S, Borck D, Haugen P, Chen M (2010) Modeling of cell-cell communication processes using Petri nets in the example of quorum sensing. Silico Biol 10:182–203. Studies in health technology and informatics, Biological petri nets, vol 162. doi:10.3233/978-1-60750-704-8-182
12. Köhler J, Baumbach J, Taubert J, Specht M, Skusa A, Rüegg A, Rawlings C, Verrier P, Philippi S (2006) Graph-based analysis and visualization of experimental results with ONDEX. Bioinformatics 22:1383–1390
13. Kormeier B, Hippe K, Töpel T, Hofestädt R (2009) CardioVINEdb: a data warehouse approach for integration of life science data in cardiovascular diseases, Im Focus das Leben. Beiträge der 39. Jahrestagung der Gesellschaft für Informatik e.V. (GI), S. Fischer et al. (Hrsg.), INFORMATIK 2009, Koellen Druck+Verlag., Bonn, 40,704–708
14. Kormeier B, Hippe K, Arrigo P, Topel T, Janowski S, Hofestädt R (2010) Reconstruction of biological networks based on life science data integration. J Integr Bioinform 7(2). doi:10.2390/biecoll-jib-2010-146
15. Lee TJ, Pouliot Y, Wagner V, Gupta P, Stringer-Calvert DW, Tenenbaum JD, Karp PD (2006) BioWarehouse: a bioinformatics database warehouse toolkit. BMC Bioinform 7:170
16. Leser U, Naumann F (2006) Informations integration. Dpunkt Verlag, Heidelberg
17. Pauling J, Rottger R, Tauch A, Azevedo V, Baumbach J (2012) CoryneRegNet 6.0–updated database content, new analysis methods and novel features focusing on community demands. Nucl Acids Res 40(Database issue):D610–D614

18. Shah SP, Huang Y, Xu T, Yuen MM, Ling J, Ouellette BF (2005) Atlas – a data warehouse for integrative bioinformatics. BMC Bioinform 6:34
19. Sheth AP, Larson JA (1990) Federated database systems for managing distributed, heterogeneous, and autonomous databases. ACM Comput Surv 22(3):183–236. 0360–0300, doi:10.1145/96602.96604, ACM, New York
20. Stevens R, Baker P, Bechhofer S, Ng G, Jacoby A, Paton NW, Goble CA, Brass A (2000) TAMBIS: transparent access to multiple bioinformatics information sources. Bioinformatics 16:184–185
21. Töpel T, Kormeier B, Klassen A, Hofestädt R (2008) BioDWH: a data warehouse kit for life science data integration. J Integr Bioinform 5(2):93. 1613-4516, http://dx.doi.org/10.2390/biecoll-jib-2008-93
22. Trissl S, Rother K, Müller H, Steinke T, Koch I, Preissner R, Frömmel C, Leser U (2005) Columba: an integrated database of proteins, structures, and annotations. BMC Bioinform 6:81

Chapter 5
Molecular Information Fusion in Ondex

Jan Taubert and Jacob Köhler

Abstract Current biological knowledge is buried in hundreds of proprietary and public life-science databases available on the World Wide Web (WWW) and millions of scientific publications. Gaining access to this knowledge can prove difficult as each database may provide different tools to query or show the data and may differ in their structure and user interface or uses a different interpretation of biological knowledge than others. Systems approaches to biological research require that existing biological knowledge (data) is made available to support on the one hand the analysis of experimental results and on the other hand the construction and enrichment of models. Data integration methods are being developed to address these issues by providing a consolidated view of molecular information fused together from multiple databases. However, a key challenge for data integration is the identification of links between closely related entries in different life sciences databases when there is no direct information that provides a reliable cross reference. Here we describe and evaluate three data integration methods to address this challenge in the context of a graph-based data integration framework (the Ondex system). We give a quantitative evaluation of their performance in two different situations: the integration and analysis of different metabolic pathways resources and the mapping of equivalent elements between the Gene Ontology and a nomenclature describing enzyme function.

Keywords Data integration • Systems Biology • Life sciences • Biological databases • Molecular information • Ontologies • Pathways • Ondex

---

J. Taubert (✉)
Wellcome Trust Genome Campus, Hinxton, Cambridgeshire CB10 1SD, UK
e-mail: taubertjan@gmail.com

J. Köhler
Dow AgroSciences LLC, 9330 Zionsville Road, Indianapolis, IN 46268, USA

5.1 Introduction

Over the last decade, biological research has changed completely. The reductionism approach of studying only a few biological entities at a time in the past is being replaced by the study of the biological system as a whole today. Systems Biology [1] seeks to understand how complex biological systems work by looking at all parts of biological systems and how they interact with each other and form the complete whole. Systems Biology can be seen as a cycle (see Fig. 5.1) consisting of the following steps:

- Having a testable hypothesis about a biological system
- Conducting experimental validation of hypothesis
- Capturing and analysis of experimental results (usually 'omics' data)
- Gain new insights (data) about a biological system from analysis results
- Refine model about a biological system to derive new hypothesis

This process requires that existing biological knowledge (data) is made available to support on the one hand the analysis of experimental results and on the other hand the construction and enrichment of models for Systems Biology.

Effective integration of biological knowledge from databases scattered around the internet and other information resources (e.g. experimental data) is recognised as a prerequisite for many aspects of Systems Biology research and has been shown to be advantageous in a wide range of use cases such as the analysis and interpretation of 'omics' data [2], biomarker discovery [3] and the analysis of metabolic pathways for drug discovery [4]. However, systems for data integration have to overcome several challenges. For example, biological data sources may contain similar or overlapping coverage, and the user of such systems is faced with the challenge of generating a consensus data set or selecting the 'best' data source. Furthermore,

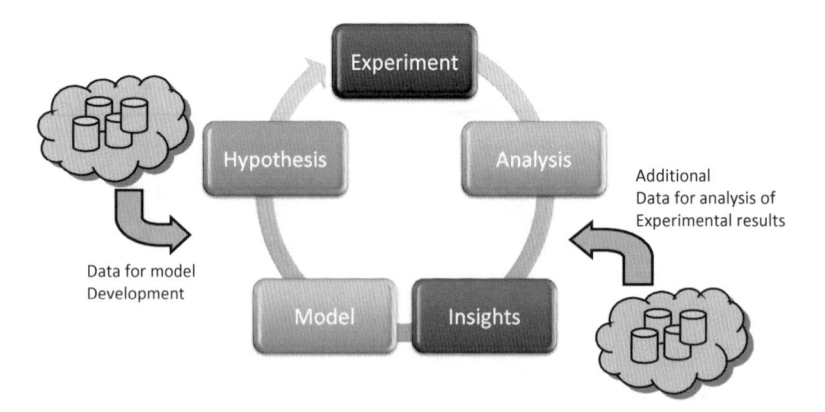

Fig. 5.1 Systems Biology cycle of experiment, analysis, insights, model and hypothesis together with requirements for large data for analysis of experimental results and model development

there are many technical challenges to data integration, like different access methods to databases, different data formats, different naming conventions and erroneous or missing data.

To address these challenges and enable effective integration of data in support of Systems Biology research, the Ondex system [2, 5–7] which is presented in this chapter was created. The Ondex system provides an integrated view across biological data sources with the aim to enable the user to gain a better understanding of biology from integrated knowledge. Ondex has been supported by BBSRC (http://www.bbsrc.ac.uk/) as part of the systems approaches to biological research initiative (SABR) and is now mainly being developed at Rothamsted Research, Manchester University and Newcastle University. The first Ondex prototype was developed at University of Bielefeld.

This book chapter is a summary and extension to previous work published in [6, 8]. It adds a new dimension to previous work by presenting integration results across time and using *Homo sapiens* as selected organism for metabolic pathway resources. We will start out by surveying different life-science data integration systems. This overview is followed by establishing a selection of challenges data integration systems are faced with and dissecting how well current systems are dealing with them. We then give a brief motivation and introduction for the Ondex system. This is followed by presenting data integration and transformation methods motivated by the stated challenges. The performance of the data integration methods is then quantitatively evaluated in two different situations: the integration and analysis of different metabolic pathways resources and the mapping of equivalent elements between the Gene Ontology and a nomenclature describing enzyme function. A brief discussion is given at the end of this book chapter.

5.1.1 Survey of Current Data Integration Systems

Several data integration systems for use in biology and related domains are in use today. Some of them use a generic approach to answer a wide range of biological questions. Others are more limited in their scope and application domain. These systems are based on principles such as link integration and hypertext navigation, data warehouses, view integration and mediator systems, workflows and mashups [9].

Software tools that solve aspects of the data integration problem are being developed for some time. The early approaches, which produced popular software such as SRS [10], use indexing methods to link documents or database entries from different databases and provide a range of text and sequence-based search and retrieval methods for users to assemble related data sets. The methods used by SRS (and related tools) address what has been described as the technical integration challenge.

More recently, data integration approaches are developed that 'drill down' into the data and seek to link objects at a more detailed level of description. Many of these approaches exploit the intuitively attractive representation of data as graphs or networks with nodes representing things and edges representing how they are related. For example, a metabolic pathway could be represented by a set of nodes identifying the metabolites linked by edges representing enzymatic reactions. Data integration systems that exploit graph-based methods include PathSys [11] or BN++ [12] and the Ondex system [13]. Both BN++ and Ondex are available as open source software.

The Visual Knowledge and BioCAD [14] software tools provide good examples for how semantic networks can be used for representing biological knowledge. The definition of the integration data structure of Ondex has been inspired by this use of semantic networks in the biology domain.

Biozon [15] is a data warehouse which includes additional derived information, such as sequence similarity or function prediction, between data entries. STRING [16] shows that multiple information sources can be combined to provide evidence for the relationship between proteins. Similar to Biozon and STRING, Ondex facilitates the information fusion of other derived information between data entities. Such information has been successfully used to improve genome annotation of *Arabidopsis thaliana* in a use case of Ondex [17].

BNDB with BN++ is the most similar system to Ondex in terms of system design and methodology. The NeAT [18] toolkit highlights how graph analysis applied to biological networks can help to reveal new insights. Furthermore it is a good example of providing such functionality via a web page.

Concluding from the presented systems and common practice in Systems Biology [5, 19], the representation of biological data as graphs or networks is a preferred choice. The complexity of the graphs or networks varies from tool to tool, for example, NeAT works with simple node and edge lists, whereas BNDB/BN++ and Ondex use a semantic-enriched graph model. Some tools like Biozon or STRING focus on aspects of providing a ready integrated knowledge base to the users. On the other hand, tools like Ondex, BNDB/BN++ or PathSys provide the user with means to assemble integrated data sets on his/her own. Visual Knowledge/BioCAD or NeAT emphasise on the biological pathways and networks analysis.

Graphical user interaction is realised in a variety of ways. Knowledge base-focused projects like Biozon or STRING tend to use a web-based interface backed by a relational database. Other data integration toolkits like BNDB/BN++ or Ondex offer a database driven backend with a dedicated front-end application and possible web service-based access. NeAT or Visual Knowledge/BioCAD loads and integrates data in an ad hoc way as part of their analysis workflows.

5.1.2 Challenges for Data Integration

Biological knowledge such as protein interactions (Fig. 5.2a), metabolic pathways (Fig. 5.2b) or biological ontologies (Fig. 5.2c) can be interpreted or understood as a network or graph. Biological databases are, however, usually implemented using table centric data structures, which do not readily allow the utilisation of

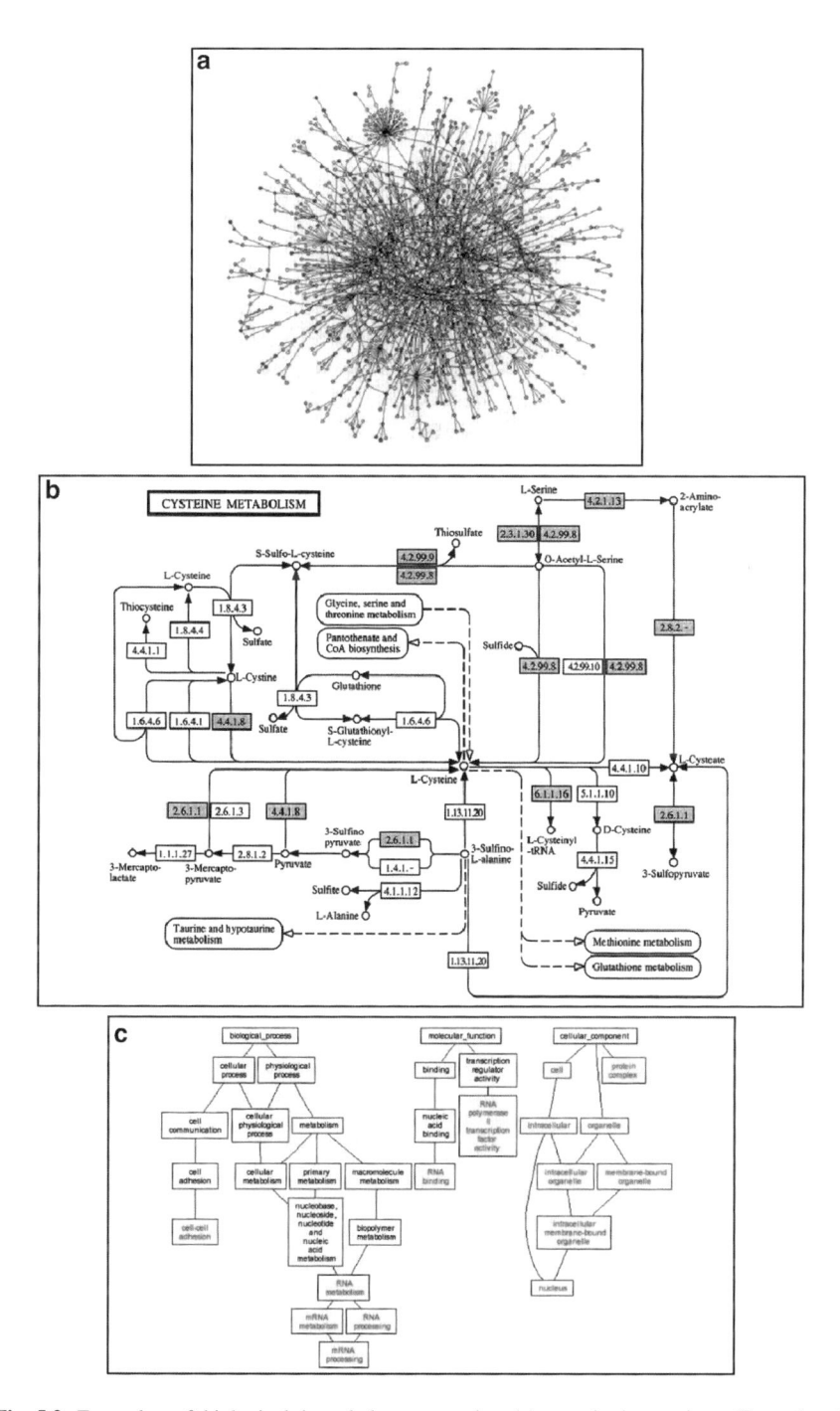

Fig. 5.2 Examples of biological knowledge as graphs: (**a**) protein interactions (Reproduced with permission from Jeong et al. [20] © Macmillan Magazines Ltd.), (**b**) metabolic pathways (Reprinted from Ogata et al. [21] with permission from Elsevier), (**c**) biological ontologies (Reprinted from Zhu et al. [22] under CC BY 2.0 licence © BioMed Central Ltd)

Table 5.1 Summarising outlined challenges for data integration systems

Challenge	Summary
First challenge	Representing biological data intuitively as a graph or network
Second challenge	Overcoming the syntactic and semantic heterogeneities between data sources
Third challenge	Provide a semantical consistent view on integrated information
Fourth challenge	Keep track of provenance during integration process
Fifth challenge	Domain-independent approach to data integration
Sixth challenge	Create a robust, usable and maintainable framework for data integration

graph analysis methods. Ondex uses a graph-based data structure which has been developed with an emphasis on providing integration of knowledge necessary for Systems Biology. Such a graph-based data structure should allow for the integration of heterogeneous data into a semantically consistent graph model and therefore support graph-based analysis algorithms and visualisation.

Biological data integration has to face the two problems of syntactic and semantic heterogeneity [23]. Syntactic heterogeneity is given by data being presented in different formats or as free text, containing spelling mistakes, wrong formatting or even missing data. Semantic heterogeneity is present in the different interpretations of data formats, symbols and names:

- Ambiguity of synonyms (exact/related), for example, Na(+)/K(+)-ATPase vs. just ATPase.
- Domain dependence of synonyms, for example, gene names in different organisms.
- Silent errors, like a typo in ENZYME Nomenclature is still valid entry (1.1.1.1 vs. 1.1.1.11).
- Unification references to other data sources can be ambiguous, for example, references to multiple splicing variants of a gene assigned to a protein.
- What is a gene, what is a protein and what is a transcript? Biological meaning is subject to interpretation and might vary.

To overcome syntactic and semantic heterogeneity in the data sources, knowledge modelling has to be adaptable for the respective domain of knowledge so that heterogeneous data sources can be transformed into a semantical consistent view. During this process it may be necessary to identify equivalent or redundant information in the data. Novel integration methods will have to be introduced to address this need. To establish trust in the integrated data, it is necessary to keep track of provenance during the whole data integration process.

Although this work has been mainly motivated by data from the life sciences, data integration is challenging in other data intensive sciences too. The integration methods should address this by being mostly domain independent. An example of a different application domain would be social networks. The methods presented in this chapter have been implemented as the core of the Ondex framework [2, 5]. One key aspect of the work on Ondex is to create a robust, usable and maintainable framework for data integration (Table 5.1).

5 Molecular Information Fusion in Ondex

Table 5.2 Challenges addressed by previous and current work

	First: data intuitively as graph or network	Second: addressing syntactic and semantic conflicts	Third: semantical consistent view	Fourth: track provenance	Fifth: domain independent	Sixth: robust, usable, maintainable framework
Visual Knowledge and BioCAD	Yes	No	Yes	No	No	Yes
Biozon	No	No	Yes	Yes	No	Yes
BNDB/BN++	Yes	Partially	Yes	No	Yes	Yes
STRING	Yes	No	Yes	Yes	No	Yes
NeAT	Yes	No	No	No	Yes	No

5.1.3 Comparison with Related Work

None of the previous presented data integration systems do address all the above-mentioned challenges as shown in Table 5.2.

The most important aspect not completely addressed by previous or related work is the second challenge of addressing syntactic and semantic heterogeneities between data sources in a systematic way. Knowledge base systems like STRING or Biozon use their own predefined database schema and load data from other data sources into this schema. During this process the mapping of source data to data objects in the system is hardwired and difficult to change. Overlapping or conflicting data between data sources usually does not get resolved. More complex systems like BNDB/BN++ provide adapters or parsers for different data sources and let the user of the system decide which selection of data source to integrate. Systems like NeAT or Visual Knowledge/BioCAD rely on the data to be in the correct format involving a larger amount of manual curation and work to be done upfront.

5.2 Motivation

Software designed for data integration in the life sciences has to address two classes of problem. It must provide a general solution to the technical (syntactic) heterogeneity, which arises from the different data formats, access methods and protocols used by different databases. More significantly, it must address the semantic heterogeneities arising from a number of sources in life-science databases. The most challenging source of semantic heterogeneity comes from the diversity and inconsistency among naming conventions for genes, gene functions, biological processes and structures among different species (or even within species). In recent years, significant progress in documenting the semantic equivalence of terms used in the naming of biological concepts and parts has been made in the development of a range of biological ontology databases which are coordinated

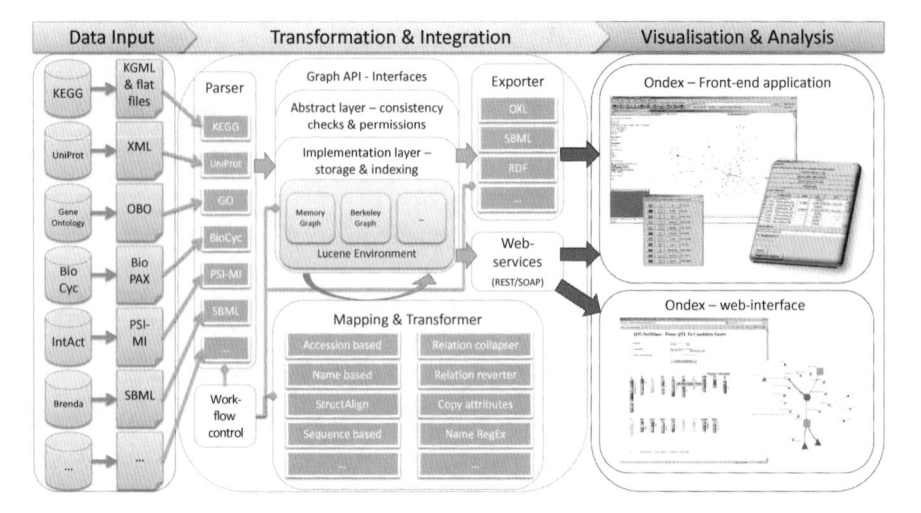

Fig. 5.3 Data integration in Ondex consists of three steps: (1) import and conversion of data sources into the data structure of Ondex (Data Input, *left*), (2) linking of equivalent or related entities of the different data sources and transformation into a semantical consistent graph (Transformation & Integration, *middle*), (3) knowledge extraction using the front-end application or web interface (Visualisation & Analysis, *right*)

under the umbrella of organisations such as the Open Biomedical Ontologies Foundry (http://www.obofoundry.org). However, the majority of biological terms still remain uncharacterised and therefore require automated methods to define equivalence relationships between them.

The integration of data in Ondex generally follows three conceptual stages as illustrated in Fig. 5.3: (1) normalising into the Ondex data structure in order to overcome predominantly technical heterogeneities between data exchange formats, (2) identifying equivalent and related entities among the imported data to overcome semantic heterogeneities at the entry level and (3) the data analysis, information filtering and knowledge extraction.

In order to make the Ondex system as extensible as possible, the second stage (middle bottom part in Fig. 5.3) has been separated both conceptually and practically. The motivations for doing this are to preserve original relationships and metadata from the original data source, make this integration step easily extensible with new methods, implement multiple methods for recognising equivalent data concepts to enhance the quality of integrated data and support reasoning methods that make use of the information generated in this step to improve the quality of integrated data.

The hypothesis here is that multiple methods for semantic data integration are necessary because of ambiguities and inconsistencies in the source data that will require different treatment depending on the source databases. In many cases, exact linking between concepts through unique names will not always be possible and therefore mappings will need to be made using inexact methods. Unless these inexact methods can be used reliably, the quality of the integrated data will be degraded.

5 Molecular Information Fusion in Ondex

To calibrate the presented data integration methods with well-structured data, the mapping of equivalent elements from the ontologies and nomenclatures extracted from the ENZYME [24] and GO [25] databases is used. To evaluate mapping methods in a more challenging integration task, the creation of an integrated data set from two important biological pathway resources, the Reactome [26] and HumanCyc [27] databases, is presented.

5.3 Methods

5.3.1 Data Import and Export

Following Fig. 5.3, the first step loads and indexes data from different sources. Ondex provides several options for loading data into the internal data warehouse, and a range of parsers have been written for commonly used data sources and exchange formats. In addition users can convert their data into an Ondex-specific XML or RDF dialect for which generic parsers are provided.

The role of all parsers is to load data from different data sources into the data structure used in the Ondex framework. In simple terms, this data structure can be seen as a graph, in which concepts are the nodes and relations are the edges. By analogy with the use of ontologies for knowledge representation in computer science, concepts are used to represent real-world objects [28]. Relations are used to represent the different ways in which concepts are connected to each other. Furthermore, concepts and relations may have additional properties and optional characteristics attached to them.

During the import process, names for concepts are lexicographically normalised by replacing non-alphanumeric characters with white spaces so that only numbers and letters are kept in the name. In addition, consistency checks are performed to identify, for example, empty or malformed concept names.

5.3.2 Data Integration Methods and Algorithms

The second step (following Fig. 5.3) links equivalent and related concepts and therefore creates relations between concepts from different data sources. Different combinations of mapping methods can be used to create links between equivalent or related concepts. Rather than immediately merging elements that are found to be equivalent, the mapping methods create a new equivalence relation between such concepts. After enough trust has been established in the results of the mapping methods by inspecting of these equivalence relations, then the information on similar elements can be fused, which is also known as molecular information fusion.

Each mapping method can be configured to create a score value reflecting the belief in a particular mapping and information about the parameters used. These scores are assigned as edge weights to the graph and form the foundation for the statistical analysis presented later. Additionally information on edges enables the user to track evidence for why two concepts were mapped by a particular mapping method.

Several constraints must be fulfilled before a mapping method creates a new link between two concepts. Under the assumption that the integrated data sources already contain all appropriate links between their own entries, new links are only created between different data sources. Biological databases often provide an NCBI taxonomy identifier for species information associated with their entries. If such identifiers are found in the graph, the mapping method ensures, in most cases, that relations are only created within the same species. In addition to species restriction, a mapping method takes the concept class of a concept into account. Only equal concept classes or specialisations of a concept class are considered to be included in a mapping pair.

5.3.2.1 Accession-Based Mapping

Most of the well-structured and managed public repositories of life-science data use accession coding systems to uniquely identify individual database entries. These codes are persistent over database versions. Cross references between databases of obviously related data (e.g. protein and DNA sequences) can generally be found using accession codes, and these can be easily exploited to link related concepts. Such concept accessions may not always present a one-to-one relationship between entries of different databases. For example, a GenBank accession found in the HumanCyc database is only unique for the coding region on the genome and not for the expressed proteins, which may exist in multiple splice variants. References presenting one-to-many relationships are call ambiguous. Concept accessions are indexed for better performance during information retrieval. Accession-based mapping by default uses only non-ambiguous concept accessions to create links between equivalent concepts, i.e. concepts that share the same references to other databases in a one-to-one relationship. This behaviour can be changed using a parameter.

Pseudocode

Let O denote the Ondex data structure consisting of a set of concepts $C(O)$ and a set of relations $R(O) \subseteq C(O) \times C(O)$. Every concept $c \in C(O)$ has a concept class $cc(c) \in CC(O)$, a data source identifier $ds(c) \in DS(O)$ and a list of concept accessions $ca(c) = \{(ca_1 \times \ldots \times ca_n) | ca_j \in CA(O)\}$. Each concept accession $ca \in CA(O)$ is a triple $ca = (ds, acc, ambiguous)$, where ds is the identifier of the data source from which the accession code acc is derived and *ambiguous* is either true or false. The bijective function *id* assigns a consecutive number $n \in \mathbb{N}$ to concepts and relations in O separately starting with 1.

```
ignore ← true or false (default)
function AccessionBasedMapping(O, ignore) {
    for all i ∈ [1..|C(O)|] do
        for all j ∈ [i..|C(O)|] do
            if ∃x ∈ ca(c_i) ∧ x ∈ ca(c_j) ∧ (¬x.ambiguous∨ignore) do
                if ds(c_i) ≠ ds(c_j) ∧ cc(c_i) = cc(c_j) do
                    O.createRelation(c_i, c_j)
}
```

Runtime Analysis

Assuming that the test if the two lists $ca(c_i)$ and $ca(c_j)$ have at least one concept accession in common takes linear time with respect to the length of the lists, for example, by using hashing strategies or ordered lists, and the average number of concept accessions on concepts is μ_{ca}, then the total runtime of accession-based mapping is $T(n) = \frac{1}{2}(n^2 + n) * \mu_{ca} \in O(n^2)$ where n is the number of concepts in the Ondex data structure.

5.3.2.2 Synonym Mapping

Entries in biological data sources often have one or more human-readable names, for example, gene names. Depending on the data source, some of these names will be exact synonyms such as the chemical name of a metabolite; others only related synonyms such as a general term for enzymatic function. Exact synonyms are especially flagged during the import process. Related synonyms are added to concepts as additional concept names. Concept names are preprocessed to strip all non-letter characters and stem special word cases before inserting them into the full-text index. Concept names are indexed for better performance and potentially fuzzy searches during information retrieval using the Apache Lucene (http://lucene.apache.org/) full-text indexing system. Fuzzy searches as supported by Lucene can be useful to overcome spelling mistakes, for example, PKM2 might be written as PK-M2 [29]. The default method for synonym mapping creates a link between two concepts if two or more concept names are matching (bidirectional best hits) to be able to cope with ambiguity of names. As a simple example of such ambiguity, the term 'mouse' shows that consideration of only one synonym is usually not enough for the disambiguation of the word, i.e. 'mouse' can mean computer mouse or the rodent *Mus musculus*. The threshold for the number of synonyms to be considered a match and an option to use only exact synonyms are parameters in the synonym mapping method.

Pseudocode

Let O denote the Ondex data structure consisting of a set of concepts $C(O)$ and a set of relations $R(O) \subseteq C(O) \times C(O)$. Every concept $c \in C(O)$ has a concept class $cc(c) \in CC(O)$, a data source identifier $ds(c) \in DS(O)$ and a list of concept names $cn(c) = \{(cn_1 \times \ldots \times cn_n) | cn_j \in CN(O)\}$. Each concept name $cn \in CN(O)$ is a tuple $cn = (name, exact)$, where *name* is the actual name of the concept and *exact* is either true or false. The bijective function *id* assigns a consecutive number $n \in \mathbb{N}$ to concepts and relations in O separately starting with 1.

```
num ← 1..N(default: 2)
exact ← true (default) or false
function SynonymMapping(O, num, exact) {
   for all i∈ [1..|C(O)|] do
      for all j∈ [i..|C(O)|] do
         if   |cn(c_i) ∩ cn(c_j)| ≥ num ∧
              (∃ x ∈ cn (c_i) ∩ cn (c_j)|x. exact ∨ ¬ exact)    do
            if ds(c_i) ≠ ds(c_j) ∧ cc(c_i) = cc(c_j) do
               O.createRelation(c_i, c_j)
}
```

Runtime Analysis

Assuming that the intersection of $cn(c_i)$ and $cn(c_j)$ can be found in linear time with respect to the size of the lists by using hashing strategies or ordered lists and the average number of concept names per concept is μ_{cn}, then the total runtime of synonym mapping is $T(n) = \frac{1}{2}(n^2 + n) * \mu_{cn} \in O(n^2)$ where n is the number of concepts in the Ondex data structure.

5.3.2.3 StructAlign Mapping

In some cases, two or more synonyms for a concept are not available in the data to be integrated. To disambiguate the meaning of a synonym shared by two concepts, the *StructAlign* mapping algorithm considers the graph neighbourhood of such concepts. A breadth-first search for a given depth (≥ 1) starting at each of the two concepts under consideration yields the respective reachability list for each concept. *StructAlign* processes these reachability lists and searches for synonym matches of concepts at each depth of the graph neighbourhood. If at any depth one or more pairs of concepts which share synonyms are found, *StructAlign* creates a link between the two concepts under consideration.

5 Molecular Information Fusion in Ondex

Pseudocode

Let O denote the Ondex data structure consisting of a set of concepts $C(O)$ and a set of relations $R(O) \subseteq C(O) \times C(O)$. Every concept $c \in C(O)$ has two additional attributes assigned: (a) a concept class $cc(c) \in CC(O)$ characterising the type of real-world entity represented by the concept (e.g. a gene) and (b) a data source identifier $ds(c) \in DS(O)$ stating the data source (e.g. HumanCyc) the concept was extracted from. Every relation $r \in R(O)$ is a tuple $r = (f, t)$ with f the 'from'-concept and t the 'to'-concept of the relation. To improve performance the algorithm is making use of indexing structures for concept names and the unique identifier returned by the bijective function *id* which assigns a consecutive number $n \in \mathbb{N}$ to concepts and relations in O separately starting with 1.

```
index ← searchable index of concept names for concepts
cutoff ← maximal depth of graph neighbourhood search
function StructAlign(O, index, cutoff) {
  matches ← new map of concepts to sets of concepts
  // search for concept name hits
  for all c∈C(O) do
    for all n∈cn(c)|n.exact do
      hits ← index.search(n.name)
      for all c'∈hits with ds(c)≠ds(c')∧cc(c)=cc(c') do
        matches[c].add(c')

  connectivity ← new map of concepts to sets of concept
  // calculate direct neighbourhood
  for all r∈R(O) with r=(f,t) do
    if ds(f)=ds(t)∧f≠t do
      connectivity[f].add(t)
      connectivity[t].add(f)
  reachability ← clone(connectivity)
  // modified breadth first search with depth cutoff
  for all i∈[1..cutoff] do
    for all (x,(y₁ ... yₙ))∈reachability do
      for all j∈[1..n] do
        reachability[x].addAll(connectivity[yᵢ])
  // look at neighbourhood of bidirectional matches
  for all (a,(b₁ ... bₙ)), (bᵢ,(c₁ ... cₘ))∈matches|a∈
    (c₁ ... cₘ) do
    na ← reachability[a]
    nb ← reachability[bᵢ]
    for all x∈na do
      if ∃y∈matches[x]|y∈nb do
        O.createRelation(a,bᵢ)
}
```

Runtime Analysis

Assuming the search for a concept name in the list of concept names takes logarithmic time with respect to the length of the list (e.g. using a self-balancing binary search tree [30]) and operations to manipulate maps and sets take constant time using hashing strategies, the runtime analysis is: Let c be the number of concepts, μ_{cn} the average number of concept names associated with a concept, r be the number of relations, μ_r the average number of relations per concept in the Ondex data structure and Δ a time constant for operations on maps and sets. The worst-case runtime of the StructAlign algorithm is then:

1. Search for concept name matches

$$T_1(c,r) = c * \mu_{cn} * \log(c * \mu_{cn}) * c * \Delta$$

2. Calculation of direct neighbourhood

$$T_2(c,r) = r * 2 * \Delta$$

3. Modified breadth-first search with depth cut-off

$$T_3(c,r) = \textit{cutoff} * c * \mu_r * \Delta$$

4. Finding bidirectional matches in neighbourhood, log(c) search time for $\exists y$

$$T_4(c,r) = c^2 * c * \Delta$$

$$T(c,r) = T_1 + T_2 + T_3 + T_4$$

$$T(c,r) = c * \mu_{cn} * \log(c * \mu_{cn}) * c * \Delta + r * 2 * \Delta$$
$$+ \textit{cutoff} * c * \mu_r * \Delta + c^2 * c * \Delta$$

Within a fully connected graph, the number of relations is $r = c * (c-1)/2$ and $\mu_r = c - 1$.

$$T(c) = \binom{c * \mu_{cn} * \log(c * \mu_{cn}) * c + c * (c-1)}{+ \textit{cutoff} * c * (c-1) + c^2 * c} * \Delta$$

$$T(c) = \left(c^2 * \mu_{cn} * \log(c * \mu_{cn}) + (1 + \textit{cutoff}) * c * (c-1) + c^3\right) * \Delta$$

$$T(c) \in O\left(c^3\right)$$

Fig. 5.4 Worked example for StructAlign. Different *shades* are used to distinguish data sources. *Node shape* represents different classes of concepts, *square* for enzymes and *circle* for metabolites. *Round arrows* show matching synonyms, whereas *vertical arrows* represent existing knowledge from data sources and *horizontal arrows* are created by StructAlign (color figure online)

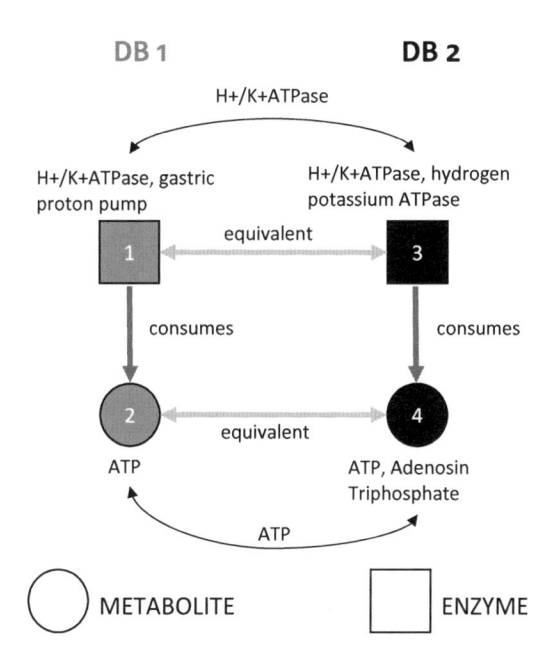

Here the average number of concept names per concept is $\mu_{cn} \ll c$. Hence the algorithm has a worst-case runtime of $O(c^3)$. Although the expected runtime on sparse graphs is $O(c^2)$ as the number of neighbours reachable for a certain depth in a sparse graph is much smaller than the number of total concepts in the graph.

Worked Example for StructAlign

Figure 5.4 shows a simple example graph of metabolites (circles) and enzymes (rectangles) originating from two data sources DB1 (left) and DB2 (right). All concepts except for concept 2 have two synonyms (exact one listed first). The 'consumes' relation (vertical arrows) is present in both data sources DB1 and DB2.

StructAlign starts to consider the first pair of concepts, here concepts 1 and 3, which share at least one exact synonym (H+/K+ATPase) and are of the same concept class (enzyme). The reachability list of concept 1 includes concept 2 and the reachability list of concept 3 includes concept 4. The undirected breadth-first search of StructAlign will find concepts 2 and 4 both being present at depth 1. As concepts 2 and 4 share at least one exact synonym (ATP) and are of the same concept class (metabolite), StructAlign collected enough evidence to create a new relation (horizontal arrows) between concepts 1 and 3. In the next step, StructAlign proceeds to the next pair of concepts 2 and 4 between DB1 and DB2, which share at least one exact synonym and will map them as being equivalent (horizontal arrows) because of the name match present between concepts 1 and 3.

5.3.2.4 Other Data Integration Methods

In addition to the mapping methods presented afore and evaluated in this study, the following selection of mapping methods shows how other information can be incorporated to deduce new relationships between concepts. This functionality is similar to that seen in Biozon [15]. A more complete list of data integration methods can be found on the Ondex web page (http://www.ondex.org).

Transitive Mapping

Transitive relationships between concepts are inferred from existing relations. For example, if concept A is identified to be equivalent to concept B and concept B is known to be equivalent to concept C, then a new equivalent relationship between concept A and concept C is created by this mapping method.

Sequence Similarity Mapping

The computation of the similarity of gene or protein sequences is achieved by exporting the sequence data into a FASTA [31] file and performing the matching using BLAST [32] or TimeLogic Decypher (http://www.timelogic.com). The results are used to create relations between concepts representing the genes or proteins. The BLAST bit score and e-Value is assigned as attributes on these relations.

External2go Mapping

The GO consortium provides reference lists of GO terms that map terms to other classification systems, for example, EC [24] enzymes or PFAM domains. The *external2go* mapping parses these lists and creates relations between entries of the GO database and entries of the other classification system.

These few examples together with the methods listed on the web page illustrate the wide range of information which is utilised by mapping methods in Ondex including simple name matches, sequence similarity search, orthology prediction, graph-pattern matching and even complex text mining-based information retrieval. Furthermore it is not difficult to add new mapping methods to Ondex.

5.3.3 Data Transformation Methods

After similar or equivalent concepts have been identified by mapping methods, the relation collapse functionality is used to merge or fuse such clusters of similar concepts connected by equivalence relations into one single concept. During

5 Molecular Information Fusion in Ondex

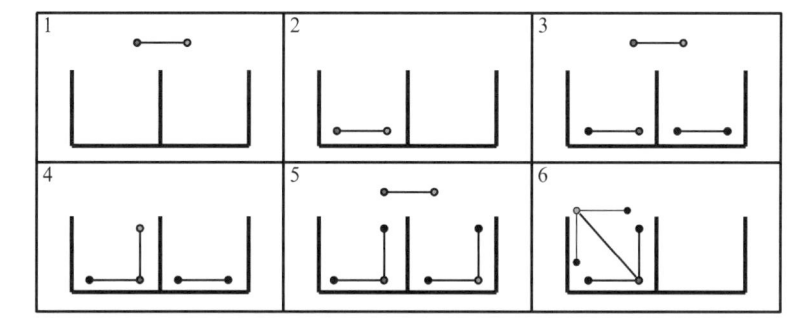

Fig. 5.5 Clustering of concepts, *1–2*; start new cluster, *3–4*; expand existing cluster, *5–6*; merge two existing clusters

this collapsing process, the molecular information of each original concept gets transferred onto the newly created fused concept, henceforth called molecular information fusion.

The collapsing of concepts consists of three main operations:

- Finding cluster of similar concepts
- Creating single collapsed concept
- Removing original concepts

Clustering of concepts, which is illustrated in Fig. 5.5, starts with iteration over all equivalence relations. For each such relation, it is determined if at least one of the two concepts connected by this relation is already a member of a cluster. If this is not the case, the relation and the two concepts are considered as the first element of a new cluster (steps 1 and 2). If one of the two concepts is already an element of an existing cluster, then the relation is added to this cluster (steps 3 and 4). If the two concepts are elements of two different clusters, these clusters are merged (steps 5 and 6).

The algorithm works with four temporary sets: nodes_open, nodes_closed, edges_open and edges_closed. The 'open' sets contain all known elements yet to explore. The 'closed' sets contain all already processed elements. The routine iterates over all concepts in the Ondex graph. For each concept all its adjacent relations are explored. If an equivalence relation is found, it is added to the edges_open set. The concept is then moved to nodes_closed, and the algorithm proceeds to explore all adjacent concepts of the elements of edges_open and moves them to edges_closed. In this fashion the algorithm switches between 'node exploration' and 'edge exploration' until no further elements to be processed are found. To avoid visiting elements which have already been analysed again, they are stored in a binary search tree so that they can be quickly re-identified. Hence each initial concept of the iteration is checked against this data structure before processing it.

The actual collapse process, which is done for every identified cluster of concepts, consists of the following steps:

- A collapse core node is created in the Ondex graph. If many nodes are collapsed into a single node, all properties of the collapsed nodes are assigned to the single representative.
- The edges going to nodes outside the current concept cluster are passed over to the collapse core node.
- All concepts of the current concept cluster are removed from the Ondex graph.

Runtime Analysis

The 'contains' and 'add' operations on the set data types in this algorithm have a runtime of $O(\log(n))$ using tree-based set data types. Let c be the number of the concepts in the Ondex graph and let μ_{cs} be the average cluster size. Then the worst-case runtime of the concept clustering algorithm is

$$O = (c * \mu_{cs} * \log(\mu_{cs}))$$

Hence the overall complexity of the algorithm is linear logarithmic.

5.3.4 Evaluation Methods

The mapping algorithms presented here can be configured using different parameters. According to the selection of the parameters, these methods yield different mapping results. To evaluate their behaviour, two different test scenarios were used: the mapping of equivalent elements in ontologies and the integration and analysis of metabolic pathways.

The evaluation of a mapping method requires the identification of a reference data set, sometimes also referred to as a 'gold standard', describing the links that really exist between data and that can be compared with those which are computed. Unfortunately, it is rare that any objective definition of a 'gold standard' can be found when working on biological data sets, and so inevitably most such evaluations require the development of expertly curated data sets. Since these are time consuming to produce, they generally only cover relatively small data subsets, and therefore the evaluation of precision and recall is inevitably somewhat limited.

In the next section, the results of mapping together two ontologies, namely, the Enzyme Commission (EC) nomenclature [24] and Gene Ontology (GO) [25], are presented. In this case, the Gene Ontology project provides a manually curated mapping to the ENZYME Nomenclature called *ec2go*. Therefore, *ec2go* has been selected as the first gold standard. The cross references between the two ontologies contained in the integrated data were also considered as the second gold standard for this scenario.

The following section also presents the results from the evaluation of a mapping created between the two metabolic pathway databases Reactome and HumanCyc. Unfortunately, a manually curated reference set is not available for this scenario. Therefore, it was necessary to rely on the cross references between the two databases that can be calculated through accession-based mapping as the nearest equivalent of a gold standard for this scenario.

5.4 Results

The mapping algorithms were evaluated using the standard measures of precision (Pr), recall (Re) and F_1-score [33]:

$$\text{Pr} = \frac{tp}{tp + fp} \quad \text{Re} = \frac{tp}{tp + fn} \quad F_1 = \frac{2 * \text{Pr} * \text{Re}}{\text{Pr} + \text{Re}}$$

The accession-based mapping algorithm (Acc) was used with default parameters, i.e. only using non-ambiguous accessions. This choice has been made to obtain a 'gold-standard' through accession-based mapping, i.e. increasing the confidence in the relations created. When evaluating the synonym mapping (Syn) and StructAlign (Struct) algorithms, parameters were varied to examine the effect of the number of synonyms that must match for a mapping to occur. This is indicated by the number in brackets after the algorithm abbreviation (e.g. Struct(1)). A second variant of each algorithm in which related synonyms of concepts were used to find a mapping was also evaluated. The use of this algorithmic variant is indicated by an asterisk suffix on the algorithm abbreviation (e.g. Syn(1)*).

5.4.1 Mapping Methods: ENZYME Nomenclature vs. Gene Ontology

The goal of this evaluation was to maximise the projection of the Enzyme Commission (EC) nomenclature onto the Gene Ontology. This would assign every EC term one or more GO terms. This evaluation has been carried out twice, once in January 2008 and a second time in the January 2013. The comparison of both results highlights the improvements made to the mapping between the two ontologies during this period.

For the first evaluation in 2008, ec2go (revision 1.67, downloaded 2008/01/21) and gene_ontology_edit.obo (revision 5.661, downloaded 2008/01/21) obtained from ftp://ftp.geneontology.org were used. Additionally enzclass.txt (last update 2007/06/19) and enzyme.dat (release of 2008/01/15) were downloaded from ftp://ftp.expasy.org.

Table 5.3 Mapping results for ENZYME Nomenclature to Gene Ontology in 2008

Method	TP, FP ec2go	TP, FP Acc	Pr, Re [%] ec2go	Pr, Re [%] Acc	F$_1$-score ec2go	F$_1$-score Acc
Ec2go	8063, 0	8049, 14	100.00, 100.00	99.83, 84.82	100.00	91.71
Acc	8049, 1441	9490, 0	84.82, 99.83	100.00, 100.00	91.71	100.00
Syn(1)	7460, 934	7462, 932	88.87, 92.52	88.90, 78.63	90.66	83.45
Syn(1)*	7605, 2581	7606, 2580	74.66, 94.32	74.67, 80.15	83.35	77.31
Syn(2)*	4734, 374	4738, 370	92.68, 58.71	92.76, 49.93	71.89	64.91
Syn(3)*	2815, 117	2816, 116	96.01, 34.91	96.04, 29.67	51.21	45.34
Struct(1)	1707, 63	1712, 58	96.44, 21.17	96.72, 18.04	34.72	30.41
Struct(1)*	1761, 279	1766, 274	86.32, 21.84	86.57, 18.61	34.86	30.63
Struct(2)	7460, 934	7462, 932	88.87, 92.52	88.90, 78.63	90.66	83.45
Struct(2)*	7605, 2581	7606, 2580	74.66, 94.32	74.67, 80.15	83.35	77.31
Struct(3)	7460, 934	7462, 932	88.87, 92.52	88.90, 78.63	90.66	83.45
Struct(3)*	7605, 2581	7606, 2580	74.66, 94.32	74.67, 80.15	83.35	77.31

Ec2go imported mapping list (1st gold standard), *Acc* accession-based mapping (2nd gold standard), *Syn* synonym mapping, *Struct* StructAlign, * allow related synonyms, *TP* true positives, *FP* false positives, *Pr* precision, *Re* recall, F$_1$-score. Synonym mapping was parameterised with a number that states how many of the names had to match to create a link between concepts. StructAlign was parameterised with the depth of the graph neighbourhood

For the second evaluation in 2013, ec2go (revision 1.487, downloaded 2012/12/22) and gene_ontology_edit.obo (daily built, downloaded 2012/12/22) have been retrieved, together with enzclass.txt (release of 2012/11/28) and enzyme.dat (release of 2012/11/28).

The data files were parsed into the Ondex data structure and the mapping algorithms applied using the Ondex pipeline. To determine the optimal parameters for this particular application case, different combination of the mapping algorithms with the variants and parameter options as described above have been systematically tested. Table 5.3 summarises the mapping results and compares the performances with the 'gold standards' data sets from ec2go and by accession mapping (Acc) generated during our analysis in 2008. Table 5.4 shows the same information for analysis results produced in 2013.

The first two rows of Tables 5.3 and 5.4 show the performance of the 'gold standard' methods tested against themselves. As can be seen by reviewing the F$_1$-scores in the subsequent rows of Tables 5.3 and 5.4, the most accurate synonym mapping requires the use of just one synonym. It does not help to search for further related synonyms (Syn(1,2,3)*). The explanation for this is that the EC nomenclature does not distinguish between exact and related synonyms. Therefore, concepts belonging to the EC nomenclature have only one preferred concept name (exact synonym) arbitrarily chosen to be the first synonym listed in the original data sources. A large number of entries in the EC nomenclature only have one synonym described, which explains the low recall of Syn(2)* and Syn(3)*.

The use of the more complex StructAlign algorithm, which uses the local graph topology to identify related concepts, has low recall when only a single synonym is

5 Molecular Information Fusion in Ondex

Table 5.4 Mapping results for ENZYME Nomenclature to Gene Ontology in 2013

Method	TP, FP ec2go	TP, FP Acc	Pr, Re [%] ec2go	Pr, Re [%] Acc	F_1-score ec2go	F_1-score Acc
Ec2go	8120, 0	8117, 3	100.00, 100.00	99.96, 77.57	100.00	87.35
Acc	8117, 2347	10464, 0	77.57, 99.96	100.00, 100.00	87.35	100.00
Syn(1)	6954, 498	7024, 428	93.32, 85.64	94.26, 67.13	89.31	78.41
Syn(1)*	7413, 2181	7538, 2056	77.27, 91.29	78.57, 72.04	83.70	75.16
Syn(2)*	4673, 537	4748, 462	89.69, 57.55	91.13, 45.37	70.11	60.58
Syn(3)*	2841, 189	2886, 144	93.76, 34.99	95.25, 27.58	50.96	42.77
Struct(1)	1449, 77	1466, 60	94.95, 17.84	96.07, 14.01	30.04	24.45
Struct(1)*	1541, 293	1562, 272	84.02, 18.98	85.17, 14.93	30.96	25.40
Struct(2)	7041, 605	7116, 530	92.09, 86.71	93.07, 68.00	89.32	78.59
Struct(2)*	7413, 2273	7538, 2148	76.53, 91.29	77.82, 72.04	83.26	74.82
Struct(3)	7041, 605	7116, 530	92.09, 86.71	93.07, 68.00	89.32	78.59
Struct(3)*	7413, 2273	7538, 2148	76.53, 91.29	77.82, 72.04	83.26	74.82

Ec2go imported mapping list (1st gold standard), *Acc* accession-based mapping (2nd gold standard), *Syn* synonym mapping, *Struct* StructAlign, * allow related synonyms, *TP* true positives, *FP* false positives, *Pr* precision, *Re* recall, F_1-score. Synonym mapping was parameterised with a number that states how many of the names had to match to create a link between concepts. StructAlign was parameterised with the depth of the graph neighbourhood

required to match and a depth cut-off of 1 is used (Struct(1) and Struct(1)*). This almost certainly results from differences in graph topology between EC nomenclature and Gene Ontology. The Gene Ontology has a more granular hierarchy, i.e. there is more than one hierarchy level between two GO terms mapped to EC terms, whereas the EC terms are only one hierarchy level apart. As the StructAlign depth cut-off search parameters are increased, more of the graph context is explored and accordingly the F_1-scores improved.

Across both tables, the highest F_1-scores come from Syn(1), Struct(2) and Struct(3), respectively. Including the related synonyms into the search (the * algorithm variants) did not improve precision. Neither did extending the graph neighbourhood search depth from Struct(2) to Struct(3) as all the neighbourhood matches had already been found within search depth 2.

During the integration of data from these data sets for this evaluation in 2008, some inconsistencies in the ec2go mapping list have been observed. The identification of such data quality issues is often a useful side effect of developing integrated data sets. The inconsistencies identified are listed in Table 5.5 and were revealed during the import of the ec2go data file after preloading the Gene Ontology and EC nomenclature into Ondex.

Presumably most of the problems are due to the previously disjoint development of both ontologies, i.e. GO references that were transferred or EC entries being deleted or vice versa. A few of the inconsistencies were possible typo errors. It remains a possibility that other 'silent' inconsistencies are still in ec2go that these integration methods would not find.

Table 5.5 Inconsistencies in ec2go in 2008

Accession	Mapping	Reason for failure
GO:0016654	1.6.4.-	Enzyme class does not exist, transferred entries
GO:0019110	1.18.99.-	Enzyme class does not exist, transferred entries
GO:0018514	1.3.1.61	Enzyme class does not exist, deleted entry
2.7.4.21	GO:0050517	GO term obsolete
GO:0047210	2.4.1.112	Enzyme class does not exist, deleted entry
1.1.1.146	GO:0033237	GO term obsolete
GO:0016777	2.7.5.-	Enzyme class does not exist, transferred entries
GO:0004712	2.7.112.1	Enzyme class does not exist, possible typo
2.7.1.151	GO:0050516	GO term obsolete

Every inconsistency was checked by hand against gene_ontology_edit.obo, enzclass.txt and enzyme.dat

A more recent analysis of data files used in 2013 revealed that the above presented inconsistencies have been corrected. The only inconsistencies identified in the newer data were:

- 1.3.5.6 to GO:0052889 (GO term is biological process, not molecular function)
- 2.5.1.46 to GO:0050983 (GO term is biological process, not molecular function)
- 2.1.1.35 to GO:0009021 (GO term obsolete)

5.4.2 Mapping Methods: Reactome vs. HumanCyc

The Reactome and HumanCyc pathway resources are both valuable for biologists interested in metabolic pathway analysis. Due to the different philosophies behind these two databases [34], however, they do have differences in their contents. Biomedical scientists wishing to work with biochemical pathway information would therefore benefit from a combined view of Reactome and HumanCyc and so this makes a realistic test. These two databases were chosen for this evaluation, because both pathway databases annotate metabolites and proteins in the pathways with standardised ChEBI [35] and UniProt [36] accessions, respectively. It is therefore possible to evaluate the precision, recall and F_1-score of the different mapping methods using accession-based mapping between these accession codes as a 'gold standard'.

For this evaluation the BioPAX [37] representations of the Reactome database (release 43 from 2012/12/10) obtained from http://www.reactome.org/download and the HumanCyc database (release 16.5 from 2012/11/06) obtained from http://humancyc.org/download.shtml were used. The Reactome database contained 1,387 metabolites and 4,650 proteins. The HumanCyc database contained 1,983 metabolites and 2,690 proteins. The evaluation results from the mapping between metabolites from these two databases are summarised in Table 5.6.

Accession-based mapping between metabolites found 856 out of 1,387 possible mappings. A closer look reveals that ChEBI identifiers are not always assigned

5 Molecular Information Fusion in Ondex

Table 5.6 Mapping results for Reactome and HumanCyc databases – metabolites

Method	TP	FP	Pr [%]	Re [%]	F_1-score
Acc	856	0	100.00	100.00	100.00
Syn(1)	218	530	29.14	25.47	27.18
Syn(1)*	468	1598	22.65	54.67	32.03
Syn(2)*	144	420	25.53	16.82	20.28
Syn(3)*	40	184	17.86	4.67	7.41
Struct(2)	238	606	28.20	27.80	28.00
Struct(2)*	430	1506	22.21	50.23	30.80
Struct(3)	238	606	28.20	27.80	28.00
Struct(3)*	430	1506	22.21	50.23	30.80

Acc accession-based mapping (gold standard), *Syn* synonym mapping, *Struct* StructAlign, * allow related synonyms, *TP* true positives, *FP* false positives, *Pr* precision, *Re* recall, F1-score. Synonym mapping was parameterised with a number that states how many of the names had to match to create a link between concepts. StructAlign was parameterised with the depth of the graph neighbourhood

Table 5.7 Mapping results for Reactome and HumanCyc databases – proteins

Method	TP	FP	Pr [%]	Re [%]	F_1-score
Acc	2826	0	100.00	100.00	100.00
Syn(1)	10	28	26.32	0.35	0.70
Syn(1)*	514	226	69.46	18.19	28.83
Syn(2)*	14	0	100.00	0.50	0.99
Struct(2)	46	36	56.10	1.63	3.16
Struct(2)*	288	112	72.00	10.19	17.85
Struct(3)	46	36	56.10	1.63	3.16
Struct(3)*	288	112	72.00	10.19	17.85

Acc accession-based mapping (gold standard), *Syn* synonym mapping, *Struct* StructAlign, * allow related synonyms, *TP* true positives, *FP* false positives, *Pr* precision, *Re* recall, F1-score. Synonym mapping was parameterised with a number that states how many of the names had to match to create a link between concepts. StructAlign was parameterised with the depth of the graph neighbourhood

to metabolite entries, most notably in HumanCyc. Therefore, the accession-based mapping does miss possible links and cannot be used naively as a gold standard for this particular application case. In this evaluation, accession-based mapping underestimates possible mappings, which leads to low precision for synonym mapping and StructAlign. A random set of the false-positive mappings returned by Syn(2)* and Struct(3) has been manually reviewed, and this revealed that a large number of the mappings made sense and metabolites shared very similar chemical names. Subject to further investigation, this example shows that relying only on accession-based data for integration might miss out some important links between data sources.

The evaluation results from the mapping between proteins from Reactome and HumanCyc are summarised in Table 5.7.

The accession-based mapping between proteins uses the UniProt accessions available in both Reactome and HumanCyc. Entries from HumanCyc can be labelled with two or more UniProt accessions representing multiple proteins involved in the same enzymatic function, whereas Reactome entries usually only have one UniProt accession. This results in one-to-many hits between Reactome and HumanCyc explaining why a total of 2,826 instead of only 2,690 mappings were found. This is a good example of how the differences in the semantics between biological data sources make it difficult to define a gold standard for evaluating integration methods.

The key finding from this evaluation based on mapping protein names is that due to different protein naming conventions in each of the two databases, name-based mapping methods cannot perform well. Manual inspection of a subset of false-negative mappings and their protein names reveals that HumanCyc is using longer names describing enzymatic functions (e.g. cytidine deaminase, cytidine aminohydrolase), whereas Reactome uses short gene names (e.g. CDA, CDD).

5.4.2.1 Visualising Results

Data integration involving large data sets can create very large networks that are densely connected. To reduce the complexity of such networks for the user, information filtering, network analysis and visualisation (see Fig. 5.3, step 3) are provided in a front-end application for Ondex [2]. The combination of data integration and graph analysis and visualisation has been shown to be valuable for a range of data integration projects in different domains, including microarray data analysis [2], support of scientific database curation [38, 39] and assessing the quality of terms and definitions in ontologies such as the Gene Ontology [40].

A particularly useful feature in the Ondex front-end is to visualise an overview of the types of data that have been imported into Ondex. This overview is called the Ondex meta-graph. It is generated as a network based on the data structure used in Ondex, which contains a type system for concepts and relations. Concepts are characterised using a class hierarchy and relations have an associated type. This information about concept classes and relation types is visualised as a graph with which the user can interact to specify semantic constraints – such as changing the visibility of concepts and relations in the visualisation and analysis of the integration data structure.

As an illustration, the integration of Reactome and HumanCyc for this evaluation results in more than 61,000 concepts and 113,000 relations. The mapping methods were run with optimal parameters identified in the previous section. After filtering down to a specific pathway using methods available in the Ondex front-end, it was possible to extract information from the integrated data as presented in Fig. 5.6.

Figure 5.6a displays parts of the *MAP kinase cascade* pathway from HumanCyc (nodes and edges in black) mapped to the corresponding entries from Reactome (indicated by bidirectional edges to blue nodes). It is now possible to visualise the differences between the two integrated pathways. Reactome contains more protein entities about a specific enzymatic function (e.g. proteins similar to *phospho-MEK*).

5 Molecular Information Fusion in Ondex 155

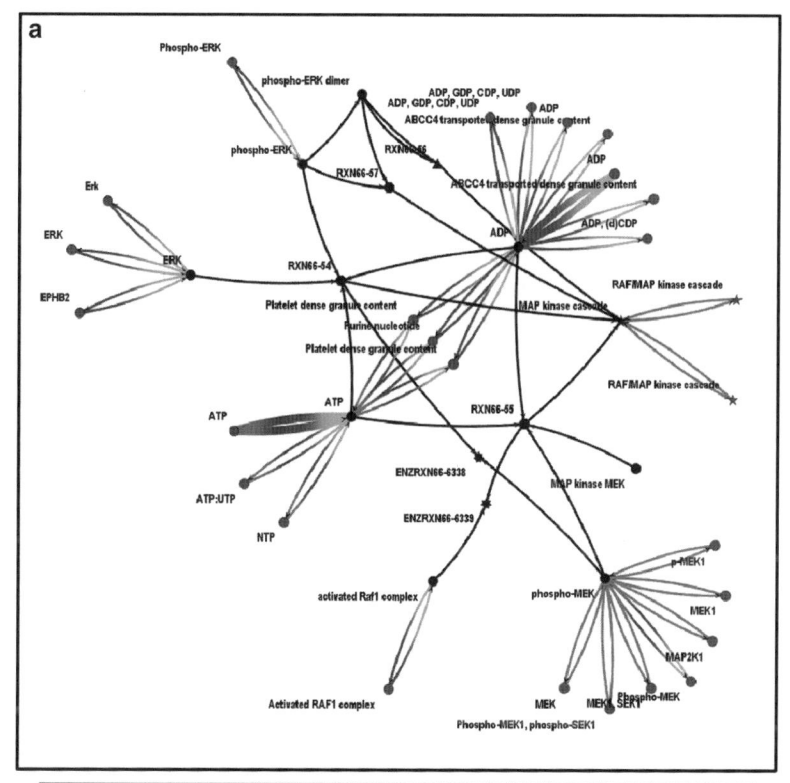

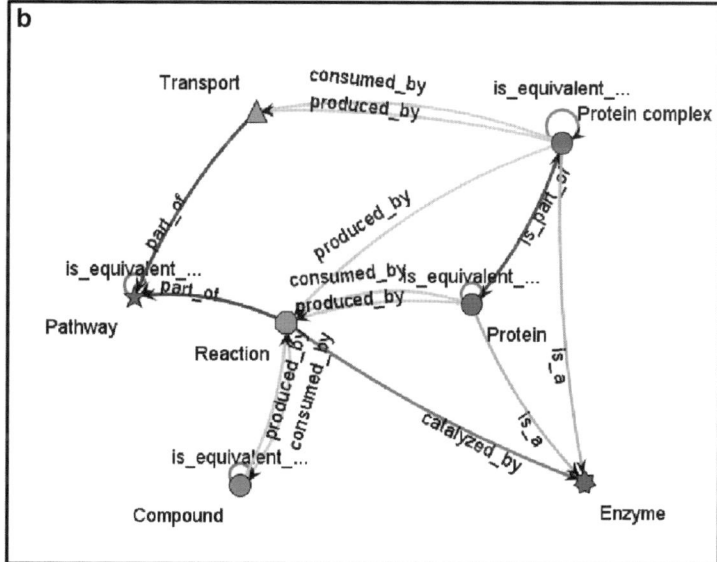

Fig. 5.6 (**a**) *MAP kinase cascade* pathway from HumanCyc with entities from Reactome. Equivalence relations are coloured by method (*red* = accession, *blue* = synonym, *green* = StructAlign) and thickness by StructAlign score. (**b**) Meta-graph providing an overview of the integrated data; *node colour* and *shape* distinguish classes; *edge colour* distinguishes different relation types (color figure online)

HumanCyc provides a larger pathway composed of more proteins than the pathway in Reactome, as the pathway concept maps to two different Reactome entries (stars, RAF/MAP kinase cascade).

The meta-graph is shown in Fig. 5.6b. This visualisation shows that the integrated data set consists of pathways (Pathway), reactions (Reaction) which are part of these pathways, metabolites (Compound) consumed or produced by the reactions, enzymes (Enzyme) catalysing the reactions and several combinations of proteins (Protein) and protein complexes (Protein complex) constituting the enzymes. The meta-graph provides the user with a useful high-level overview of the conceptual schema for this integrated data.

The last step to complete the molecular information fusion of the data presented in Fig. 5.6a would be to select the best equivalence relations and use the relation collapse data transformation to merge similar concept nodes together. To reduce the number of false-positive mappings, one would choose only such equivalence relations which are found by a combination of data integration methods (different edge colours) and at the same time carry a high confidence score (edge thickness) assigned by the data integration methods.

5.5 Discussion

Alternative methods for creating cross references (mappings) between information in different but related data sources have been presented. This is an essential component in the integration of data having different technical and semantic structures. Two realistic evaluation cases were used to quantify the performance of a range of different methods for mapping between the concept names and synonyms used in these databases. A quantitative evaluation of these methods shows that a graph-based algorithm (StructAlign) and mapping through synonyms can perform as well as using accession codes. In the particular application case of linking chemical compound names between pathway databases, the StructAlign and synonym-based algorithms outperformed the most direct mapping through accession codes by identifying more elements that were indirectly linked. Manual inspection of the false-positive mappings showed that both StructAlign and synonym mapping methods can be used where accession codes are not available to provide links between equivalent data source concepts. The combination of all three mapping methods yields the most complete projection between different data sources. This is an important result, because it is not always possible to find suitable accession code systems that provide the direct cross references between databases once you move outside the closely related data sources that deal with biological sequences and their functional annotations.

A similar approach to StructAlign called 'SubTree Match' has been described in [41] for aligning ontologies. This work extends this idea into a more general approach for data integration for biological networks and, furthermore, presents a formal evaluation in terms of precision and recall.

A particular challenge in this evaluation has been to identify suitable 'gold standard' data sets against which to assess the success of the algorithms developed. The results presented here are therefore not definitive, but represent the best quantitative comparison that could be achieved in the circumstances. Therefore, these results represent a pragmatic evaluation of the relative performance of the different approaches to concept name matching for data integration of life-science data sources.

Acknowledgements We would like to thank all current and previous contributors to the Ondex system (see www.ondex.org). The main part of this work has been carried out at Rothamsted Research. Rothamsted Research receives grant in aid from the Biotechnology and Biological Sciences Research Council (BBSRC). This work was supported by BBSRC SABR award BB/F006039/1 and TSB project TP 5082–33372. JT also would like to thank EMBL-EBI for allowing time to write this chapter.

WWW Link List (In Order of First Occurrence)

Name of resource	Brief description	WWW link
Ondex system	Data integration, visualisation and analysis framework for life-science data	http://www.ondex.org
BBSRC	Biotechnology and Biological Sciences Research Council in the United Kingdom	http://www.bbsrc.ac.uk
SRS	Sequence Retrieval System for biological data	http://www.instem.com/solutions/srs.html
PathSys	Graph-based system for creating a combined database of biological pathways, gene regulatory networks and protein interaction maps	http://biologicalnetworks.net/PathSys/
BN++ and BiNA	Biological data warehouse combined with biological network analyser	http://www.bina.unipax.info/
BioCAD	Integrated software for biosystem reverse engineering	http://biosoft.kaist.ac.kr/
Biozon	Unified biological knowledge resource with emphasis on protein and DNA characterisation and classification	http://www.biozon.org
STRING	Database of known and predicted protein interactions	http://string-db.org/
NeAT	Network Analysis Tools	http://rsat.bigre.ulb.ac.be/rsat/index_neat.html
OBO	Open Biomedical Ontologies Foundry	http://www.obofoundry.org
ENZYME (EC)	Nomenclature Committee of the International Union of Biochemistry and Molecular Biology	http://www.chem.qmul.ac.uk/iubmb/enzyme/

(continued)

(continued)

Name of resource	Brief description	WWW link
GO	The Gene Ontology	http://www.geneontology.org/
Reactome	Curated knowledgebase of biological pathways in humans	http://www.reactome.org
HumanCyc	Encyclopedia of Homo sapiens Genes and Metabolism	http://humancyc.org/
NCBI Taxonomy	Provides a taxonomy browser, taxonomy resources and other information	http://www.ncbi.nlm.nih.gov/taxonomy
GenBank	GenBank is the NIH genetic sequence database	http://www.ncbi.nlm.nih.gov/genbank/
Apache Lucene	Open source full-text indexing system	http://lucene.apache.org
BLAST	The Basic Local Alignment Search Tool	http://blast.ncbi.nlm.nih.gov
Decypher	Hardware accelerated sequence aligner	http://www.timelogic.com
PFAM	Large collection of protein families	http://pfam.sanger.ac.uk
Ec2go	Mapping file from EC to GO	http://www.geneontology.org/external2go/ec2go
ChEBI	Chemical Entities of Biological Interest	http://www.ebi.ac.uk/chebi
UniProt	Universal Protein Resource is a catalog of information on proteins	http://www.uniprot.org

References

1. Biotechnology and Biological Sciences Research Council (2007) Systems biology. http://www.bbsrc.ac.uk/publications/topic/systems-biology.aspx
2. Köhler J, Baumbach J, Taubert J, Specht M, Skusa A, Ruegg A, Rawlings C, Verrier P, Philippi S (2006) Graph-based analysis and visualization of experimental results with ONDEX. Bioinformatics 22(11):1383–1390
3. Gaylord M, Calley J, Qiang H, Su EW, Liao B (2006) A flexible integration and visualisation system for biomarker discovery. Appl Bioinformatics 5(4):219–223
4. Fischer HP (2005) Towards quantitative biology: integration of biological information to elucidate disease pathways and to guide drug discovery. Biotechnol Annu Rev 11:1–68
5. Köhler J, Rawlings C, Verrier P, Mitchell R, Skusa A, Ruegg A, Philippi S (2005) Linking experimental results, biological networks and sequence analysis methods using Ontologies and Generalised Data Structures. In Silico Biol 5(1):33–44
6. Taubert J, Hindle M, Lysenko A, Weile J, Köhler J, Rawlings CJ (2009) Linking life sciences data using graph-based mapping. Paper presented at the proceedings of the 6th international workshop on data integration in the life sciences, Manchester, UK
7. Taubert J, Sieren KP, Hindle M, Hoekman B, Winnenburg R, Philippi S, Rawlings C, Köhler J (2007) The OXL format for the exchange of integrated datasets. J Integr Bioinform 4(3):63
8. Taubert J (2011) ONDEX - a data integration framework for the life sciences. Bielefeld University, Bielefeld
9. Goble C, Stevens R (2008) State of the nation in data integration for bioinformatics. J Biomed Inform 41(5):687–693. doi:S1532-0464(08)00017-8 [pii] 10.1016/j.jbi.2008.01.008
10. Etzold T, Ulyanov A, Argos P (1996) SRS: information retrieval system for molecular biology data banks. Methods Enzymol 266:114–128

11. Baitaluk M, Qian X, Godbole S, Raval A, Ray A, Gupta A (2006) PathSys: integrating molecular interaction graphs for systems biology. BMC Bioinformatics 7:55
12. Küntzer J, Blum T, Gerasch A, Backes C, Hildebrandt A, Kaufmann M, Kohlbacher O, Lenhof H-P (2006) BN++ − a Biological Information System. J Integr Bioinform 3(2):34. doi:10.2390/biecoll-jib-2006-34
13. Smith B, Ceusters W, Klagges B, Kohler J, Kumar A, Lomax J, Mungall C, Neuhaus F, Rector AL, Rosse C (2005) Relations in biomedical ontologies. Genome Biol 6(5):R46
14. Lee D, Kim S, Kim Y (2007) BioCAD: an information fusion platform for bio-network inference and analysis. BMC Bioinformatics 8(Suppl 9):S2. doi:1471-2105-8-S9-S2 [pii] 10.1186/1471-2105-8-S9-S2
15. Birkland A, Yona G (2006) BIOZON: a system for unification, management and analysis of heterogeneous biological data. BMC Bioinformatics 7:70. doi:1471-2105-7-70 [pii] 10.1186/1471-2105-7-70
16. Jensen LJ, Kuhn M, Stark M, Chaffron S, Creevey C, Muller J, Doerks T, Julien P, Roth A, Simonovic M, Bork P, von Mering C (2009) STRING 8 – a global view on proteins and their functional interactions in 630 organisms. Nucleic Acids Res 37(Database issue):D412–D416. doi:gkn760 [pii] 10.1093/nar/gkn760
17. Pesch R, Lysenko A, Hindle M, Hassani-Pak K, Thiele R, Rawlings C, Köhler J, Taubert J (2008) Graph-based sequence annotation using a data integration approach. J Integr Bioinform 5(2):94. doi:10.2390/biecoll-jib-2008-94
18. Brohee S, Faust K, Lima-Mendez G, Sand O, Janky R, Vanderstocken G, Deville Y, van Helden J (2008) NeAT: a toolbox for the analysis of biological networks, clusters, classes and pathways. Nucleic Acids Res 36(Web Server issue):W444–W451. doi:gkn336 [pii] 10.1093/nar/gkn336
19. Dwyer T, Rolletschek H, Schreiber F (2004) Representing experimental biological data in metabolic networks. Paper presented at the proceedings of the second conference on Asia-Pacific bioinformatics, vol 29, Dunedin, New Zealand
20. Jeong H, Mason SP, Barabasi AL, Oltvai ZN (2001) Lethality and centrality in protein networks. Nature 411(6833):41–42. doi:10.1038/35075138
21. Ogata H, Goto S, Fujibuchi W, Kanehisa M (1998) Computation with the KEGG pathway database. Biosystems 47(1–2):119–128
22. Zhu H, Cabrera RM, Wlodarczyk BJ, Bozinov D, Wang D, Schwartz RJ, Finnell RH (2007) Differentially expressed genes in embryonic cardiac tissues of mice lacking Folr1 gene activity. BMC Dev Biol 7:128. doi:10.1186/1471-213X-7-128
23. Gardner SP (2005) Ontologies and semantic data integration. Drug Discov Today 10(14):1001–1007. doi:S1359-6446(05)03504-X [pii] 10.1016/S1359-6446(05)03504-X
24. Bairoch A (2000) The ENZYME database in 2000. Nucleic Acids Res 28(1):304–305
25. Ashburner M, Ball CA, Blake JA, Botstein D, Butler H, Cherry JM, Davis AP, Dolinski K, Dwight SS, Eppig JT, Harris MA, Hill DP, Issel-Tarver L, Kasarskis A, Lewis S, Matese JC, Richardson JE, Ringwald M, Rubin GM, Sherlock G (2000) Gene ontology: tool for the unification of biology. The gene ontology consortium. Nat Genet 25(1):25–29. doi:10.1038/75556
26. Jupe S, Akkerman JW, Soranzo N, Ouwehand WH (2012) Reactome – a curated knowledgebase of biological pathways: megakaryocytes and platelets. J Thromb Haemost. doi:10.1111/j.1538-7836.2012.04930.x
27. Caspi R, Altman T, Dreher K, Fulcher CA, Subhraveti P, Keseler IM, Kothari A, Krummenacker M, Latendresse M, Mueller LA, Ong Q, Paley S, Pujar A, Shearer AG, Travers M, Weerasinghe D, Zhang P, Karp PD (2012) The MetaCyc database of metabolic pathways and enzymes and the BioCyc collection of pathway/genome databases. Nucleic Acids Res 40(Database issue):D742–D753. doi:10.1093/nar/gkr1014
28. Smith B (2004) Beyond concepts: ontology as reality representation. In: Varzi A, Vieu L (eds) Proceedings of FOIS. IOS Press, Amsterdam

29. Schuemie MJ, Mons B, Weeber M, Kors JA (2007) Evaluation of techniques for increasing recall in a dictionary approach to gene and protein name identification. J Biomed Inform 40(3):316–324. doi:S1532-0464(06)00097-9 [pii] 10.1016/j.jbi.2006.09.002
30. Knuth D (1997) Section 6.2.3: Balanced trees. In: The art of computer programming, vol 3, Sorting and searching, 2nd edn. Addison-Wesley, Reading, 1998. ISBN 0-201-89685-0
31. Pearson WR (1990) Rapid and sensitive sequence comparison with FASTP and FASTA. Methods Enzymol 183:63–98
32. Altschul SF, Madden TL, Schaffer AA, Zhang J, Zhang Z, Miller W, Lipman DJ (1997) Gapped BLAST and PSI-BLAST: a new generation of protein database search programs. Nucleic Acids Res 25(17):3389–3402. doi: 10.1093/nar/25.17.3389
33. Goutte C, Gaussier E (2005) A probabilistic interpretation of precision, recall and F-score, with implication for evaluation. In: Losada DE, Fernandez-Luna JM (eds) European Colloquium on IR Research (ECIR'05), 2005, Springer Berlin Heidelberg, pp 345–359. http://dx.doi.org/10.1007/978-3-540-31865-1_25
34. Stobbe MD, Houten SM, Jansen GA, van Kampen AH, Moerland PD (2011) Critical assessment of human metabolic pathway databases: a stepping stone for future integration. BMC Syst Biol 5:165. doi:10.1186/1752-0509-5-165
35. Degtyarenko K, de Matos P, Ennis M, Hastings J, Zbinden M, McNaught A, Alcantara R, Darsow M, Guedj M, Ashburner M (2008) ChEBI: a database and ontology for chemical entities of biological interest. Nucleic Acids Res 36(Database issue):D344–D350. doi:10.1093/nar/gkm791
36. Apweiler R, Bairoch A, Wu CH, Barker WC, Boeckmann B, Ferro S, Gasteiger E, Huang H, Lopez R, Magrane M, Martin MJ, Natale DA, O'Donovan C, Redaschi N, Yeh LS (2004) UniProt: the universal protein knowledgebase. Nucleic Acids Res 32(Database issue):D115–D119. doi:10.1093/nar/gkh13132/suppl_1/D115 [pii]
37. Bader G, Cary M (2005) BioPAX – biological pathways exchange language. BioPAX workgroup. http://www.biopax.org/release/biopax-level2-documentation.pdf
38. Baldwin TK, Winnenburg R, Urban M, Rawlings C, Köhler J, Hammond-Kosack KE (2006) PHI-base provides insights into generic and novel themes of pathogenicity. Mol Plant Microbe Interact 19(12):1451–1462
39. Winnenburg R, Baldwin TK, Urban M, Rawlings C, Köhler J, Hammond-Kosack KE (2006) PHI-base: a new database for pathogen host interactions. Nucleic Acids Res 34(Database issue):D459–D464
40. Köhler J, Munn K, Rüegg A, Skusa A, Smith B (2006) Quality control for terms and definitions in ontologies and taxonomies. BMC Bioinformatics 7:212
41. Zhang L, Gu J-G (2005) Ontology based semantic mapping architecture. In: Fourth international conference on machine learning and cybernetics. IEEE

Chapter 6
Text Mining on PubMed

Timofey V. Ivanisenko, Pavel S. Demenkov, and Vladimir A. Ivanisenko

Abstract A technology of linguistic analysis with the use of computer methods is called a text mining.

Computer tools based on this technology can provide a wide range of tasks, including:

1. The task of finding a relevant literature with the user-specified criteria and determination of the correspondence between single article or manually specified picks of articles and researching area of knowledge or a set of predesignated areas
2. The task of identification and extraction of names of biological objects that can be found in the raw text (e.g., genes, proteins, metabolites) with extra information on them, such as the type of object and names of its synonyms
3. The task of establishment of relationships between objects that had been automatically recognized in text with the representation of the obtained data in a form convenient for the further analysis, for example, in the form of associative networks

Keywords Text mining • Associative genetic networks • Automated PubMed analysis • Knowledge extraction system

6.1 Systems for the Automated Search of Literature

Systems for the thematic search of the literature are extremely important in almost any kind of scientific research. Their main task is automated determination of the level of relevance between electronic publications and information of interest for

T.V. Ivanisenko (✉) • P.S. Demenkov • V.A. Ivanisenko
Institute of Cytology and Genetics SB RAS, Laboratory of the Computer Proteomics, Novosibirsk, Russia
e-mail: itv@bionet.nsc.ru

specialists. The most common techniques for the development of such systems are the use of logic and vector models as well as mining techniques; often they are combined in order to improve the search.

6.1.1 Logical Query Model

Logical queries allow to perform search of documents with a user-specified strings of keywords associated by logical operators AND/OR/NOT by comparison of user-specified queries with all available documents. In case of full or partial string matches, considering logical operators in the query, the document will be defined as satisfying to the query or not. For example, the query «*p53 AND open-angle glaucoma*» will display all documents containing a name of the protein "*p53*" and "*open-angle glaucoma*" disease at the same time the query «*p53 NOT cancer*» will show only documents that involve "*p53*," but not involve a "*cancer*" disease. The advantage of this method is its easy implementation. At the same time, its main drawbacks are the lack of features for the formation of complex queries that can allow, for example, to consider the relationships between objects, as well as excessive search redundancy [1].

6.1.2 Vector Query Model

Vector query algorithm was proposed by Joyce and Needham [2]. It is based on the idea that similar documents should meet to the simular requirements.

The algorithm is based on a representation of each document as a mathematical vector of terms in which each term is corresponding to the frequency number of its matches in the text, such vectors are related to control vector that is formed with a user-specified query, and as a result the establishment of the extent to which articles specified to the provided subject area takes place. This approach offers the feature of the combination of similar documents into clusters, which can significantly improve the time and the quality of search. The first query vector algorithm has been implemented by Salton in SMART (Salton's Magical Automatic Retriever of Text) search engine [3].

6.1.3 Mining Methods

These methods include:
- Methods of the statistical correlations intended for the formation of rules for establishment of relations between documents and prespecified categories [4].

- Clustering methods based on different semantic attributes of the document set with the use of linguistic and mathematical methods without a priori knowledge; as a result of such analysis, a taxonomy of documents or visual map, providing effective coverage of large amounts of data, is created [5].
- Methods for analysis of the relationships for identification of descriptors (key phrases) in the documents that provide flexible navigation in text [6].
- Methods for the identification of facts designed for the extraction of knowledge, in order to improve the classification, retrieval, and clusterization of documents [7].

6.1.4 Existing Search Systems

The Entrez system [8] (http://www.ncbi.nlm.nih.gov/sites/gquery) allows to make search of information on biological databases supported by NCBI, such as PubMed, GenBank, Structure, and Genome. It is based on a model of vector and logical queries as well as mining techniques.

Muller and his colleagues developed a search engine, Textpresso [9] (http://www.textpresso.org), specialized on *Caenorhabditis elegans* that includes over 3,800 of full-text articles and 16,000 of abstracts. It is based on the modified vector method, containing articles that were previously divided on separate sentences as well as on terms appropriate to *C. elegans* and stored at the database; the search queries are divided into words, and such approach allowed authors to improve the quality of search in comparison with classical method of vector queries.

PubMatrix [10] (http://pubmatrix.grc.nia.nih.gov) is a system that allows to make search on the PubMed database by comparing a user-specified sample of terms, such as gene or protein names with a set of their functions. As a result, the system provides a list of abstracts of scientific publications containing links between these genes or proteins and their functions.

6.2 Systems for Automated Identification of Biological Objects in Texts

Krauthammer and Nenadic distinguished three stages of automatic recognition of biological objects [11]:

- Extraction of names, synonyms, and abbreviations in the unstructured text
- Identification and establishment of relationships between objects
- Representation of obtained information in the formalized form

There are three main approaches for automated identification of the names of biological objects in text:

- With the use of rules and templates
- With the use of statistical and machine learning methods
- With the use of thematic vocabularies

6.2.1 Methods for Identification of the Biological Objects with the Use of Rules and Templates

These methods are based on the use of a set of regular expressions (rules or patterns) that normally are formed manually by specialists [12] and intended to identification of terms according to their syntactic and semantic features. Ananiadou and McNaught in their work concluded that systems implemented by use of these methods can get the better-quality results in comparison to other approaches [13]. The main disadvantage of methods based on templates is a poor quality in analysis of complex sentences.

6.2.2 Recognition of Biological Entities with the Use of Statistical Algorithms and Machine Learning Methods

The use of statistical approaches allows the identification of terms based on the frequency values of their occurrence in text; these methods are effective in solving the problem of keywords. Systems based on machine learning methods are designed to search for specified classes of terms in the text and allow to do direct identification of objects with their classification with the help of "training samples." These samples are used to "train" method and allow them to produce high-quality object recognition and classification on a specific biological problem. The main problems of machine learning methods are poor availability of training samples and the high need of a large amount of high-quality data [13].

Collier and colleagues used a hidden Markov model and automated analysis of orthographic word features for the extraction of the terms related to the ten predesignated classes [14]. The results of this system were highly dependent from the quality of samples. Thus, for a class of proteins, F-score value was 75.9 %, while F-score value for RNA was much less due to their low representation in the training set. Similar results (F-score of 75 %) occurred in Morgan and colleagues' analysis of gene names for Drosophila genus (small flies) [15]. They used a hidden Markov model in conjunction with contextual analysis and simple spelling rules.

Kazama and colleagues used the method of the support vector machine with a GENIA training set [16, 17]. The so-called B-I-O tags were used for the annotation: the B tag allowed to identify preterm structures, the I tag contrasted the words forming the part of the term, and the O tag was used for words going after terms. Tags were supplemented with information related to different classes of molecular-genetic objects. For instance, tag B protein was associated with the words that were situated in front of the names of proteins. The F-score value for this method was 50 %.

6.2.3 Recognition of Biological Objects with Dictionaries

These methods are based on the use of thematical words for the search for biological object names by the comparing of text with terms from the dictionary. The advantage of such approach over other methods is fast term classification by types with their reference to the various databases. The main disadvantages are the inability to recognize the novel names and a high degree of false-positive results related to short and nonunique names [18].

The BioThesaurus web-based system [19] (http://pir.georgetown.edu/pirwww/iprolink/biothesaurus.shtml) was designed for the establishment of interactions between genes and proteins in unstructured text. The system was based on the use of vocabulary compiled from the different databases: UniProt [20]; NCBI resources devoted to genes and proteins [21], including Entrez Gene, RefSeq, and GenPept; and genomic databases of model organisms such as MGD [22], SGD [23], RGD [24], FlyBase [25], and WormBase [26] and some other sources. The total volume of the dictionary was about 2.8 million of unique gene and protein names.

6.2.4 Recognition of Biological Objects with the Combining of Different Methods

For today most of modern systems designed for the identification of names of biological objects in texts are combining several different approaches. For example, popular is a combination of methods based on patterns with machine learning. This allows to achieve more higher values of completeness and accuracy. Tsuruoka and Tsujii used the search with the dictionary along with machine learning methods [27]. On the first step (recognition phase), the text was scanned using a dictionary for protein name candidates. The problem of spelling variation was solved with an approximate string-matching technique. On the second step (filtering phase), each candidate was checked if it is a name of a protein or not with a machine learning method. The classifier was trained on an annotated corpus GENIA [28] and used the context of the term and the term itself as the features for the classification. Only "accepted" candidates were recognized as names of proteins. The F-measure (the harmonic value of the precision and recall values) for this system was 70.2 %.

Hakenberg with colleagues developed a GNAT system [29] (http://cbioc.eas.asu.edu/gnat/) for the identification of the names of genes from various organisms in the texts of abstracts of scientific publications. For the identification of gene names, dictionaries (for each of the 25 organisms, a separate dictionary was compiled) and machine learning methods were used. The search of noncanonical forms of names was done using automates with the ending number of states, while the identification of canonical names was done with the help of dictionaries based on Entrez, GO, UniProt, and other databases. The F-score of the system was 81.4 % (the precision and recall were 90.8 % and 73.8 %, respectively).

6.3 Systems for the Recognition of Interactions Between Biological Objects

For the solution of task of automated extraction of information about the molecular and genetic interactions between biological objects from the literature, the following methods are widely used:

- Methods based on the co-occurrence of objects in the text
- Methods based on a set of rules and patterns (shallow parsing)
- Methods based on a deep syntactic analysis of the separated sentences (full or deep parsing)

The co-occurrence method is based on a calculation of the frequency of co-occurrence of object names in the text. It is assumed that the more two objects can be mentioned in the same text, the more likely they are related with each other. The main advantages of these methods are the easy implementation and high value of recall. But on the other side, the precision of such method is not very high and this method does not allow the identification of type of relationships between objects. Coremine Medical (http://www.coremine.com/medical) and FACTA [30] (http://text0.mib.man.ac.uk/software/facta/main.html) are examples of such systems. At the BRENDA database (http://www.brenda-enzymes.org), co-occurrence method was used for the extraction of data about associations between diseases and enzymes [31].

The shallow parsing is based on the extraction of information from texts with the use of partial relations between words in a sentence using a set of specific patterns and rules. A SUISEKI (System for Information Extraction on Interactions) [32] designed for the automated analysis of the syntactic structure of phrases and other developments for the extraction of protein interactions is based on this method. The core of the system is the number of rules that allow capturing different language constructions that are commonly used to describe interactions. The rules are implemented as frames of the form "[protein/gene] binds/associates/... [protein/gene]" as well as the form describing specific relations, such as "[noun indicating interaction] of [protein/gene] with [protein/gene]." The Chilibot [33] (http://www.chilibot.net) extracts sentences from abstracts of scientific publications related to a pair or a list of genes, proteins, or keywords and uses shallow parsing for the classification of the extracted sentences as noninteractive, interactive, or simple abstract co-occurrence.

Information extraction systems based on the full-sentence parsing approach tend to be more precise as they deal with the structure of an entire sentence, and variations of the full parsing-based approach have been applied for biomedical information retrieval. However, full parsers are significantly slower and require more memory than shallow analyses because they have to deal with general syntactic ambiguity and handle the full set of possible structures of whole sentences.

The full (deep) parsing is based on the language description with the help of formal grammars. Such approach is usually more accurate than shallow parsing as it is working with the structure of an entire sentence. On the other hand, the main disadvantages are the full dependence from the quality and fullness of the training set and high requirements to memory. The MedScan system [34] (http://www.elsevier.com/online-tools/pathway-studio/training-support#faqs) from Pathway Studio used a full syntactic parser for the analysis of the semantic and lexical structure of sentences and search of interactions between various biological objects, including small molecules, genes, proteins, protein functional classes, diseases, and cell processes.

6.4 The ANDSystem Tool

The ANDSystem tool incorporates methods for automated extraction of knowledge from the PubMed abstracts of scientific publications and factographic databases [35]. The ANDSystem consists of three main modules: module of linguistic text analysis and extraction of knowledge from text; the ANDCell database, containing the results of knowledge extraction from PubMed in the form of associative networks; and the ANDVisio tool that provides a graphical interface for ANDCell, intended for the graphical visualization and analysis of associative gene networks comprising relationships between biological processes, diseases, and molecular-genetic objects (proteins, genes, metabolites). The vertices of such networks are molecular-genetic objects, diseases, and processes while the edges between the vertices represent types of associations. Considered are the following objects: genes, proteins, microRNAs, metabolites, molecular processes and pathways, cellular components, and diseases (Fig. 6.1).

The following types of relationships are established between molecular-genetic objects: association, interactions, co-expression, treatment, catalytic reactions, conversion of molecules, degradation of a protein, regulation of gene expression, regulation of activity or function, regulation of transport, regulation of stability or degradation, and regulation of molecular-biological processes and diseases.

Algorithms for extraction of knowledge from text implemented in ANDSystem are based on the use of dictionaries and templates [36]. A thesaurus of genes was compiled with the use of the NCBI gene database; for the protein dictionary, a Swiss-Prot database was used; a list of diseases was extracted from the PharmGKB; for the metabolites, a ChEBI database was analyzed; biological processes and cellular components were obtained from Gene Ontology; and for microRNA, miRBase

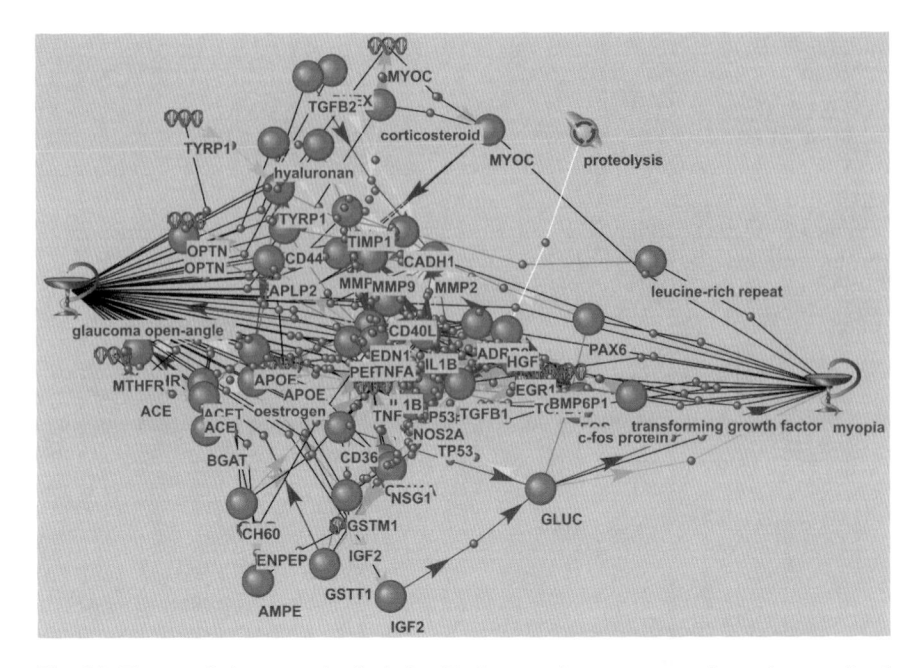

Fig. 6.1 The associative network of relationship between human genes and proteins associated with open-angle glaucoma and myopia generated with ANDVisio

was used. The extraction of relationships between described biological objects from text was done with a help of about 4,000 manually created templates. The obtained knowledge base now consists of over five million facts about relationships between diseases, molecular-genetic objects, and biological processes.

With the ANDVisio, an associative network describing relationship between human genes and proteins associated with open-angle glaucoma and myopia diseases [37]. The built network contains 15 genes and 50 proteins that are associated with myopia and open-angle glaucoma at the same time and over 400 relationships between them (Fig. 6.1). It identified 26 pathways between myopia and open-angle glaucoma containing the most important objects and relationships, including SMAD3, PAX6, IPO13, GCR, NOE3, MYOC proteins, and the OLFM3 gene.

References

1. Shatkay H, Wilbur WJ (2000) Finding themes in medline documents: probabilistic similarity search. In: Hoppenbro J, Souza Lima T, Papazoglou M, Sheth A (eds) Proceedings IEEE advances in digital libraries 2000, Washington DC, May 2000, pp 183–192
2. Joyce T, Needham RM (1997) The thesaurus approach to information retrieval. American documentation (1958) 9:192–197. In: Sparck Jones K, Willet P (eds) Readings in information retrieval. Morgan Kaufmann Publishers Inc, California (1997), pp 15–20

3. Salton G (1968) Automatic information organization and retrieval. McGraw Hill, New York
4. Sebastiani F (1999) Machine learning in automated text categorization. Technical report IEI-B4-31-1999, Istituto di Elaborazione dell'Informazione. CNR, Pisa
5. Кириченко КМ, Герасимов МБ (2001) Обзор методов кластеризации текстовых документов. Материалы международной конференции Диалог, т 2, Аксаково, 2001
6. Гаврилова ТА, Хорошевский ВФ (2000) Базы знаний интеллектуальных систем. Учебник, Питер, Санкт-Петербург, 2000
7. Ильин Н, Киселёв С, Танков С, Рябышкин В (2006) Технологии извлечения знаний из текста, Открытые системы, 6, 2006
8. Schuler G, Epstein J, Ohkawa H, Kans J (1996) Entrez: molecular biology database and retrieval system. Methods Enzymol 266:141–162
9. Muller HM, Kenny EE, Sternberg PW (2004) Textpresso: an ontology-based information retrieval and extraction system for biological literature. PLoS Biol 2:309
10. Becker K et al (2003) PubMatrix: a tool for multiplex literature mining. BMC Bioinforma 4:61
11. Krauthammer M, Nenadic G (2004) Term identification in the biomedical literature. J Biomed Inform 37:512–526
12. Krallinger M, Morgan A, Smith L, Leitner F, Tanabe L, Wilbur J, Hirschman L, Valencia A (2008) Evaluation of text mining systems for biology: overview of the Second BioCreative community challenge. Genome Biol 9(2):1
13. Ananiadou S, McNaught J (eds) (2006) Text mining for biology and biomedicine. Artech House, Norwood
14. Collier N, Nobata C, Tsujii J (2000) Extracting the names of genes and gene products with a hidden Markov model. In: Proceedings of COLING 2000, Saarbruecken, pp 201–207
15. Morgan A, Yeh A, Hirschman L, Colosimo M (2003) Gene name extraction using FlyBase resources. In: Proceedings of NLP in biomedicine. ACL 2003, Sapporo, pp 1–8
16. Kazama J, Makino T, Ohta Y, Tsujii J (2002) Tuning support vector machines for biomedical named entity recognition. In: ACL-02 workshop on natural language processing in biomedical applications, Pennsylvania, July 2002
17. Kim JD, Ohta T, Tateisi Y, Tsujii J (2003) GENIA corpus – a semantically annotated corpus for bio-textmining. Bioinformatics 19(1):180–182
18. Cohen KB, Hunter L (2005) Natural language processing and systems biology. In: Dubitzky W, Azuaje F (eds) Artificial intelligence and systems biology. Springer, Dordrecht
19. Liu H, Hu ZZ, Zhang J, Wu C (2006) BioThesaurus: a web-based thesaurus of protein and gene names. Bioinformatics 22:103–105
20. Bairoch A, Apweiler R, Wu CH et al (2007) The Universal Protein Resource (UniProt). Nucleic Acids Res 35:193–197
21. Wheeler D, Church D, Federhen S et al (2003) Database resources of the National Center for Biotechnology. Nucleic Acids Res 31:28–33
22. Eppig JT et al (2005) The Mouse Genome Database (MGD): from genes to mice — a community resource for mouse biology. Nucleic Acids Res 33:471–475
23. Christie KR et al (2004) Saccharomyces Genome Database (SGD) provides tools to identify and analyze sequences from Saccharomyces cerevisiae and related sequences from other organisms. Nucleic Acids Res 32:311–314
24. De la Cruz N et al (2005) The Rat Genome Database (RGD): developments towards a phenome database. Nucleic Acids Res 33:485–491
25. Drysdale RA, Crosby MA (2005) FlyBase: genes and gene models. Nucleic Acids Res 33:390–395
26. Chen N et al (2005) WormBase: a comprehensive data resource for Caenorhabditis biology and genomics. Nucleic Acids Res 33:383–389
27. Tsuruoka Y, Tsujii J (2003) Boosting precision and recall of dictionary-based protein name recognition. In: Ananiadou S, Tsujii J (eds) Proceedings of the ACL 2003 workshop on natural language processing in biomedicine, Stroudsburg, July 2003, vol 13. Association for Computational Linguistics, Stroudsburg, pp 41–48

28. Ohta T, Tateishi Y, Mima H, Tsujii J (2002) Genia corpus: an annotated research abstract corpus in molecular biology domain. In: Proceedings of the human language technology conference, San Diego, March 2002
29. Hakenberg J et al (2008) Inter-species normalization of gene mentions with Gnat. Bioinformatics 24:126–132
30. Tsuruoka Y, Tsujii J, Ananiadou S (2008) FACTA: a text search engine for finding associated biomedical concepts. Oxford J 24(21):2559–2560
31. Scheer M, Grote A, Chang A et al (2011) BRENDA, the enzyme information system in 2011. Nucleic Acids Res 39:670–676
32. Blaschke C, Valencia A (2001) The potential use of SUISEKI as a protein interaction discovery tool. Genome Inform 12:123–134
33. Chen H, Sharp BM (2004) Content-rich biological network constructed by mining PubMed abstracts. BMC Bioinforma 5:147
34. Nikitin A, Egorov S, Daraselia N, Mazo I (2003) Pathway studio – the analysis and navigation of molecular networks. Bioinformatics 19(16):2155–2157
35. Demenkov PS, Ivanisenko TV, Kolchanov NA, Ivanisenko VA (2012) ANDVisio: a new tool for graphic visualization and analysis of literature mined associative gene networks in the ANDSystem. Silico Biol 11(3):149–161
36. Demenkov PS, Aman EE, Ivanisenko VA (2008) Associative network discovery (AND) – the computer system for automated reconstruction networks of associative knowledge about molecular-genetic interactions. Comput Technol 13(2):15–19
37. Podkolodnaya OA, Yarkova EE, Demenkov PS, Konovalova OS, Ivanisenko VA, Kolchanov NA (2011) Application of the ANDCell computer system to reconstruction and analysis of associative networks describing potential relationships between myopia and glaucoma. Russ J Genet 1(1):21–28

Part III
Network Visualization, Modeling and Analysis

Chapter 7
Network Visualization for Integrative Bioinformatics

Andreas Kerren and Falk Schreiber

Abstract Approaches to investigate biological processes have been of strong interest in the past few years and are the focus of several research areas like systems biology. Biological networks as representations of such processes are crucial for an extensive understanding of living beings. Due to their size and complexity, their growth and continuous change, as well as their compilation from databases on demand, researchers very often request novel network visualization, interaction, and exploration techniques. In this chapter, we first provide background information that is needed for the interactive visual analysis of various biological networks. Fields such as (information) visualization, visual analytics, and automatic layout of networks are highlighted and illustrated by a number of examples. Then, the state of the art in network visualization for the life sciences is presented together with a discussion of standards for the graphical representation of cellular networks and biological processes.

Keywords Biological networks • Visualization • Graph drawing • Visual analytics • Interaction • Exploration • SBGN • Visualization tools

---

A. Kerren (✉)
Department of Computer Science, Linnaeus University, Vejdes Plats 7, SE-351 95 Växjö, Sweden
e-mail: andreas.kerren@lnu.se

F. Schreiber
Martin Luther University Halle-Wittenberg, Von-Seckendorff-Platz 1, D-06120 Halle, Germany

IPK Gatersleben, Corrensstrasse 3, D-06466 Gatersleben, Germany
e-mail: schreibe@ipk-gatersleben.de

7.1 Introduction

Many biological processes are represented as networks. Examples are networks from the area of molecular biology, such as metabolic networks, protein interaction networks, and gene regulatory networks, but also from other areas of the life sciences such as ecological networks, phylogenetic networks, neuronal networks, chemical structures, and infection networks. Network modeling, analysis, and visualization are important steps towards a systems biological understanding of organisms and organism communities. The graphical depiction of such networks supports the understanding of the underlying processes and is essential to make sense of much of the complex biological data that is now being generated.

A picture of a network is called a *network diagram* or a *network map*; see Fig. 7.1 for an SBGN map of a metabolic pathway. A network diagram representing

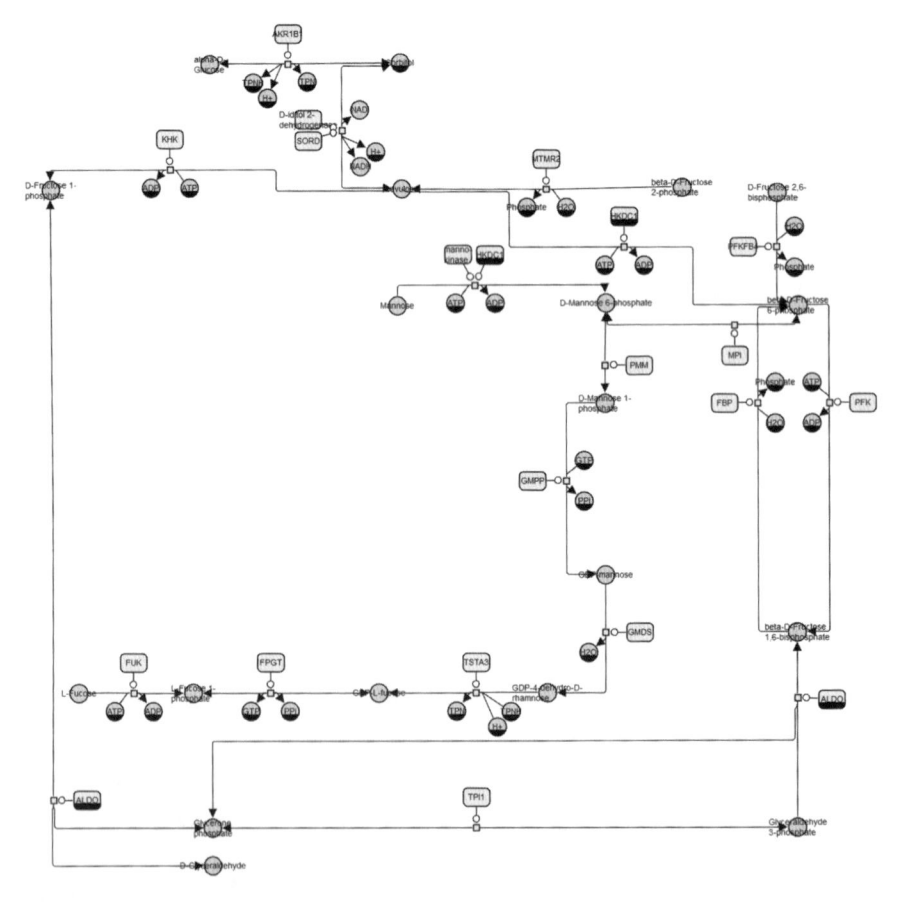

Fig. 7.1 A map of a metabolic pathway shown in the SBGN standard [88], derived from *KEGG* [61], computed and displayed by *Vanted* [110]

biological processes consists of a set of elements (called *nodes* or *vertices*) and their connections or interactions (called *edges*). These elements and connections often have a defined appearance and are placed in a specific layout. Due to the size and complexity of such networks, methods for their automatic visualization and interactive exploration are desired.

Network diagrams or maps have been produced manually for a long time. Examples are textbooks on biochemistry [8,96], biological network posters [94,99], and some electronic information systems such as *ExPASy* [4] and *KEGG* [61]. The drawings in these resources have been created manually long before their use and provide only a restricted view of the data. These maps represent the knowledge at the time of their generation and are static, hence cannot be changed by an end user. Therefore, this type of biological network visualization is often called *static visualization*.

Because of the size and complexity of biological networks, their steady growth and continuous change, as well as the compilation of user-specific networks from databases, novel automatic visualization, interaction, and exploration methods are desired. The generation of a network map on demand is called *dynamic visualization*. Such visualizations are automatically created by the end user from up-to-date data. Their advantages are, inter alia, that they can be modified to provide particular views at the data and often navigation and exploration methods are supported in interactive systems.

This review gives a brief introduction into (information) visualization, visual analytics, and automatic layout of networks, presents the state of the art in automatic network visualization for the life sciences, and standards for the graphical representation of cellular networks and biological processes. It is structured in two main parts as follows: Sect. 7.2 provides information about the foundations from computer science in general and looks into the subareas of *information visualization*, *graph drawing* (network visualization), and *visual analytics* in particular. Section 7.3 takes a closer look at the visualization of biological networks and discusses methods, some important tools, and the SBGN standard. It looks into the application and extension of computer science methods for the special requirements of the life sciences.

7.2 Background

The effective visualization of biological networks is influenced by research from many different fields. In the past, such networks were simply considered as large graphs (or hypergraphs), and a suitable visual representation was restricted to finding an appropriate (static) graph layout. Nowadays, research in the visualization of large and complex networks is more focused on interactive exploration and analysis that includes the consideration of additional data that might be attached to various graph elements or that might be the basis for the construction of biochemical networks. The process of such a data collection and storage will heavily increase in the future. This is especially true in systems biology where, for example, the huge amount of *omics data automatically generated by high-throughput technologies [3, 39] lead to the challenge of interpreting all of these data sets in context of networks. The fundamental problem today is to transform the data—which is typically not preprocessed, erratic, stored in idiosyncratic formats, sometimes uncertain, and often composed of various types (multidimensional, time dependent, geospatial, etc.)—into information and make it useful/available/analyzable to analysts. Often, this challenge is called the *information overload problem*. Positive effects of such a transformation are then to discover something that is interesting (like patterns or outliers) or to monitor a huge data set in real time [70].

Because of this general view on the problem, we provide a more general background section. First, we discuss the field of information visualization in the next subsection. We highlight the most important definitions/aims and present a brief high-level overview of visual representations and interaction techniques. Then, we outline the field of graph drawing and discuss the most often used layout algorithms. Finally, a relatively new field, called visual analytics, is introduced. Due to page limitations, we cannot give a comprehensive overview of all aspects of the aforementioned research fields. Instead, we present a selection of fundamental ideas/approaches and refer to the literature including surveys.

7.2.1 Information Visualization

Information visualization (InfoVis) is a research area which focuses on the use of interactive visualization techniques to help people understand and analyze data. While related fields such as scientific visualization involve the presentation of data that has some physical or geometric correspondence, information visualization centers on abstract information without such correspondences, i.e., information that cannot be mapped into the physical world in most cases. Examples of such abstract data are symbolic, tabular, networked, hierarchical, or textual information sources. The ever-increasing amount of data generated or made available every day amplifies the urgent need for InfoVis tools. To give the field a firm base, InfoVis combines several aspects of different research areas, such as scientific visualization, human-computer interaction, data mining, information design, cognitive psychology, visual perception, cartography, graph drawing, and computer graphics [73, 74].

7.2.1.1 The Importance of Human Visual Perception and Visual Metaphors

Human information processing and the human capability of information reception have to be adequately taken into account when developing visualization tools. This should be reflected in an appropriate user interface design, a clean requirement analysis and modeling, and perhaps most important an efficient interaction between the human analyst and the computer. Discussing the different features of our eye, the various process models of human visual perception (incl. preattentive perception and features) or our capabilities of pattern recognition would go beyond the scope of this background section. There are many good textbooks that deal with these topics in context of visualizations: we recommend the books of Ware [141], Kerren et al. [74], and Ward et al. [140].

Edward Tufte, one of the leaders in the field of visual data exploration, describes in his illustrated textbooks [131–133] how information can be prepared so that the visual representation depicts both the data and the data context. The use of suitable visual metaphors assists our brain in its endeavor to connect new information received through the visual input channels to existing information stored in short- or long-term memory [72]. Tufte inspired many InfoVis researchers in their ambition to develop novel visual representations for the data sets under consideration (the process of representing a concrete data set by an appropriate visual structure is called "visual mapping") as well as interaction techniques which support a better understanding of the data.

7.2.1.2 Visual Representations

Visual mappings explain how data models can be expressed using visual metaphors and be converted into corresponding visual representations which are suitable for interaction. This is typically done in the 2D space, because 3D representations usually introduce unnecessary clutter and navigation problems. We highlight the most important visualization techniques for basic data types in the following paragraphs. Of course there are other types of data that have to be considered. We refer to the literature if the reader is interested to get more information, such as [27, 102] for geo-spatial data, [2] for time-series data, or [41, 126, 140] for a comprehensive discussion of visual representations in general.

Visualization Techniques for Multivariate Data

Multivariate (or multidimensional) data sets can mostly be described as data tables with n data objects and m attributes/features, i.e., for each object exists an attribute vector with m dimensions. The attribute values can be classified into nominal, ordinal, or quantitative. In practice, we often have a large amount of data objects and many attributes with different types. Finding a suitable visual representation is thus challenging, and the right choice might depend on further parameters like application domain, integration into a larger visualization environment, or support of specific interaction techniques. In general, visual mappings for multivariate data can roughly be categorized as follows:

Point-based approaches: This class of techniques projects n-dimensional objects from the data space to a lower-dimensional—typically 2D—display space [140].

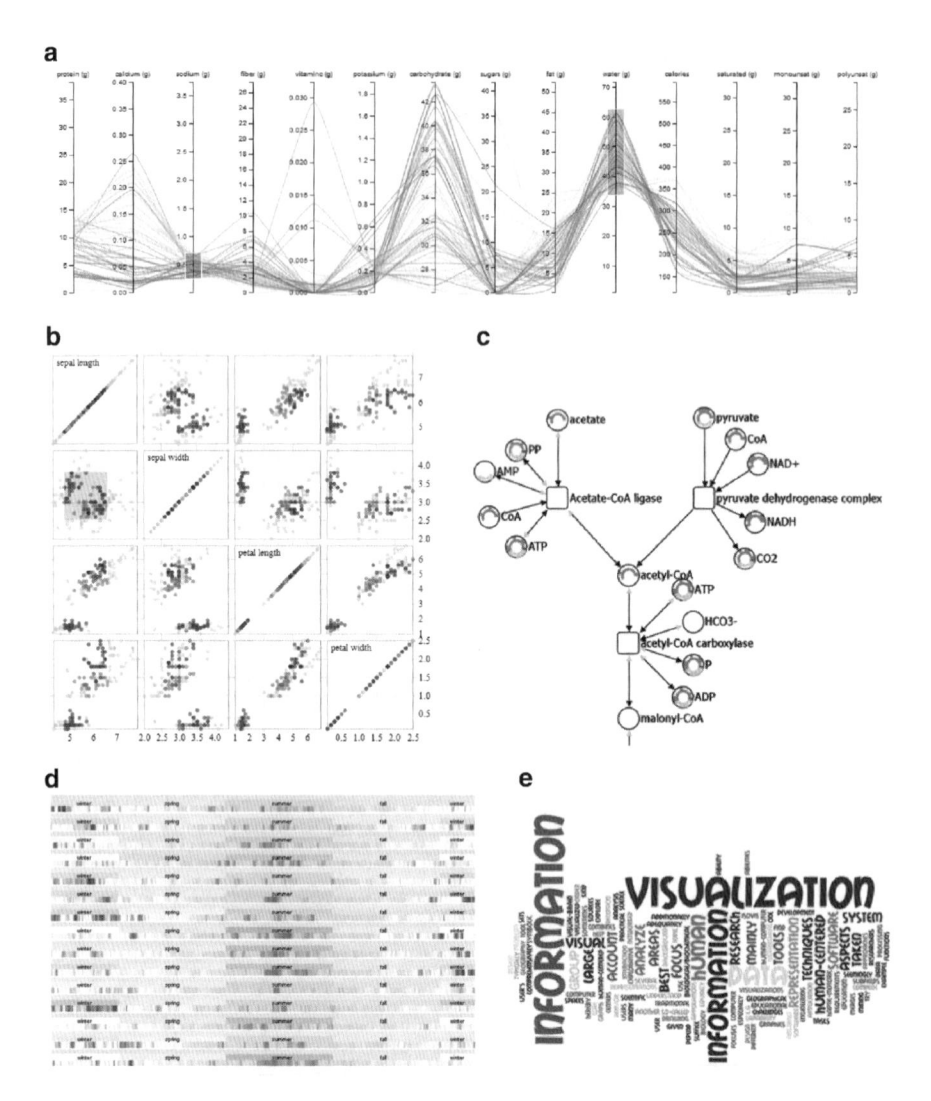

Fig. 7.2 Some examples of often used visualization techniques. The screenshots in (**a**) and (**b**) were produced with D3 [22]. (**a**) Parallel coordinates that visualize a nutrient content data set with more than 1,000 data objects and 14 attributes (available online [31]). Note that the visible polylines were interactively selected in the 3rd and 10th axes. (**b**) A scatterplot matrix showing data from the Iris data set (available online [11]). Also in this case, the colored points indicate data selected by the user (see the *grey*-colored selection in the plot of the first column, second row). (**c**) Small icons/glyphs are embedded into the graph nodes of a metabolic network. In this case, they indicate reachable nodes in other (color-coded) pathways [60]. (**d**) A pixel-based approach to visualize weather data of a city. The rows represent years, and the temperatures (color-coded from *blue* over *white* to *red*) of each day are ordered from *left* to *right* [90]. (**e**) Sample tag cloud of a text document which is related to information visualization (generated with Wordle [32]) (Color figure online)

There are different variations: *scatterplot matrices*, for instance, consist of a grid of 2D scatterplots each showing a possible pair of dimensions/attributes [19]; see Fig. 7.2b for an example. Dimensional reduction techniques, such as multidimensional scaling (*MDS*) [92, 145], principal component analysis (*PCA*) [53], or self-organizing maps (*SOMs*) [80], project n-dimensional data records into 2D/3D directly. The idea is to preserve properties of the multivariate data space during the projection, i.e., similar data objects in data space should also be similar in display space which is represented by neighborhood. Note that absolute positions in the display space are less important, in contrast to relative positions.

Axis-based approaches: Here, a multidimensional data object is usually represented by a polyline, and its attribute values are marked on coordinate axes which can be arranged in various ways. Thus, the user can read the attribute values from the intersections between the coordinate axes and the polyline. The most prominent examples are *parallel coordinate systems* [49] (cf. Fig. 7.2a) or *star plots* [16] (also called Kiviat diagrams).

Icon-based approaches: Icon- or glyph-based approaches are coherent graphical entities that represent the attribute values of a data record by modification of the entity's visual features, such as line thickness, size, color, and orientation. There are many different realizations, such as *stick figures* [106], *Chernoff faces* [18], or *shape coding* [7]. A variant of so-called rose diagrams [100] is shown in Fig. 7.2c.

Pixel-based approaches: Such approaches try to maximize the available display space by mapping attribute values to single pixels. There is only one degree of freedom to represent such a value by a pixel: its color. Therefore, the challenge in the development of pixel-based representations is to arrange the used pixels on the screen in a meaningful way. Well-known examples are *recursive patterns* [65] or the *VisDB* tool [66] for the analysis of databases. Figure 7.2d exemplifies the idea in context of the visualization of weather data collected over time.

Visualization Techniques for Hierarchical Data and Networks

Networks and trees are in the center of our interest in this chapter. Therefore, we provide an own Sect. 7.2.2 for a deeper discussion of suitable visualization possibilities for these data types and focus there on traditional node-link approaches. For the sake of completeness, we want to note that there are also so-called space-filling methods that try to solve some conceptual problems of node-link diagrams, such as the high space consumption and difficult inclusion of many (and complex) attributes into the drawing. *Treemaps* fall into this category in which the hierarchy is recursively mapped to rectangular areas [52]. Other examples are *Beamtrees* [134], *sunburst* approaches [108], or *network matrices* [1].

Visualization Techniques for Text and Documents

Today, the availability of texts and documents is overwhelming, and people want to actively deal with them to solve specific problems. Typical questions are as follows: what documents contain a text about a specific topic? Or are there similar documents to those that I already have? Information visualization is capable of supporting the aforementioned tasks in several ways.

Text visualization: First, we focus on approaches to the visualization of a single text document. *Tag Clouds* provide information about the frequency of words contained in a text [63]. The approach uses different font sizes for each word in the text to indicate how often a certain word is used in comparison with the other words as shown in Fig. 7.2e. Several extensions and related approaches exist, such as *Wordle* or *ManiWorlde* [77, 138]. *SparkClouds* extend the original tag cloud idea with a temporal variable by so-called sparklines [87]. Thus, trends can easily be identified and analyzed. An approach for visual literary analysis is called *Literature Fingerprinting* [67]. It supports the visual comparison of texts by calculating features (e.g., word/sentence length or measurement of vocabulary richness) for different hierarchy levels and by creating characteristic fingerprints of the texts.

Document visualization: Collections of text documents can be structured to some extent (software packages, wikis, etc.) or relatively unstructured (e-mails, patents, etc.). Early approaches, e.g., *Lifestreams* [34], simply arranged documents according to specific attribute values such as time tags. More recent works analyze the documents by metrics, such as similarity, and perform cluster analyses or compute SOMs. Conceptually similar (by looking at the resulting visual representation) is *ThemeScapes* [147] that follows a natural landscape metaphor. Single documents are categorized and then mapped to a document map as topic areas, whereas the documents themselves are shown as small dots. "Mountains" in the landscape represent document concentrations in a thematic environment (density), height lines connect concept domains, etc. There are many more recent approaches that make use of the same metaphor, such as [104]. In order to carry out comparisons of text documents using tag clouds, *Parallel Tag Clouds* [20] arrange tags on vertical lines for each document. Identical words are then highlighted by connection lines.

7.2.1.3 Interaction Techniques

Interaction techniques in information visualization are mechanisms "for modifying what the users see and how they see it" [140]. There are many taxonomies of interaction techniques in the literature which help to better understand the design space of interaction; a nice overview is provided by Yi et al. [148]. In the following, we present a simplified and shortened classification of interaction methods for information visualization from our paper [70] which is based on [43] of its own:

Data and view specification: This category focuses on the data space and how the data is visually represented (corresponds to data transformations and visual mappings in the InfoVis Reference Model [14]):

- *Encode/visualize:* Users can choose the visual representation of the data records including graphical features, such as color and shape. Visual representations typically depend on the data types as discussed in Sect. 7.2.1.2.
- *Reconfigure:* Some interaction techniques allow the user to map specific attributes to graphical entities. An example is the mapping of attributes in a multivariate data set to different axes in a scatterplot.
- *Filter:* This technique is of great importance as it allows the user to interactively reduce the data shown in a view. Popular methods are *dynamic queries* by using range sliders [146] or picking a set of nodes in a network visualization for further analyses by performing a "lasso" selection [44].
- *Sort:* Ordering of records according to their values is a fundamental operation in the visual analysis process. This is, for example, important in network analysis where nodes might be sorted based on specific centrality values [150].

View manipulation: Our second category addresses interacting with visual representations (view transformations in the InfoVis Reference Model).

- *Select:* Selection is often used in advance of a filter operation. The aim is to select an individual object or a set of objects in order to highlight, manipulate, or filter them out. Examples include putting a placemark on a virtual map to highlight a spatial area or the specification of attribute ranges in parallel coordinate systems as seen in Fig. 7.2a.
- *Navigate/explore:* This important class of interaction techniques typically modify the level of detail in visualizations following the mantra *overview first, zoom and filter, and details on demand* [121]. Well-known approaches are *focus and context* [111], *overview and detail* [51], *zooming and panning* [137], and *semantic zooming* [127].
- *Coordinate/connect:* Linking a set of views or windows together to enable the user to discover related items. Brushing and linking techniques (e.g., *histogram brushing* [89]) are used in almost all information visualizations, such as in [59].
- *Organize:* Large visualization systems often consist of several windows and workspaces that have to be organized on the screen. Adding and removing views can be confusing to the analyst. Some systems help the user to better overview and to preserve his/her mental map by grouping of views or by assigning specific places where they have to appear [50, 91].

Note that it is possible and also common practice to combine the aforementioned techniques. The given literature references only point to selected example works and make no claim to be complete.

7.2.2 Graph Drawing and Network Visualization

In this subsection, we distinguish between *graphs* and *multivariate networks*. A (simple) graph $G = (V, E)$ consists of a finite set of vertices (or nodes) V and a set of edges $E \subseteq \{(u, v) | u, v \in V, u \neq v\}$, whereas a multivariate network N consists of an underlying graph G plus additional attributes that are attached to the nodes and/or edges. To describe the fundamental ideas of graph visualization algorithms more efficiently, we have to provide some definitions:

- An edge $e = (u, v)$ with $u = v$ is called a *self-loop*.
- If an edge e exists several times in E, then it is called a *multiple edge*.
- A *simple graph* has no self-loops and no multiple edges. Here, we assume that all graphs are simple graphs for the sake of convenience.
- The *neighbors* of a node v are its adjacent nodes.
- The *degree* of a node v is the number of its neighbors.
- A *directed graph* (or digraph) is a graph with directed edges, i.e., (u, v) are ordered pairs of nodes.
- A directed graph is called *acyclic* if it has no directed cycles, i.e., there is no directed path where the same node is visited twice.
- A graph is *connected* if there is a path between u and v for each pair (u, v) of nodes.
- A graph is *planar* if it can be drawn in the 2D plane without intersections of edges (*edge crossings*).

7.2.2.1 Traditional Graph Drawing (GD)

Graph drawing algorithms compute a 2D/3D layout of the nodes and the edges, mainly based on so-called node-link diagrams [141]. They play a fundamental role in network visualization. Particular graph layout algorithms can give an insight into the topological structure of a network if properly chosen and implemented. The graph readability is affected by quantitative measurements called *aesthetic criteria* [24], such as:

- Minimization of edge crossings
- Minimization of the drawing area
- Displaying the symmetries of the graph topology
- Constraining edge lengths
- Constraining the number of edge bends
- Maximization of the resolution

Thus, graph drawing generally deals with the ways of drawing graphs according to the set of predefined aesthetic criteria [17]. A problem is that these criteria are often contradictory, and problems which aim to optimize the criteria are often NP-hard. Therefore, many GD algorithms are heuristics. Note that we only focus

7 Network Visualization for Integrative Bioinformatics

on traditional GD approaches in this subsection. There are further possibilities to represent graphs, such as matrix representations [1] or hybridizations between both approaches [44] (cf. Sect. 7.2.1.2).

In the following paragraphs, a selection of drawing approaches is presented. These are layout methods for trees, force-based layout techniques, and hierarchical drawings. There are many more approaches not discussed here, for instance, orthogonal layouts [29], visualization of hypergraphs [9], or dynamic layouts for graphs that change over time [25] (a possible application of dynamic approaches is visualizing the evolution of biochemical networks [112], for instance). Implementing good graph drawing algorithms is usually complicated and time-consuming. Therefore, a number of different open source libraries were developed, such as *JUNG* [105] and many others, that allow to simply call predefined methods for the computation of a specific graph layout.

Tree Drawings

Trees are a special case of directed (acyclic) graphs that usually have a distinguished node called the *root* of the tree. We can regard a tree as a digraph with all edges oriented away from the root. A binary tree is a rooted tree where each node has at most two children (we assume here that binary trees are ordered). The graph drawing community developed a lot of different layout methods for binary and general trees. In this context, there is another set of more specified aesthetic criteria especially for (binary) trees:

- Nodes at the same level of the tree should lie along a straight line, and the straight lines defining the levels should be parallel.
- A left subtree should be positioned to the left of its parent node and a right subtree to the right.
- A parent node should be centered over its subtrees.
- Two isomorphic subtrees should be drawn equally. Graph isomorphism means that there is a bijection between two graphs, so that any two nodes u and v are adjacent in the first graph if and only if their bijections are adjacent in the second graph.
- A tree and its mirror image should produce drawings that are reflections of one another.
- Integer coordinates should be preferred which leads to a grid drawing at the end.

Many tree layout algorithms use a divide and conquer strategy, such as the well-known Reingold/Tilford algorithm for binary trees [107]. In a postorder traversal of the tree, the following simple steps are executed:

1. Draw the left subtree.
2. Draw the right subtree.
3. Combine both drawings with a specific minimum distance.

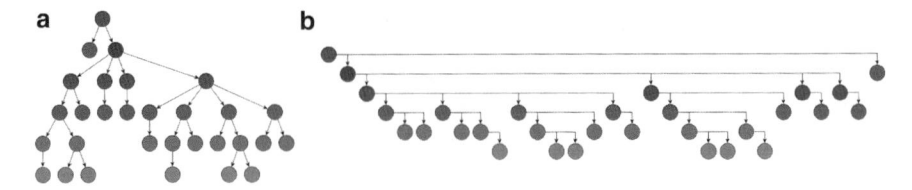

Fig. 7.3 Two sample tree layouts that were computed and displayed by the *yED* graph editor [149]. The identical input tree has 30 nodes and 29 edges. (**a**) A standard tree layout for general trees. (**b**) A so-called HV-drawing in which the layout algorithm switches between the *horizontal* and *vertical* orientation

4. Place the root of both subtrees at the next upper level exactly in the center of its subtrees.
5. In case the parent node has only one subtree, place the root in a specific horizontal distance.

Reingold/Tilford runs in linear time and can relatively easily be extended for the layout of general trees [13, 139]. Of course, there are further possibilities of drawing trees with the help of node-link diagrams, such as radial layouts, H-trees, or HV-trees. We refer the reader to the standard literature [24, 64]. Figure 7.3 shows two example layouts computed with the *yED* tool [149].

Force-Based Drawings

Force-based layout techniques use a physical analogy to draw graphs and are widely used in practice. This is because of several reasons: the physical metaphor makes them easy to understand and to code, the results are suitable for many application fields, they are easy to extend with additional constraints, and the process of obtaining an equilibrium state (see below) can be animated which looks pretty nice. A simple version of a force-based layout algorithm using spring and electrical repulsion forces is introduced in the following. Here, the edges between nodes are modeled as springs, and the nodes can be considered as charged particles that repel each other. For the x-component of the force vector on a node v, the following holds (y-component analogous):

$$\sum_{(u,v) \in E} (\text{sti}_{uv}(d_{uv} - l_{uv})) \hat{x}_{uv} + \sum_{(u,v) \in V \times V} \frac{\text{rep}_{uv}}{d_{uv}^2} \hat{x}_{uv} \quad (7.1)$$

Here, $\hat{x}_{uv}$ denotes the unit vector of $(x_v - x_u)$. d_{uv} is the Euclidean distance between u and v, l_{uv} is the zero-energy (natural) length of the spring between u and v (i.e., no force if $d_{uv} = l_{uv}$), $\text{sti}_{uv} \in [0, 1]$ is the stiffness of the spring between u and v (i.e., the larger this parameter the more the tendency for d_{uv} to be close to l_{uv}), and finally rep_{uv} is the strength of the electrical repulsion between the two nodes. In Eq. 7.1,

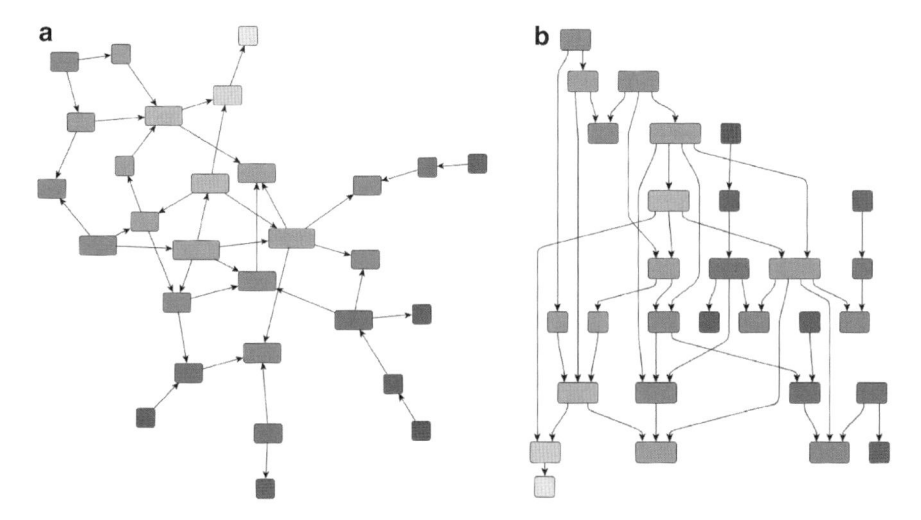

Fig. 7.4 Two sample graph layouts that were computed and displayed by the *yED* graph editor [149]. The identical input digraph has 29 nodes and 39 edges. (**a**) Result of a force-based layout algorithm. (**b**) Layered (or hierarchical) drawing

the first sum represents the spring force between two nodes u and v connected with an edge and the second sum the repulsion force between v and other nodes. Both forces together build a complete force system for all graph elements. Depending on the underlying physical model, the repulsion forces avoid that nodes are getting too close, and the spring forces provide a uniform edge length, for instance. In the current formula, Hook's law is used to specify the spring force between two nodes, i.e., if the distance between the two nodes is larger than the natural length of the spring, then the nodes attract each other. And the strength of the attraction is proportional to the difference between distance and natural length.

A simple algorithm that computes a final graph layout consists of a loop which firstly computes the forces of all nodes and then moves each node a bit into the direction of its force vector computed in Eq. 7.1. At the beginning, all nodes are positioned randomly. The loop is left if the sum of all forces together is small enough (equilibrium state) or after a specific number of iterations. This strategy works for undirected and directed graphs, with and without cycles, cf. Fig. 7.4a.

Layered (Hierarchical) Drawings of Directed Graphs

A general aim for the layout of a directed graph is to compute a so-called *monotone* drawing in which all edges point into the same direction. Such a monotone drawing has some advantages in the interpretation of the digraph's topology [47]. Obviously, the input digraph must be acyclic in that case, otherwise we would get edges that flow backwards (called *feedback edges*). In practice this apparent hard condition is not really a problem, because we can use such a drawing method for general directed graphs if we change the direction of a minimal number of the feedback edges. This step is known as *cycle removal*. By doing so, we get a directed acyclic graph (DAG) that is drawn by using a method for computing monotone layouts, such as a layered drawing as explained in this paragraph. If the final layout is ready, we simply reverse the feedback edges again.

Many people prefer a hierarchical structure of the final graph layout, i.e., the nodes of the graph are arranged on vertical or horizontal, parallel layers in the 2D plane. Often, such a structure is already given by the input data. For instance, if someone wants to visualize hyperlinks (edges) between the HTML pages (nodes) of a website, then usually the pages are already hierarchically organized. In the following, we briefly present a standard technique for layered drawings that is based on the fundamental work of Sugiyama et al. [129].

The basic idea is very simple and intuitive; it has three phases. In the first phase, the nodes of the graph are assigned to a number of layers (we can skip this phase if there is already a layering in the input graph). This layer assignment problem is NP-complete if we want to minimize the height and the width of the final layering. A further complication occurs if edges span over several layers: then we have to introduce the so-called dummy nodes that lie on the spanned layers, i.e., a long edge is thus subdivided by the dummy nodes. This strategy causes modified edges which only reach from one layer to the next one (the digraph is called *proper* in such cases) and is needed for the second phase. After the layer assignment, we have to eliminate the number of edge crossings. This is done by reordering the graph nodes and the dummy nodes within each layer. With the help of the dummy nodes, the algorithm gets control over the edge positioning, and in consequence, it is possible to avoid crossings of edges that span over several layers. Minimizing edge crossings in a proper layered digraph is NP-complete, even if there are only two layers. Note that the node positions (x-coordinates) on the layers are relative only up to now (the y-coordinates of the nodes are already specified by the node layers if we assume to have horizontal layers). The final phase is the real coordinate assignment of all nodes on the layers, i.e., we assign concrete x-coordinates for each (normal and dummy) node. Also this task leads to an optimization problem that can be solved, for instance, by linear programming (LP). Constraints of the LP are then the fixed orderings in the layers, and the target function is specified by the straightness of the edges. As a final step, we remove the dummy nodes and obtain the wished layered drawing as shown in Fig. 7.4b.

7.2.3 Multivariate Network Visualization

Good drawing algorithms as described in the previous subsection will not solely solve the problem of visualizing multivariate networks. There are several reasons for this statement. First, the most traditional graph drawings do not scale well, i.e., they are not able to represent huge data sets with many thousands of nodes and/or edges. Second, additional multivariate data cannot be intuitively embedded into a standard drawing. The InfoVis community tried to address those issues by visualization approaches that provide filtering and interaction possibilities in order to reduce the number of graph elements under consideration as well as by methods to visually analyze attributes in context of the underlying graph topology. Several approaches can be found in the literature that attempt to offer solutions for the problem of visualizing multivariate networks: *multiple and coordinated views, integrated approaches, semantic substrates, attribute-driven layouts,* and *hybrid approaches* [57]. We will discuss these concepts in the following paragraphs:

Multiple and coordinated views: This category of solutions aims to combine several views and present them together. Coordinated views allow the use of the most powerful visualization techniques for each specific view and data set [41, 109]. As an application example, we highlight the work of Shannon et al. [120] who realized this idea in the network visualization domain. They use two distinct views: one view shows a parallel coordinate approach for the visual representation of the network attributes and the other view displays a node-link drawing of a graph. Their tool is equipped with a variety of visualization and interaction techniques; both views are coordinated by linking and brushing [126] techniques. The drawback of multiple views is that they split the displayed data because of the spatial separation of the visual elements.

Integrated approaches: To provide a combined picture, attributes and the underlying graph can be displayed in one single view. "Integrated views can save space on a display and may decrease the time a user needs to find out relations; all data is displayed in one place" [41]. One example is described in Borisjuk et al. [10] work on the visualization of experimental data in relation of a metabolic network. The authors used a straightforward approach by employing small diagrams instead of representing the nodes as simple circles or rectangles. Each diagram, e.g., a bar chart, shows experimental data that is related to the regarded node. This approach provides a view to all available information, but the embedding of the visualizations into the nodes causes the nodes to grow in size. This issue may affect the readability of the network due to the overlaps that may appear when the number of nodes and the attributes is high [71]. Thus, it does not scale well. However, the problem of space usage and clutter introduced by such approaches can be avoided by using focus and context techniques (cf. Sect. 7.2.1). *Magic lenses* are one of several possibilities that are able to interactively visualize the node attributes within the same view as exemplified in Fig. 7.5.

Semantic substrates: In order to further avoid clutter in multivariate network visualizations, some researchers realized the idea of so-called semantic substrates that "are non-overlapping regions in which node placement is based on node attributes": Shneiderman and Aris [122] introduced this idea and combined it with sliders to control the edge visibility and thus to ensure comprehensibility of the edges' end nodes. One conceptual drawback of such approaches is that the underlying graph topology is not (completely) visible.

Attribute-driven layouts: Those layouts use the display of the network elements to present insight about the attached multivariate data instead of visualizing the graph

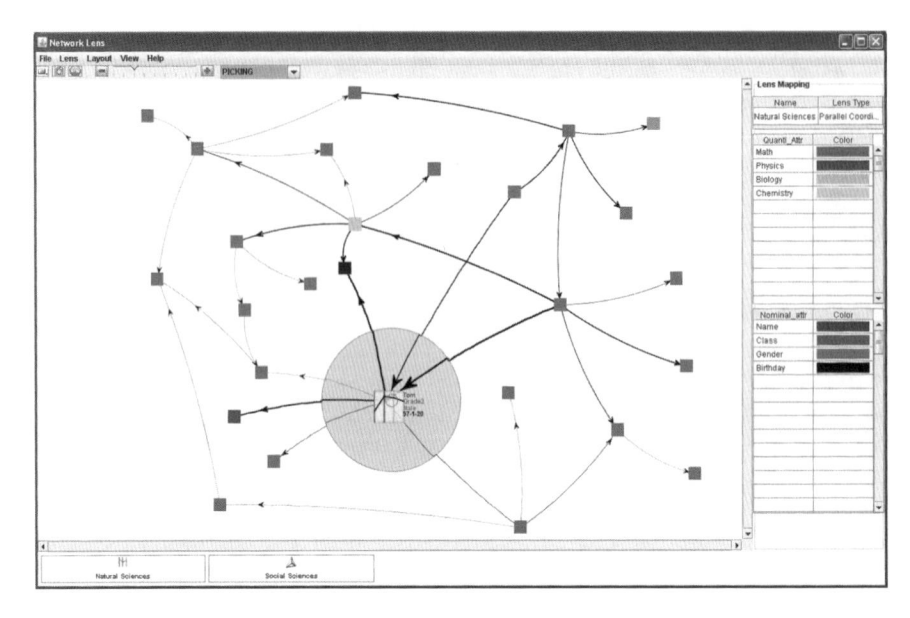

Fig. 7.5 Overview of the *Network Lens* tool [58]. The graphical user interface is divided into three distinctive parts: the main network visualization area, the lens information area on the *right-hand side*, and the *bottom* part where user-produced lenses are preserved. It offers a way to visualize additional network attributes (displayed inside of the circular lens), while preserving the overall network topology and context. The lens in the screenshot covers one node only and shows a small parallel coordinate diagram with four quantitative as well as four nominal attributes belonging to that node. The user is able to move the lens with the mouse or to translate the graph behind the lens

topology itself. While being similar to semantic substrates, this technique does not necessarily place the nodes into specific regions. Instead, it uses calculations based on node attributes to control the placement of a node in the graph layout. An example is *PivotGraph* [142] which uses a grid layout to show the relationship between (node) attributes and links.

Hybrid approaches: They combine at least two of the previously discussed techniques. The most common combinations are multiple coordinated views with any of the integrated approaches. For instance, Rohrschneider et al. [112] integrate additional attributes of a biological network inside the nodes and edges; see Fig. 7.6. The authors also use other visual metaphors for creating multiple coordinated views to show time-related data of the network.

7.2.4 Visual Analytics

Visual analytics (VA) "is the science of analytical reasoning facilitated by interactive visual interfaces" [130]. A crucial property of this research field is that computational methods of data analysis are combined with interactive visualization

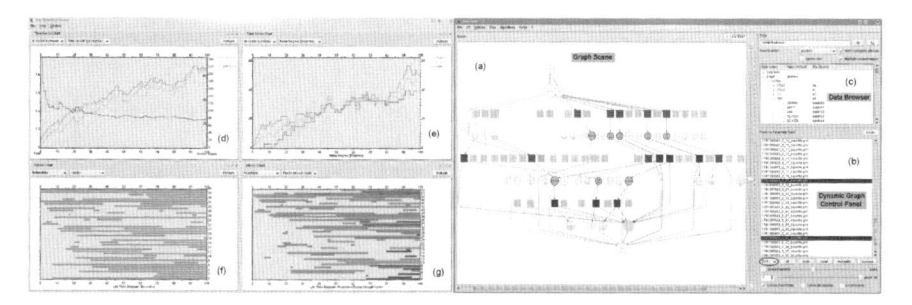

Fig. 7.6 The screenshot shows a tool for the visual analysis of dynamic metabolic networks [112]. On the *left-hand side*, two time-series charts of selected attributes display attribute dynamics over time. Interval charts represent the dynamic topology of the graph in terms of life times of metabolites, enzymes, and reactions. On the *right*, the graph scene shows the set union graph (= the super graph that summarizes all nodes/edges of the individual graphs that appear over time) with the applied node coloring scheme which supports distinguishing between older and newer nodes

techniques in order to analyze data more efficiently. *Automatic data analysis* covers various aspects from data storage and organization to automatic analysis algorithms, such as support vector machines, neural networks, and PCA. It might be classified among others into data management, data mining, and machine learning. For many data analysis problems, fully automated analysis methods only work for well-defined and well-understood problems, i.e., there has to exist a model of the underlying problem [68]. Otherwise, traditional data mining techniques will not work. Even if a model exists, then the results of the automated analyses have to be sufficiently communicated to and interpreted by analysts. Here, *interactive visualizations* come into the play as they are able to support the analyst to discover (possibly unexpected) patterns, trends, or relationships in the data. Interaction techniques (as presented in Sect. 7.2.1.3) are of particular importance to visually analyze large volumes of data. Interaction allows, among other things, to explore "unknown" data collections following Shneiderman's mantra of information visualization [121] or to build hypotheses with the help of "What if?" questions and to verify them visually or with algorithmic methods. The need to combine interactive visualization with computational analysis methods is obvious and opens novel possibilities to address the information overload problem. A more detailed discussion on VA can be found in [68, 69, 130].

As an example from the field of visual network analysis, we have selected the *ViNCent* tool [75, 150] that combines exploratory data visualization with automatic analysis techniques, such as computing a variety of centrality values for network nodes as well as hierarchical clustering or node reordering based on centrality values. Automatic and interactive approaches are seamlessly integrated in one single analysis framework which provides insight into the importance of an individual node or groups of nodes and allows quantifying the network structure; see Fig. 7.7.

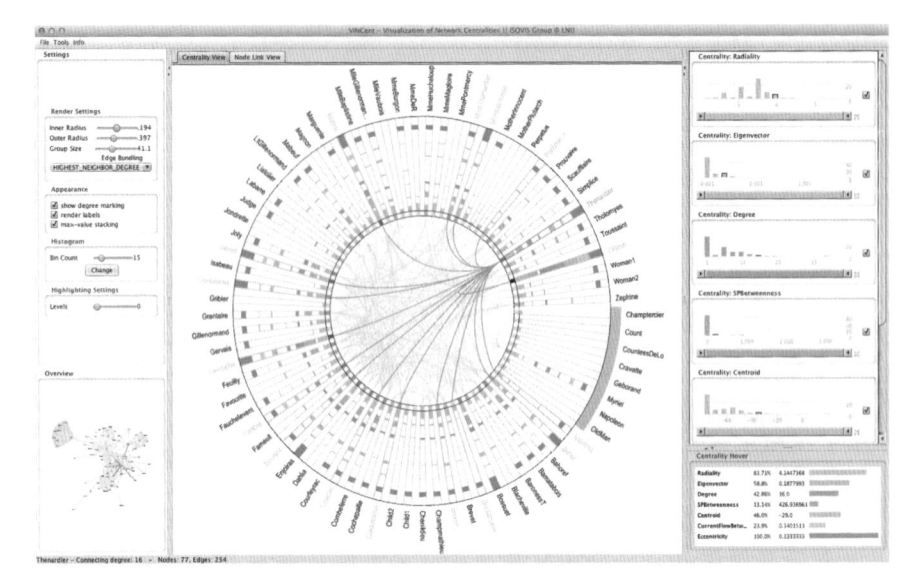

Fig. 7.7 Overview of the *ViNCent* user interface [150]. The *center* shows the radial centrality view of the input network. The *right side* displays the corresponding histograms of the network centralities as well as detailed values of the network centralities for the currently hovered node. Histograms can be used to filter the views. The *left panel* allows changing the render settings and displays an overview of the respective node-link layout of the network. A node group has been manually selected and is shown as a *light-blue* stripe along the outer *circle* in the centrality view as well as in the overview (*bottom left*) by using a background region of the same color (Color figure online)

7.3 Visualization of Biological Networks

Visual representations of biological networks are widely used in the life sciences. Examples are shown in textbooks, on pathway posters, in databases, and by a large number of tools for the analysis and visualization of biological processes. Well-known software tools are listed in Sect. 7.3.1.2. Software tools often use established layout methods as described in Sect. 7.2.2 to visualize biological networks automatically. Sometimes those algorithms are modified, for example, by adding extra forces to force-based approaches. However, often these methods do not or only partly take into account specific requirements for the visualization of a particular biological network, and hence these visualizations are usually difficult to understand, especially if large networks are visualized.

In the following subsections, we will introduce some typical solutions for common networks from molecular biology, discuss domain-adapted solutions for particular networks, list major tools for the visualization of biological networks, and finally discuss the *Systems Biology Graphical Notation* (SBGN) as the graphical standard for biological networks.

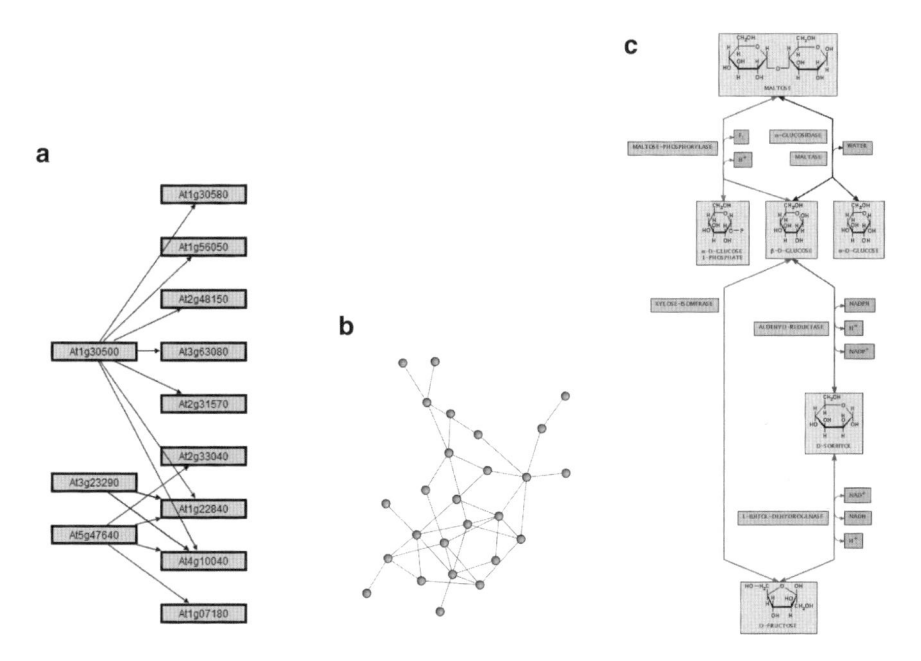

7 Network Visualization for Integrative Bioinformatics 191

Fig. 7.8 Three sample layouts of biological networks. (**a**) and (**b**) were computed and displayed by the *Vanted* system [110]; (**c**) was computed by *BioPath* [33]. (**a**) A gene regulatory network (nodes represent genes, edges represent regulation, and labels show gene names). (**b**) A protein interaction network (nodes represent proteins; edges represent interaction). (**c**) A metabolic network (nodes represent metabolites, enzymes, and reactions; edges represent consumption and production)

7.3.1 Methods

7.3.1.1 Early Approaches

Driven by the emerging availability of biological networks from databases in the mid-1990s, several groups started to either use existing graph drawing algorithms or design extensions to these algorithms to automatically visualize biological networks. In the following, we present such early work for the three major types of networks from molecular biology.

Signal Transduction and Gene Regulatory Networks

These networks represent regulation or directed interaction between biological entities (such as genes) and are usually modeled as directed graphs; see Fig. 7.8a. There are two widely used methods to visualize such networks: force-based and layered drawings. Several systems provide force-based graph drawing methods for the visualization of these networks, for example, *PATIKA* [23] and *GeNet* [118]. These tools typically use well-known force-based algorithms such as Eades' algorithm [28], often based on existing layout libraries and systems like *Pajek* [5] or *yFiles* [144]. There are some improvements of the general force-based method to consider application-specific requirements such as the representation of subcellular locations. One example is implemented in the *PATIKA* system.

Signal transduction and gene regulatory networks are directed graphs and, for example, the visualization of the main direction is important to understand the flow of information through the network. Therefore, layered drawing methods are often employed for the computation of maps of these networks. Some tools using this layout method are *TransPath* [85] and *BioConductor* [15]. Often layout libraries for layered drawings such as *dot* [84] are used.

Protein Interaction Networks

These networks represent proteins and their interactions and are modeled as undirected graphs; see Fig. 7.8b. Several systems which employ force-based graph drawing methods for their visualization have been presented, for instance [12,42,98, 119]. Also some work on interactive exploration of protein interaction networks has been done, for example, by combining circular and force-based layouts and smooth transitions between subsequent drawings using animation [35].

Metabolic Networks

These networks represent the transformation of metabolites into each other and are usually modeled as directed graphs; see Fig. 7.8c. There are two common approaches to visualizing metabolic networks: force-based and layered drawing methods. Several network analysis tools support force-based layouts, for example, *BioJAKE* [113], *Cytoscape* [119], *PathwayAssist* [101], and *VisANT* [45]. Frequently they visualize not only metabolic but also other types of biological networks. However, force-based approaches mostly do not meet common application-specific requirements. Such requirements are, inter alia, different sizes of nodes, the special placement of co-substances and enzymes, and the general direction of pathways.

Layered drawings are often used as they emphasis the main direction in the network. Tools supporting layered drawings are largely based on existing software libraries. Such solutions show the main direction within networks and partly deal with different node sizes. However, there is no specific placement of co-substances or special pathways such as cycles. Examples are *PathFinder* [40] (which uses the *VCG* library [114]) and *BioMiner* [123] (which employs *yFiles* [144]). The earliest approach to our knowledge is from Karp and Paley, where the complete network is separated into parts such as trees, paths, and circles, and the parts are laid out separately [62]. Although not a layered drawing algorithm as described in Sect. 7.2.2, it results in an overall layout with some layered structure. Extended layered drawings consider cyclic structures within the network or show pathways of different topology using different layouts, such as the algorithm by Becker and Rojas [6]. An advanced layered drawing algorithm for metabolic networks considering all relevant visualization requirements has been presented in [115].

7.3.1.2 Current Approaches and Tools

There are many challenges in current research of biological network visualization and visual analytics, such as visual analysis of integrated and correlated data, visual comparison of networks, integrated and overlapping networks, graphical representation of paths and flows, and hierarchical networks; see [3, 39]. Consequently, this field has become very research active and, for example, several special algorithms have been presented in the last few years concerning the layout of biological networks. Among them are grid-based methods [81], clustered circular layouts [38], and constraint-based methods [116]. The quality of these specialized layout algorithms is often much better than just applying standard methods, an example is shown in Fig. 7.1.

A broad range of more than 170 tools for the modeling, analysis, and visualization of biological networks is nowadays available on the Internet. These tools change often rapidly, new tools emerge, and old tools obtain new features or are not longer maintained. Therefore, only a small set of some important tools will be listed here. Other reviews are available, for example, Suderman and Hallett in 2007 compared more than 35 tools regarding network and data visualization [128]; Kono et al. compared tools for pathway representation, mapping and editing, and data exchange in 2009 [83]; and Gehlenborg et al. looked at visualization tools for interaction networks and biological pathways in 2010 [39].

The following tools may be of interest to the reader. As the functionality of the tools changes rapidly over time, we do not provide a feature list but encourage the reader to visit the respective tool websites given below:

- *BiNa* [86]   (http://bit.ly/y6ix9i)
- *BioUML* [82]   (http://bit.ly/yIETIt)
- *CellDesigner* [36, 37]   (http://bit.ly/A0FQiF)
- *CellMicrocosmos* [125]   (http://bit.ly/WJ8cnE)
- *Cytoscape* [119, 124]   (http://bit.ly/wY2sbG)
- *Omix* [26]   (http://bit.ly/zL52vB)
- *Ondex* [78, Chap. 5]   (http://bit.ly/AetZjz)

- *Pathway Projector* [83] (http://bit.ly/zo5x2M)
- *PathVisio* [135] (http://bit.ly/zunwxW)
- *Vanted* [54, 110] (http://bit.ly/Aigr0T)
- *VisAnt* [45, 46] (http://bit.ly/agZBni)

7.3.2 SBGN Standard

Biological networks shown in books, articles, and online resources are often difficult to understand as the same biological concept can be shown by using different graphical representations. Therefore, it is time-consuming to get familiar with the graphical notation used, but this also carries the danger of misinterpretation. Consequently, particularly for molecular-biological networks such as gene regulatory, signal transduction, protein interaction, and metabolic networks, there were several attempts to define a uniform representation. This includes Kitano's Process Diagrams [76], Kohn's Molecular Interaction Maps [79], and Michal's representation of metabolic pathways [95]. However, a single map type is often not enough to adequately illustrate the complexity of biological processes, and none of the mentioned attempts has asserted itself as a widely used standard.

Since 2006, there is a new initiative which partly builds on earlier standardization attempts and is closely connected with the successful exchange format *SBML* (System Biology Markup Language) [48]: *SBGN*—the System Biology Graphical Notation [88]. Additional material can be found under http://sbgn.org, and formal specifications are available [93, 97, 103]; see the previously mentioned website for the latest version of the specification.

SBGN supports three corresponding views or maps on a biological process: *process description* which describes elements (cellular building blocks like molecules, and nucleic acid sequences but also other information like observable events) and interactions between these elements; *entity relationship* which presents the interaction between biological entities and the influence of entities on other elements; and *activity flow* which focuses on the flow of information from one activity to another. These different language types enable to show different aspects of biological processes. A process description contains, for example, a molecule often several times in different states, e.g., phosphorylated or unphosphorylated, while both other map types show in each case only one occurrence of such a molecule. Figure 7.9 shows two molecular-biological networks in SBGN notation.

There are several tools supporting SBGN, including *CellDesigner* [36], *EPE* (Edinburgh Pathway Editor) [30], *PathVisio* [135], and *SBGN-ED* [21] (an extension of *Vanted* [110]). A comparison has been done by Junker et al. [56]. There is also SBGN support for tool developers [136].

7 Network Visualization for Integrative Bioinformatics

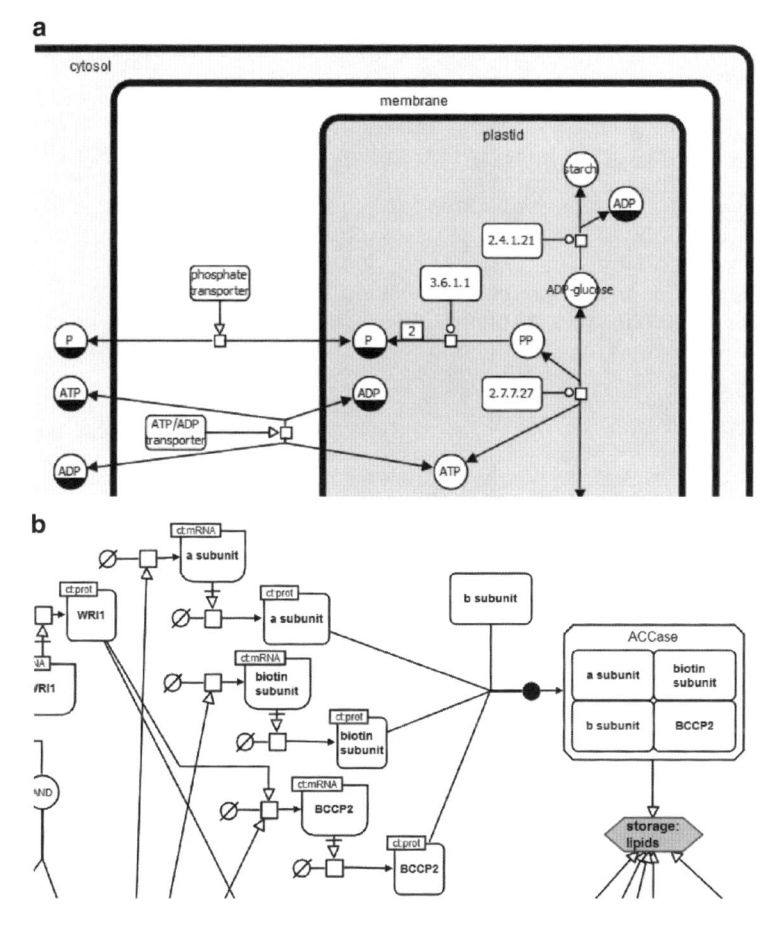

Fig. 7.9 Two examples of SBGN maps. (**a**) Part of a metabolic pathway in SBGN notation (pathway derived from *MetaCrop* [117], an information system based on *Meta-All* [143]). (**b**) Part of a gene regulatory network in SBGN notation (derived from *RIMAS* [55])

References

1. Abello J, van Ham F (2004) Matrix zoom: a visual interface to semi-external graphs. In: Proceedings of the IEEE symposium on information visualization, Austin. IEEE Computer Society, Los Alamitos, Texas, pp 183–190
2. Aigner W, Miksch S, Schumann H, Tominski C (2011) Visualization of time-oriented data. Springer, London/New York
3. Albrecht M, Kerren A, Klein K, Kohlbacher O, Mutzel P, Paul W, Schreiber F, Wybrow M (2010) On open problems in biological network visualization. In: Proceedings of the international symposium on graph drawing (GD '09), Chicago. LNCS, vol 5849. Springer, pp 256–267

4. Appel RD, Bairoch A, Hochstrasser DF (1994) A new generation of information retrieval tools for biologists: the example of the ExPASy WWW server. Trends Biochem Sci 19:258–260
5. Batagelj V, Mrvar A (2004) Pajek – analysis and visualization of large networks. In: Jünger M, Mutzel P (eds) Graph drawing software. Springer, Berlin/New York, pp 77–103
6. Becker MY, Rojas I (2001) A graph layout algorithm for drawing metabolic pathways. Bioinformatics 17(5):461–467
7. Beddow J (1990) Shape coding of multidimensional data on a microcomputer display. In: Proceedings of the 1st conference on visualization '90, VIS '90, San Francisco. IEEE Computer Society Press, Los Alamitos, pp 238–246
8. Berg JM, Tymoczko JL, Stryer L (2002) Biochemistry. W H Freeman, New York
9. Berge C (1989) Hypergraphs: the theory of finite sets. North-Holland, Amsterdam
10. Borisjuk L, Hajirezaei MR, Klukas C, Rolletschek H, Schreiber F (2005) Integrating data from biological experiments into metabolic networks with the DBE information system. Silico Biol 5(2):93–102
11. Bostock M Edgar Anderson's Iris data set scatter plot matrix. http://mbostock.github.com/d3/talk/20111116/iris-splom.html. Last accessed 13 Mar 2013
12. Breitkreutz BJ, Stark C, Tyers M (2003) Osprey: a network visualization system. Genome Biol 4(3):R22
13. Buchheim C, Jünger M, Leipert S (2002) Improving walker's algorithm to run in linear time. In: Revised papers from the 10th international symposium on graph drawing, GD '02, Irvine. Springer, London, pp 344–353
14. Card S, Mackinlay J, Shneiderman B (eds) (1999) Readings in information visualization: using vision to think. Morgan Kaufmann, San Francisco
15. Carey VJ, Gentry J, Whalen E, Gentleman R (2005) Network structures and algorithms in BioConductor. Bioinformatics 21(1):135–136
16. Chambers JM, Cleveland WS, Kleiner B, Tukey PA (1983) Graphical methods for data analysis. Wadsworth, Belmont
17. Chen C (2004) Information visualization: beyond the horizon, 2nd edn. Springer, London/Berlin/Heidelberg
18. Chernoff H (1973) The use of faces to represent points in k-dimensional space graphically. J Am Stat Assoc 68:361–368
19. Cleveland WC, McGill ME (1988) Dynamic graphics for statistics. CRC, Boca Raton
20. Collins C, Viegas F, Wattenberg M (2009) Parallel tag clouds to explore and analyze faceted text corpora. In: Proceedings of the IEEE symposium on visual analytics science and technology (VAST '09), Atlantic City. IEEE Computer Society, pp 91–98
21. Czauderna T, Klukas C, Schreiber F (2010) Editing, validating and translating of SBGN maps. Bioinformatics 26(18):2340–2341
22. D3. Data-driven documents. http://d3js.org. Last accessed 13 Mar 2013
23. Demir E, Babur O, Dogrusöz U, Gürsoy A, Nisanci G, Çetin Atalay R, Ozturk M (2002) PATIKA: an integrated visual environment for collaborative construction and analysis of cellular pathways. Bioinformatics 18(7):996–1003
24. Di Battista G, Eades P, Tamassia R, Tollis IG (1999) Graph drawing: algorithms for the visualization of graphs. Prentice Hall, Upper Saddle River
25. Diehl S, Görg C, Kerren A (2001) Preserving the mental map using foresighted layout. In: Ebert DS, Favre JM, Peikert R (eds) Data visualization 2001, Eurographics, Ascona. Springer, Vienna, pp 175–184
26. Droste P, Miebach S, Niedenführ S, Wiechert W, Nöh K (2011) Visualizing multi-omics data in metabolic networks with the software Omix: a case study. Biosystems 105(2):154–161
27. Dykes J, MacEachren AM, Kraak MJ (2005) Exploring geovisualization. Pergamon, Oxford
28. Eades P (1984) A heuristic for graph drawing. Congressus Numerantium 42:149–160
29. Eiglsperger M, Fekete SP, Klau GW (2001) Orthogonal graph drawing. In: Kaufmann M, Wagner D (eds) Drawing graphs. Lecture notes in computer science, vol 2025. Springer, Berlin/Heidelberg, pp 121–171

30. EPE. EPE Edinburgh PAthway editor. http://epe.sourceforge.net/SourceForge/EPE.html. Last accessed 02 Aug 2012
31. ExposeData.com. Nutrient contents – parallel coordinates. http://exposedata.com/parallel/. Last accessed 13 Mar 2013
32. Feinberg J. Wordle. http://www.wordle.net. Last accessed 13 Mar 2013
33. Forster M, Pick A, Raitner M, Schreiber F, Brandenburg FJ (2002) The system architecture of the BioPath system. Silico Biol 2(3):415–426
34. Freeman E, Fertig S (1995) Lifestreams: organizing your electronic life. In: AAAI fall symposium on AI applications in knowledge navigation and retrieval, Cambridge. Association for the Advancement of Artificial Intelligence, pp 38–44
35. Friedrich C, Schreiber F (2003) Visualization and navigation methods for typed protein-protein interaction networks. Appl Bioinform 2(3 Suppl):19–24
36. Funahashi A, Morohashi M, Kitano H (2003) CellDesigner: a process diagram editor for gene-regulatory and biochemical networks. Biosilico 1(5):159–162
37. Funahashi A, Matsuoka Y, Jouraku A, Kitano H, Kikuchi N (2006) CellDesigner: a modeling tool for biochemical networks. In: Proceedings of the 38th conference on winter simulation, winter simulation conference, Monterey, pp 1707–1712
38. Fung D, Wilkins M, Hart D, Hong S (2010) Using the clustered circular layout as an informative method for visualizing protein-protein interaction networks. Proteomics 10(14):2723–2727
39. Gehlenborg N, O'Donoghue SI, Baliga NS, Goesmann A, Hibbs MA, Kitano H, Kohlbacher O, Neuweger H, Schneider R, Tenenbaum D, Gavin AC (2010) Visualization of omics data for systems biology. Nat Methods 7:S56–S68
40. Goesmann A, Haubrock M, Meyer F, Kalinowski J, Giegerich R (2002) PathFinder: reconstruction and dynamic visualization of metabolic pathways. Bioinformatics 18(1):124–129
41. Görg C, Pohl M, Qeli E, Xu K (2007) Visual representations. In: Kerren A, Ebert A, Meyer J (eds) Human-centered visualization environments. LNCS, tutorial, vol 4417. Springer, Berlin, pp 163–230
42. Han K, Ju BH, Park JH (2002) InterViewer: dynamic visualization of protein-protein interactions. In: Kobourov SG, Goodrich MT (eds) Proceedings of the international symposium on graph drawing (GD '02), Irvine. LNCS, vol 2528. Springer, pp 364–365
43. Heer J, Shneiderman B (2012) Interactive dynamics for visual analysis. Commun ACM 55(4):45–54
44. Henry N, Fekete JD, McGuffin MJ (2007) Nodetrix: a hybrid visualization of social networks. IEEE Trans Vis Comput Graph 13:1302–1309
45. Hu Z, Mellor J, Wu J, DeLisi C (2004) VisANT: an online visualization and analysis tool for biological interaction data. BMC Bioinform 5(1):e17
46. Hu Z, Hung JH, Wang Y, Chang YC, Huang CL, Huyck M, DeLisi C (2009) VisANT 3.5: multi-scale network visualization, analysis and inference based on the gene ontology. Nucl Acids Res 37(Web Server issue):W115–W121
47. Huang W, Eades P, Hong SH (2009) A graph reading behavior: geodesic-path tendency. In: Proceedings of the IEEE Pacific visualization symposium, 2009 (PacificVis '09), Beijing, pp 137–144
48. Hucka M, Finney A, Sauro HM, Bolouri H, Doyle JC, Kitano H, Arkin AP, Bornstein BJ, Bray D, Cornish-Bowden A, Cuellar AA, Dronov S, Gilles ED, Ginkel M, Gor V, Goryanin II, Hedley WJ, Hodgman TC, Hofmeyr JH, Hunter PJ, Juty NS, Kasberger JL, Kremling A, Kummer U, Le Novere N, Loew LM, Lucio D, Mendes P, Minch E, Mjolsness ED, Nakayama Y, Nelson MR, Nielsen PF, Sakurada T, Schaff JC, Shapiro BE, Shimizu TS, Spence HD, Stelling J, Takahashi K, Tomita M, Wagner J, Wang J (2003) The systems biology markup language (SBML): a medium for representation and exchange of biochemical network models. Bioinformatics 19:524–531
49. Inselberg A, Dimsdale B (1990) Parallel coordinates: a tool for visualizing multi-dimensional geometry. In: Proceedings of the IEEE conference on visualization (Vis '90), San Francisco. IEEE Computer Society, pp 361–378

50. Javed W, Elmqvist N (2012) Exploring the design space of composite visualization. In: Proceedings of the IEEE Pacific symposium on visualization (PacificVis '12), Songdo. IEEE Computer Society Press, pp 1–8
51. Jerding DF, Stasko JT (1998) The information mural: a technique for displaying and navigating large information spaces. IEEE Trans Vis Comput Graph 4(3):257–271
52. Johnson B, Shneiderman B (1991) Tree-maps: a space-filling approach to the visualization of hierarchical information structures. In: Proceedings of the 2nd conference on visualization (Vis '91), San Diego. IEEE Computer Society Press, Los Alamitos, pp 284–291
53. Jolliffe I (2002) Principal component analysis. Springer, New York
54. Junker BH, Klukas C, Schreiber F (2006) VANTED: a system for advanced data analysis and visualization in the context of biological networks. BMC Bioinform 7:109
55. Junker A, Hartmann A, Schreiber F, Bäumlein H (2010) An engineer's view on regulation of seed development. Trends Plant Sci 15(6):303–307
56. Junker A, Rohn H, Czauderna T, Klukas C, Hartmann A, Schreiber F (2012) Creating interactive, web-based and data-enriched maps using the systems biology graphical notation. Nat Protoc 7:579–593
57. Jusufi I (2012) Towards the visualization of multivariate biochemical networks. Licentiate thesis, Linnaeus University
58. Jusufi I, Dingjie Y, Kerren A (2010) The network lens: interactive exploration of multivariate networks using visual filtering. In: Proceedings of the 14th international conference on information visualisation (IV '10), London. IEEE Computer Society Press, pp 35–42
59. Jusufi I, Kerren A, Aleksakhin V, Schreiber F (2012) Visualization of mappings between the gene ontology and cluster trees. In: Proceedings of the SPIE 2012 conference on visualization and data analysis (VDA '12), IS&T/SPIE, Burlingame. SPIE, vol 8294, pp 8294–20
60. Jusufi I, Klukas C, Kerren A, Schreiber F (2012) Guiding the interactive exploration of metabolic pathway interconnections. Inf Vis 11(2):136–150
61. Kanehisa M, Goto S, Kawashima S, Nakaya A (2002) The KEGG databases at GenomeNet. Nucl Acids Res 30(1):42–46
62. Karp PD, Paley SM (1994) Automated drawing of metabolic pathways. In: Lim H, Cantor C, Bobbins R (eds) Proceedings of the international conference on bioinformatics and genome research, Tallahassee, pp 225–238
63. Kaser O, Lemire D (2007) Tag-cloud drawing: algorithms for cloud visualization. In: Proceedings of tagging and metadata for social information organization (WWW '07), Banff
64. Kaufmann M, Wagner D (1999) Drawing graphs: methods and models. Lecture notes in computer science, tutorial, vol 2025. Springer, Berlin/Heidelberg
65. Keim DA (2002) Information visualization and visual data mining. IEEE Trans Vis Comput Graph 7(1):1–8
66. Keim D, Kriegel HP (1994) Visdb: database exploration using multidimensional visualization. IEEE Comput Graph Appl 14(5):40–49
67. Keim D, Oelke D (2007) Literature fingerprinting: a new method for visual literary analysis. In: Proceedings of the IEEE symposium on visual analytics science and technology (VAST '07), Sacramento. IEEE Computer Society Press, pp 115–122
68. Keim D, Andrienko G, Fekete JD, Görg C, Kohlhammer J, Melançon G (2008) Visual analytics: definition, process, and challenges. In: Kerren A, Stasko JT, Fekete JD, North C (eds) Information visualization: human-centered issues and perspectives. Lecture notes in computer science, vol 4950. Springer, Berlin/Heidelberg, pp 154–175
69. Keim D, Kohlhammer J, Ellis G, Mansmann F (eds) (2010) Mastering the information age – solving problems with visual analytics. Eurographics Digital Library, Goslar
70. Kerren A, Schreiber F (2012) Toward the role of interaction in visual analytics. In: Proceedings of the winter simulation conference, winter simulation conference, WSC '12, Berlin, pp 420:1–420:13
71. Kerren A, Ebert A, Meyer J (eds) (2007) Human-centered visualization environments. LNCS, tutorial, vol 4417. Springer, Berlin

72. Kerren A, Ebert A, Meyer J (2007) Introduction to human-centered visualization environments. In: Kerren A, Ebert A, Meyer J (eds) Human-centered visualization environments. LNCS, tutorial, vol 4417. Springer, Berlin, pp 1–9
73. Kerren A, Stasko JT, Fekete JD, North C (2007) Workshop report: information visualization human-centered issues in visual representation, interaction, and evaluation. Inf Vis 6(3):189–196
74. Kerren A, Stasko JT, Fekete JD, North C (eds) (2008) Information visualization: human-centered issues and perspectives. Lecture notes in computer science, vol 4950. Springer, Berlin/Heidelberg
75. Kerren A, Köstinger H, Zimmer B (2012) Vincent – visualisation of network centralities. In: Proceedings of the international conference on information visualization theory and applications (IVAPP '12), INSTICC, Rome, pp 703–712
76. Kitano H (2003) A graphical notation for biochemical networks. Biosilico 1(5):169–176
77. Koh K, Lee B, Kim B, Seo J (2010) Maniwordle: providing flexible control over wordle. IEEE Trans Vis Comput Graph 16:1190–1197
78. Köhler J, Baumbach J, Taubert J, Specht M, Skusa A, Rüegg A, Rawlings C, Verrier P, Philippi S (2006) Graph-based analysis and visualization of experimental results with ONDEX. Bioinformatics 22(11):1383–1390
79. Kohn KW, Aladjem MI (2006) Circuit diagrams for biological networks. Mol Syst Biol 2:e2006.0002
80. Kohonen T, Schroeder MR, Huang TS (eds) (2001) Self-organizing maps, 3rd edn. Springer, New York/Secaucus
81. Kojima K, Nagasaki M, Jeong E, Kato M, Miyano S (2007) An efficient grid layout algorithm for biological networks utilizing various biological attributes. BMC Bioinform 8:76
82. Kolpakov FA (2002) BioUML – framework for visual modeling and simulation of biological systems. In: Proceedings of the international conference on bioinformatics of genome regulation and structure, Novosibirsk. Springer, pp 130–133
83. Kono N, Arakawa K, Ogawa R, Kido N, Oshita K, Ikegami K, Tamaki S, Tomit M (2009) Pathway projector: web-based zoomable pathway browser using KEGG Atlas and Google maps API. PLoS ONE 4(11):e7710
84. Koutsofios E, North S (1995) Drawing graphs with dot. Technical report, AT&T Bell Laboratories, Murray Hill
85. Krull M, Voss N, Choi C, Pistor S, Potapov A, Wingender E (2003) TRANSPATH: an integrated database on signal transduction and a tool for array analysis. Nucl Acids Res 31(1):97–100
86. Küntzer J, Backes C, Blum T, Gerasch A, Kaufmann M, Kohlbacher O, Lenhof HP (2007) BNDB – the biochemical network database. BMC Bioinform 8:367
87. Lee B, Riche N, Karlson A, Carpendale S (2010) Sparkclouds: visualizing trends in tag clouds. IEEE Trans Vis Comput Graph 16(6):1182–1189
88. Le Novère N, Hucka M, Mi H, Moodie S, Schreiber F, Sorokin A, Demir E, Wegner K, Aladjem MI, Wimalaratne SM, Bergman FT, Gauges R, Ghazal P, Kawaji H, Li L, Matsuoka Y, Villéger A, Boyd SE, Calzone L, Courtot M, Dogrusoz U, Freeman TC, Funahashi A, Ghosh S, Jouraku A, Kim S, Kolpakov F, Luna A, Sahle S, Schmidt E, Watterson S, Wu G, Goryanin I, Kell DB, Sander C, Sauro H, Snoep JL, Kohn K, Kitano H (2009) The systems biology graphical notation. Nat Biotechnol 27(8):735–741
89. Li Q, Bao X, Song C, Zhang J, North C (2003) Dynamic query sliders vs. brushing histograms. In: CHI '03 extended abstracts on human factors in computing systems, CHI EA '03, Fort Lauderdale. ACM, New York, pp 834–835
90. Liu J (2012) Visualization of weather data: temperature trend visualization. Bachelor's thesis, School of Computer Science, Physics and Mathematics, Linnaeus University, Växjö
91. MacNeil S, Elmqvist N (2013) Visualization mosaics for multivariate visual exploration. Comput Graph Forum 32:38–50
92. Mardia KV (1979) Multivariate analysis. Academic, London/New York

93. Mi H, Schreiber F, Novère NL, Moodie S, Sorokin A (2009) Systems biology graphical notation: activity flow language level 1. Nat Preced. doi:10.1038/npre.2009.3724.1
94. Michal G (1993) Biochemical pathways (Poster). Boehringer Mannheim, Mannheim
95. Michal G (1998) On representation of metabolic pathways. BioSystems 47:1–7
96. Michal G (1999) Biochemical pathways. Spektrum Akademischer Verlag, Heidelberg
97. Moodie S, Novère NL, Sorokin A, Mi H, Schreiber F (2009) Systems biology graphical notation: process description language level 1. Nat Preced. doi:10.1038/npre.2009.3721.1
98. Mrowka R (2001) A Java applet for visualizing protein-protein interaction. Bioinformatics 17(7):669–670
99. Nicholson DE (1997) Metabolic pathways map (Poster). Sigma Chemical Co., St. Louis
100. Nightingale F (1858) Notes on matters affecting the health, efficiency, and hospital administration of the British Army. Harrison & Sons, London
101. Nikitin A, Egorov S, Daraselia N, Mazo I (2003) Pathway studio – the analysis and navigation of molecular networks. Bioinformatics 19(16):2155–2157
102. Nöllenburg M (2007) Geographic visualization. In: Kerren A, Ebert A, Meyer J (eds) Human-centered visualization environments. LNCS, tutorial, vol 4417. Springer, Berlin, pp 257–294
103. Novère NL, Moodie S, Sorokin A, Schreiber F, Mi H (2009) Systems biology graphical notation: entity relationship language level 1. Nat Preced. doi:10.1038/npre.2009.3719.1
104. Oesterling P, Scheuermann G, Teresniak S, Heyer G, Koch S, Ertl T, Weber G (2010) Two-stage framework for a topology-based projection and visualization of classified document collections. In: Proceedings of the IEEE symposium on visual analytics science and technology (VAST '10), Salt Lake City. IEEE Computer Society, pp 91–98
105. O'Madadhain J, Fisher D, Nelson T. JUNG – Java universal network/graph framework. http://jung.sourceforge.net/. Last accessed 27 Jan 2013
106. Pickett RM, Grinstein GG (1988) Iconographic displays for visualizing multidimensional data. In: Proceedings of the 1988 IEEE international conference on systems, man, and cybernetics, Beijing, vol 1, pp 514–519
107. Reingold EM, Tilford JS (1981) Tidier drawing of trees. IEEE Trans Softw Eng 7(2):223–228
108. Richard JS, Catrambone R, Guzdial M, Mcdonald K (2000) An evaluation of space-filling information visualizations for depicting hierarchical structures. Int J Hum Comput Stud 53:663–694
109. Roberts JC (2004) Exploratory visualization with multiple linked views. In: MacEachren A, Kraak MJ, Dykes J (eds) Exploring geovisualization. Elseviers, Amsterdam
110. Rohn H, Junker A, Hartmann A, Grafahrend-Belau E, Treutler H, Klapperstück M, Czauderna T, Klukas C, Schreiber F (2012) VANTED v2: a framework for systems biology applications. BMC Syst Biol 6:139
111. Rohrschneider M, Heine C, Reichenbach A, Kerren A, Scheuermann G (2010) A novel grid-based visualization approach for metabolic networks with advanced focus & context view. In: Proceedings of the international symposium on graph drawing (GD '09), Chicago. LNCS, vol 5849. Springer, pp 268–279
112. Rohrschneider M, Ullrich A, Kerren A, Stadler PF, Scheuermann G (2010) Visual network analysis of dynamic metabolic pathways. In: Proceedings of the 6th international conference on advances in visual computing – volume Part I (ISVC '10), Las Vegas. Springer, Berlin/Heidelberg, pp 316–327
113. Salamonsen W, Mok KY, Kolatkar P, Subbiah S (1999) BioJAKE: a tool for the creation, visualization and manipulation of metabolic pathways. In: Proceedings of the Pacific symposium on biocomputing, Big Island, pp 392–400
114. Sander G (1994) Graph layout through the VCG tool. In: Tamassia R, Tollis IG (eds) Proceedings of the DIMACS international workshop on graph drawing (GD '94), Princeton. Springer, pp 194–205
115. Schreiber F (2002) High quality visualization of biochemical pathways in BioPath. Silico Biol 2(2):59–73
116. Schreiber F, Dwyer T, Marriott K, Wybrow M (2009) A generic algorithm for layout of biological networks. BMC Bioinform 10:375

117. Schreiber F, Colmsee C, Czauderna T, Grafahrend-Belau E, Hartmann A, Junker A, Junker BH, Klapperstück M, Scholz U, Weise S (2012) MetaCrop 2.0: managing and exploring information about crop plant metabolism. Nucl Acids Res 40(1):D1173–D1177
118. Serov VN, Spirov AV, Samsonova MG (1998) Graphical interface to the genetic network database GeNet. Bioinformatics 14(6):546–547
119. Shannon P, Markiel A, Ozier O, Baliga NS, Wang JT, Ramage D, Amin N, Schwikowski B, Ideker T (2003) Cytoscape: a software environment for integrated models of biomolecular interaction networks. Genome Res 13(11):2498–2504
120. Shannon R, Holland T, Quigley A (2008) Multivariate graph drawing using parallel coordinate visualisations. Technical report 2008-6, School of Computer Science and Informatics, University College Dublin
121. Shneiderman B (1996) The eyes have it: a task by data type taxonomy for information visualizations. In: Proceedings of the IEEE symposium on visual languages (VL '96), Boulder. IEEE Computer Society, pp 336–343
122. Shneiderman B, Aris A (2006) Network visualization by semantic substrates. IEEE Trans Vis Comput Graph 12:733–740
123. Sirava M, Schäfer T, Eiglsperger M, Kaufmann M, Kohlbacher O, Bornberg-Bauer E, Lenhof HP (2002) BioMiner – modeling, analyzing, and visualizing biochemical pathways and networks. Bioinformatics 18(Suppl. 2):S219–S230
124. Smoot ME, Ono K, Ruscheinski J, Wang PL, Ideker T (2011) Cytoscape 2.8: new features for data integration and network visualization. Bioinformatics 27(3):431–432
125. Sommer B, Künsemöller J, Sand N, Husemann A, Rumming M, Kormeier B (2010) Cellmicrocosmos 4.1 – an interactive approach to integrating spatially localized metabolic networks into a virtual 3d cell environment. In: Fred ALN, Filipe J, Gamboa H (eds) Proceedings of the first international conference on bioinformatics (BIOINFORMATICS '10), Valencia, pp 90–95
126. Spence R (2007) Information visualization: design for interaction, 2nd edn. Prentice Hall, Harlow
127. Stasko J, Muthukumarasamy J (1996) Visualizing program executions on large data sets. In: Proceedings of the IEEE symposium on visual languages (VL '96), Boulder. IEEE Computer Society, pp 166–173
128. Suderman M, Hallett MT (2007) Tools for visually exploring biological networks. Bioinformatics 23(20):2651–2659
129. Sugiyama K, Tagawa S, Toda M (1981) Methods for visual understanding of hierarchical system structures. IEEE Trans Syst Man Cybern SMC-11(2):109–125
130. Thomas JJ, Cook KA (2006) A visual analytics agenda. IEEE Comput Graph Appl 26(1):10–13
131. Tufte ER (1990) Envisioning information. Graphics Press, Cheshire
132. Tufte ER (1997) Visual explanations: images and quantities, evidence and narrative. Graphic Press, Cheshire
133. Tufte ER (2001) The visual display of quantitative information, 2nd edn. Graphics Press, Cheshire
134. van Ham F, van Wijk JJ (2002) Beamtrees: compact visualization of large hierarchies. In: Proceedings of the IEEE symposium on information visualization (InfoVis '02), Boston. IEEE Computer Society, pp 93–100
135. van Iersel MP, Kelder T, Pico AR, Hanspers K, Coort S, Conklin BR, Evelo C (2008) Presenting and exploring biological pathways with PathVisio. BMC Bioinform 9:399. 1–9
136. van Iersel MP, Villéger A, Czauderna T, Boyd SE, Bergmann FT, Luna A, Demir E, Sorokin AA, Dogrusöz U, Matsuoka Y, Funahashi A, Aladjem MI, Mi H, Moodie SL, Kitano H, Novère NL, Schreiber F (2012) Software support for SBGN maps: SBGN-ML and LibSBGN. Bioinformatics 28(15):2016–2021
137. Van Wijk JJ, Nuij WAA (2003) Smooth and efficient zooming and panning. In: Proceedings of the IEEE conference on information visualization (InfoVis '03), Seattle. IEEE Computer Society, Washington, DC, pp 15–22

138. Viegas FB, Wattenberg M, Feinberg J (2009) Participatory visualization with wordle. IEEE Trans Vis Comput Graph 15:1137–1144
139. Walker JQ (1990) A node-positioning algorithm for general trees. Softw Pract Exp 20(7):685–705
140. Ward M, Grinstein G, Keim DA (2010) Interactive data visualization: foundations, techniques, and application. A.K. Peters, Natick
141. Ware C (2004) Information visualization: perception for design, 2nd edn. Morgan Kaufmann, San Francisco
142. Wattenberg M (2006) Visual exploration of multivariate graphs. In: Proceedings of the SIGCHI conference on human factors in computing systems (CHI '06), Montreal. ACM, New York, pp 811–819
143. Weise S, Grosse I, Klukas C, Koschützki D, Scholz U, Schreiber F, Junker BH (2006) Meta-All: a system for managing metabolic pathway information. BMC Bioinform 7:465
144. Wiese R, Eiglsperger M, Kaufmann M (2001) yFiles: visualization and automatic layout of graphs. In: Mutzel P, Jünger M, Leipert S (eds) Proceedings of the international symposium on graph drawing (GD '01), Vienna. LNCS, vol 2265. Springer, pp 453–454
145. Williams M, Munzner T (2004) Steerable, progressive multidimensional scaling. In: Proceedings of the IEEE symposium on information visualization (InfoVis '04), Austin. IEEE Computer Society Press, pp 57–64
146. Williamson C, Shneiderman B (1992) The dynamic homefinder: evaluating dynamic queries in a real-estate information exploration system. In: Proceedings of the international ACM conference on research and development in information retrieval (SIGIR '92), Copenhagen. ACM, New York, pp 338–346
147. Wise J, Thomas J, Pennock K, Lantrip D, Pottier M, Schur A, Crow V (1995) Visualizing the non-visual: spatial analysis and interaction with information from text documents. In: Proceedings of the IEEE symposium on information visualization (InfoVis '95), Atlanta. IEEE Computer Society, pp 51–58
148. Yi JS, Kang YA, Stasko J, Jacko J (2007) Toward a deeper understanding of the role of interaction in information visualization. IEEE Trans Vis Comput Graph 13(6):1224–1231
149. yWorks. yEd graph editor. http://www.yworks.com/en/products_yed_about.html. Last accessed 02 Aug 2012
150. Zimmer B, Jusufi I, Kerren A (2012) Analyzing multiple network centralities with ViNCent. In: Proceedings of SIGRAD 2012: interactive visual analysis of data, Växjö, 29–30 Nov 2012. Number 81 in Linköping electronic conference proceedings. Linköping University Electronic Press, pp 87–90

Chapter 8
Biological Network Modeling and Analysis

Sebastian Jan Janowski, Barbara Kaltschmidt, and Christian Kaltschmidt

Abstract Each scientist needs to be aware of the complexity of cellular life and the modeling possibilities to be able to reconstruct, analyze, and simulate biological systems. Bioinformatics modeling, analysis, and simulation are highly interdisciplinary disciplines using techniques and concepts from computer science, statistics, mathematics, chemistry, biology, biochemistry, genetics, and physics, among others. Without knowledge about these research topics, it is almost impossible to produce good theoretical models, which can be used for hypothesis testing. Therefore, this chapter gives an impression of what can be modeled from the bioinformatics and biological point of view and introduces into biological networks, common analysis techniques from graph theory, and possibilities to reconstruct, simulate, and share biological networks based on database content.

S.J. Janowski (✉)
Faculty of Technology, Department of Bioinformatics, Bielefeld University,
Universitätsstraße 25, D-33501 Bielefeld, Germany
e-mail: janowski@cebitec.uni-bielefeld.de

B. Kaltschmidt
Faculty of Biology, Molecular Neurobiology, Bielefeld University, Universitätsstraße 25,
D-33501 Bielefeld, Germany
e-mail: barbara.kaltschmidt@uni-bielefeld.de

C. Kaltschmidt
Faculty of Biology, Cell Biology, Bielefeld University, Universitätsstraße 25,
D-33501 Bielefeld, Germany
e-mail: c.kaltschmidt@uni-bielefeld.de

M. Chen and R. Hofestädt (eds.), *Approaches in Integrative Bioinformatics: Towards the Virtual Cell*, DOI 10.1007/978-3-642-41281-3_8,
© Springer-Verlag Berlin Heidelberg 2014

8.1 What Can or Should Be Modeled?

What is cellular life? The simplest answer from the biological point of view is the following: anything that contains DNA or RNA [1], shows self-organization, and has evolved over time as described by Manfred Eigen [2]. Motivated to seek a theory to understand life, many decades ago researchers embarked on the study of biological systems [3, 4]. Their main goal is not to imitate life but rather to understand the universal logic and properties of living systems. Cellular functions which do not rely on simple enumeration of molecular components and processes, such as transcription, translation, and modifications, are carried out constantly. These components never act as one independent element. Thus, present-day cellular biology is challenged to reconstruct coupled dynamical models with many differing elements and strongly interacting systems. Therefore, scientists endeavor to provide a new look at data on the present organisms to validate or reject hypotheses.

The main task for modern biology is to trace phenotypical properties back to specific molecules. Therefore, theoretical models are constructed, consisting of the formation of switching rules that obligate cell features. With modern systems biology and bioinformatics, those theoretical models are pictured. Therefore, natural sciences produce a holistic view of different levels of organizations. Using causal relations, theoretical models are constructed using several different switching rules. Through the turning on and off of one or more genes, as controlled by one or more molecules, the properties and dynamics of a cell can change. This can result in different cell behavior, where the concentration of some other molecule is altered, with the effect of turning on or off some other genes [1, 5].

Thus, to model and investigate cellular life, several different key components of real-life systems have to be considered. The central dogma of molecular biology stated by Francis Crick in 1958 describes the basic information flow in cells with the following sentence: "DNA makes RNA, which in turn makes Proteins" [6, 7]. In general, this statement is correct, whereas it is very simplified. Nowadays, natural science has investigated many processes and functions in detail, such as transcription, translation, and posttranslational modification, among others, which extend this stated dogma. The investigation of other regulatory processes, such as microRNA fine regulation, is still in their beginning phases. Table 8.1 gives an example of specific cell-type characteristics and dynamics to show the variety of living organisms [8].

Although all these presented aspects have to be considered in the modeling of a biological system and put into relationship with the biological dogma, it is neither recommended nor practical to model all aspects. Too many unknown parameters will come up, with the danger being that a fitted model will match to nearly anything. Fitted parameters can be even misleading or become meaningless. Furthermore, the larger the model, the longer it will take to determine parameters and to analyze properties of interest. Therefore, each model has to be limited to a practical size and linked to clear scientific questions.

Table 8.1 Biological cell characteristics for *E. coli*, yeast (*S. cerevisiae*), and mammalian (human fibroblast) based on [8]

Property	*E. coli*	Yeast (*S. cerevisiae*)	Mammalian (human fibroblast)
Cell volume	$\sim 1\,\mu m^3$	$\sim 1{,}000\,\mu m^3$	$\sim 10{,}000\,\mu m^3$
Proteins/cell	$\sim 4 \times 10^6$	$\sim 4 \times 10^9$	$\sim 4 \times 10^{10}$
Genes	$\sim 4{,}500$	$\sim 6{,}600$	$\sim 30{,}000$
Size of regulator binding site	~ 10 bp	~ 10 bp	~ 10 bp
Size of promoter	~ 100 bp	$\sim 1{,}000$ bp	$\sim 10^4$ to 10^5 bp
Size of gene	$\sim 1{,}000$ bp	$\sim 1{,}000$ bp	$\sim 10^4$ to 10^6 bp (with introns)
Diffusion time of protein across cell	~ 0.1 s $D = 10\,\mu m^2/s$	~ 10 s	~ 100 s
Diffusion time of small molecule across cell	~ 0.1 ms $D = 1{,}000\,\mu m^2/s$	~ 10 ms	~ 0.1 s
Time to transcribe a gene	~ 1 min (80 bp/s)	~ 1 min	~ 30 min (including mRNA processing)
Time to translate a protein	~ 2 min (40 aa/s)	~ 2 min	~ 30 min (including mRNA nuclear export)

One possibility to limit model size is by using biological networks. These networks can be restricted to only one -omic level, such as metabolomics or proteomics. The main advantage of biological networks is that they can be used to answer scientific questions with the focus on important regulatory elements, rather than building up whole systems.

8.2 Biological Networks

Cellular life is mostly a network of interacting elements. To visually represent and analyze the various interactions and relationships, biological systems can be modeled as biological networks, which are based on mathematical graphs (see Definition 1).

Definition 1. A graph is an ordered pair $G = (V, E)$:

- Comprising of a set V of vertices and a set E of edges, where each edge is assigned to two (not necessarily disjunct) vertices.
- The order of a graph is $|V|$, comprised of the number of vertices.
- The size of a graph is $|E|$, comprised of the number of edges.
- The degree of a vertex is the number of edges that connect to it and are defined by $N_G(v)$ or $N(v)$.

The objects, represented by nodes, are called "vertices" and the links, represented by directed or undirected arrows, are called "edges." In general, the smallest level of details is the molecular level, describing DNA, RNA, proteins, and metabolites interacting with each other. Thus, nodes can be any kind of biological compounds belonging to such a system. Edges are used to represent biological relations and processes, such as activation, inhibition, and expression, among others. To model all system elements, information flow, and dynamics, different biological networks were introduced as described in the following:

- **Transcription networks (or gene regulation networks)**
  Transcriptional networks control the gene expression within cells in time, space, and amplitude [9]. Usually these kinds of networks describe how one gene is controlled by the product of another gene. Therefore, the highly interconnected processes are modeled with a directed graph, in which nodes represent gene, transcription factors, and/or proteins and edges indicate mechanisms, such as transcription, DNA binding, protein synthesis, and degradation, among others. Furthermore, the synthesis of RNA, posttranscriptional events, mRNA turnover, and translation can also be considered. However, as these kinds of networks model a wide range of biological processes, they play a major role in protein-protein interaction networks, signal transduction networks, metabolic networks, and others, which are described in the following.
- **Protein interaction networks**
  In terms of the degree of regulation, it becomes apparent that a protein can never be investigated in isolation. Moreover, it has to be examined in the context of other proteins and their interacting network, in the so-called protein-protein interaction networks. The majority of biological processes within a cell are controlled and mediated by proteins [1, 5]. They interact with other molecules, such as low-molecular-weight compounds, lipids, and nucleic acids to ensure transcription, translation, splicing, mechanical strength, transport, immunity, signal transduction, growth, development, and many other processes. The types of interactions range from transient interactions, occurring for a limited time, such as they appear in protein kinases, protein phosphates, and others, up to static interactions, such as the transfer of biosynthetic intermediates between catalytic sites without the diffusion into the enzyme's surrounding. A further important aspect of protein-protein interaction is the signal transmissions from the external environment to specific locations within the cells.

  However, such protein-protein interaction networks enable the scientist to investigate protein functions, system dynamics, and biological mechanisms [9–15]. Reconstructing these kinds of networks, unknown proteins can be grouped into known biological context and important proteins into functional groups, subnetworks, and motifs identified and examined in detail. This kind of analysis has become so important and powerful that it already contributes to new therapeutic strategies [13, 16, 17].

- **Signal transduction networks**
  Signal transduction networks are of special interest in biological and medical sciences as many diseases are related to disturbances in signaling networks [18]. In general, signal transduction links intracellular processes to the extracellular environment of a cell. The general aim is to model and describe cellular functions in response to external stimuli. Therefore, information transmission is modeled, starting with the binding of extracellular ligands to receptors and resulting in cell response that triggers a cascade of signal transduction reactions. The sequence of reactions involved mainly relies on reversible chemical modifications and complex formations, such as phosphorylation. The final targets of the processes are transcription factors and metabolic enzymes. In summary, signal transduction pathways transform a set of inputs into a set of outputs.

  In contrast with other networks, such as protein-protein interaction networks, signaling networks are basically directed. From the topological point of view, the networks involve many different motifs, such as positive and negative feedback loops. One of the most prominent examples is the negative feedback loop of the transcription factor NF-κB [19, 20].

- **Metabolic networks**
  Metabolic networks have a fundamental importance in biochemistry and biotechnology, as many scientists modify or alter metabolic networks to produce fine chemicals, antibiotics, industrial enzymes, antibodies, etc. Furthermore, metabolic networks are used in biomedicine enabling a better understanding of metabolic mechanisms and for controlling infections. Therefore, scientists examine differences, synergies, and other interactions between human beings and pathogens. In general, the main goal of metabolic networks is the modeling of cellular processes, such as the uptaking and digesting of substrates from the environment, energy generation, growth, and cell survival, among others. Many of these networks are available online in databases, such as KEGG [21], EcoCyc [22], and BioCyc [23]. The networks refer to metabolites (amino acids, glucose, polysaccharides, glycans, etc.) and their biochemical reactions.

- **Correlation networks**
  Correlation networks represent statistical associations between variables derived from experiments, such as derived from whole genome arrays, mass spectrometry, and enzyme-based proteomic experiments, among others [9]. The global analysis approach is to give a broad overview of the state of the organism. Due to technological advances in systems biology, experimental approaches are able to provide qualitative and quantitative information, which can be used for comprehensive insights into biological systems.

  Usually the resulting datasets are mainly independent variable-unit entries. However, based on the experimentally measured values, correlations can be determined from either the probability point of view or the strength of variable units. The first approach measures if two values have a connection by coincidence or if there seems to be a real link. Therefore, correlation coefficients are calculated expressing the connection probability. The accuracy of this approach mainly depends on the sample size of the experiment. Examining a large number of samples increases the probabilities for finding real connections and, moreover, increases the probability of identifying whether weak connections are true. The second approach only considers connection from the strength of variable units, instead of the sampling size. However, an experimental validation based upon the results is the best way to confirm a predicted correlation.

- **Neuronal networks**
  In neuronal networks artificial neurons are connected to each other. The aim is to reconstruct systems as they appear in real life [24,25]. Thus, connections between neurons are modeled with neuronal summation, in which potentials and electric gap junctions define firing strategies and signal transduction from one neuron to another. In neuronal networks, neurons only respond to a subset of mostly simple stimuli given by their neighbors, whereas, in real systems, the information flow is based on inhibitory postsynaptic potentials and excitatory postsynaptic potentials. The modeling and analysis of neuronal networks has attracted wide interest in life sciences. For example, the subject of one application field is to model systems which are able to learn complex patterns and therefore build a kind of artificial intelligence.

- **Phylogenetic networks**
  Phylogenetic networks describe the evolution and relationship between different organisms. Usually, phylogenetic reconstructions are presented by trees rather than networks, in which branch points represent the evolutionary separation of two organisms. However, trees do not consider vertical and horizontal gene-transfer events. Thus, phylogenetic networks describe evolutionary processes in more detail. Kunin et al. give one prominent example of such a phylogenetic network in their article "The net of life: Reconstructing the microbial phylogenetic network" [26].

- **Ecological networks**
  Ecological networks typically present food webs. Food webs are limited representations of real ecosystems describing ecological communities focusing on trophic interactions between consumers and resources ("what eats what") [27–29]. In general, two trophic categories exist, called trophic levels. The first ones are the autotrophs, which produce organic matter from inorganic substances. The second level, the heterotrophs, obtains organic matter by feeding on autotrophs and other heterotrophs. It is a unified system of exchange, adopted to analyze interrelationships between community structure, stability, and ecosystem processes.

  The analysis of food webs has shown that the evolution of realistic food web structures can be explained on the basis of simple rules regarding population abundance and species occurrence. For example, ecologists and mathematics have figured out early on that the structure of food webs consists of nonrandom properties, such as scaling laws. By examining a predator-prey model (resource-consumer, plant-herbivore, parasite-host), it becomes obvious that the size of one species is crucial to the stability of the whole system [30].

8 Biological Network Modeling and Analysis

However, food webs are an important representation for the prediction of ecological events. They are mainly used to understand biological systems and moreover to protect them from outside influences, such as climate change, foreign wild species, and the narrowing of the habitat.

Summarized, the presented biological networks are able to capture all -omic levels and, furthermore, able to model ecological events and other correlations. With these advantages bioinformatics and systems biology have a set of powerful integrated frameworks to present, integrate, and visualize knowledge. Furthermore, graph theory comes with powerful approaches to analyze those networks as described in the following.

8.3 Biological Network Analysis Based on Graph Theory

As mentioned in the previous section, graphs or networks can be used to model many types of biological relations, biological processes, and biological questions. Furthermore, geometry and topology can give important clues about organization and information flow within a system. Graph analysis can determine structural properties of a network. Furthermore, graph theory can analyze vertex degrees, path lengths, diameter, and many other structural properties.

In general, graphs can have different types as presented in Fig. 8.1. In a **directed graph** an edge between the vertices u and v is represented by the ordered pair (u, v) [31]. Visually the ordered pair represents the direction of the arrowhead. However, there is a big difference between directed and undirected graphs for a given number

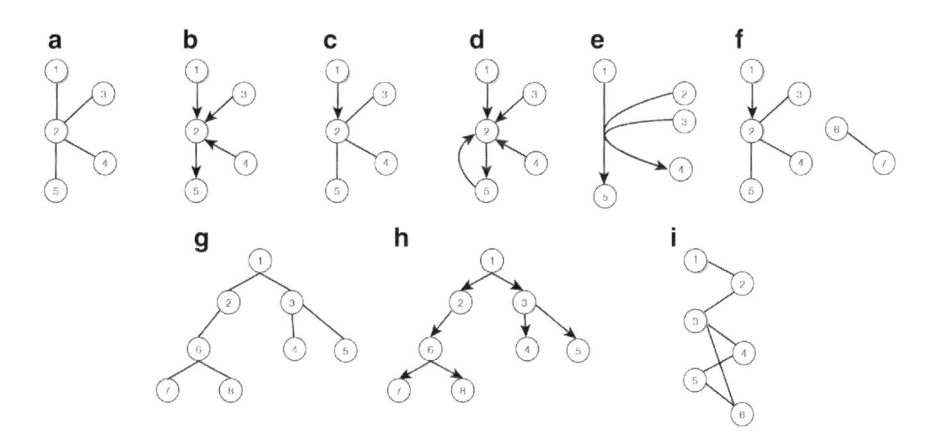

Fig. 8.1 Different graph types as they may appear in biological networks: (**a**) undirected, (**b**) directed, (**c**) mixed, (**d**) multigraph, (**e**) hyper-graph, (**f**) unconnected graphs, (**g**) tree, (**h**) rooted tree, and (**i**) bipartite graph

of vertices. The **amount of directed graphs** $N_{\text{dir}}(V)$ with V vertices is much higher than the amount of possible undirected graphs $N_{\text{undir}}(V)$ [9]:

$$\frac{N_{\text{dir}}(V)}{N_{\text{undir}}(V)} = 2^{\frac{(V^2-1)}{2}} \tag{8.1}$$

A **mixed graph** has both directed and undirected pairs. In the biological context it can represent protein-protein interaction networks, where some interactions are undirected, such as protein-complex bindings, and some interactions, such as activation, phosphorylation, and other processes are directed. A **multigraph** contains multiple edges, where two or more edges are incident to the same two vertices. A **hyper-graph** is characterized by more than two elements, which are connected to one interaction. Hyper-graphs are often used to model metabolic networks where several substances are used in one reaction to produce another substance.

A graph is **bipartite** if there is a partition of its vertex set $V = S \cup T$, such that each edge in E has exactly one end vertex in S and one end vertex in T. A **tree** is an undirected, acyclic graph, where vertices with only one edge are called leaves. All other vertices are inner vertices. The depth of such a tree is the length of the path from the root to a vertex. The height is the maximal depth. A rooted tree is often regarded as a directed graph [31].

A **subgraph** $G' = (V', E')$ of the graph $G = (V, E)$ is a graph where $V' \in V$ and $E' \in E$ [31]. The **density** of a graph is given by

$$\frac{2\,|\,E\,|}{|\,V\,|\,(|\,V\,|-1)} \tag{8.2}$$

This definition indicates how dense or connected a graph is determining vertex degrees [32].

Two graphs G and G' are **isomorphic** $G \simeq G'$, if there exist a bijection $\varphi : V - > V'$ between the vertex sets of G and G', such that any two vertices u and v of G are adjacent in G if and only if (u) and (v) are adjacent in G', based on $xy \in E \Leftrightarrow \varphi(x)\varphi(y) \in E' \; \forall x, y \in V$ [31].

Global network properties are topological entities, such as distance, average path length, and diameter. A **path** is a sequence $(v_0, e_1, v_1, e_2, \ldots, v_{k-1}, e_k, v_k)$ of vertices and edges. The **length of a path** is given by its number of edges. The **distance** between two vertices is given by $d_G(u, v)$. A **shortest path** between two vertices is a path with minimal length d_{ij}. The **average path** length is defined by $d = \langle d_{ij} \rangle$. The **diameter** is defined by $d_m = \max(d_{ij})$, which represents the maximum path length. The correlation between edges and vertices is given by $\varepsilon(G) := |E|/|V|$ [31, 32].

An **Eulerian path** is a path which contains every edge exactly once. A graph is an **Eulerian graph** if it contains an Eulerian path [31]. A path in an undirected graph that visits each vertex exactly once is called a **Hamiltonian path**. A graph that contains a Hamiltonian path is a **Hamilton graph** [31].

Going further into detail, vertex degrees and other topological indices are described in the following, which serve as a base for centrality measurements. Network centralities are a common method to determine important elements within a system. In the social sciences it is a common task to model relationships with graphs and, based on that, to identify people that are more influential than others. Similar questions can also be asked of biological networks.

A **centrality** is defined by the function $\mathscr{C} : V \mapsto \mathbb{R}$ on a directed or undirected graph $G = (V, E)$, which assigns a real number to every vertex (vertex degree). If one vertex is more central than another one, then $\mathscr{C}(v_1) > \mathscr{C}(v_2)$ is given [33].

A **vertex degree** $\delta_G(v) = \delta(v)$ is the number of edges $|E(v)|$ incident to the vertex, with loops counted twice. The **minimum degree** is characterized by $\delta(G) := \min\{d(v) \mid v \in V\}$, the **maximum degree** by $\Delta(G) := \max\{d(v) \mid v \in V\}$, and the **average degree** by:

$$d(G) := \sum_{v \in V} \frac{d(v)}{|V|} \tag{8.3}$$

The relation between the degrees is given by $\delta(G) \leq d(G) \leq \Delta(G)$ [9, 31, 32].

However, centrality measurements are only comparable inside the same network, and some measurements can only be applied on connected networks. One of the first centrality measurements is the **degree centrality**, defined by

$$\mathscr{C}_{\deg}(v) := |e | e \in E \wedge v \in e| \tag{8.4}$$

This measurement counts the number of edges connected to a vertex. In several studies, this measurement was used to identify essential elements within a biological network. A study on *Saccharomyces cerevisiae* revealed that proteins with a high degree centrality are more essential in comparison to others [34]. Other studies described similar findings with degree centralities as described by Hahn et al. [35].

The **average neighbor degree** is defined by Junker and Schreiber [9]

$$k_{i,nn} = \frac{1}{k_i} \sum_{j=1}^{N_v} A_{ij} k_j \tag{8.5}$$

for each vertex n_i over all vertices N. A is the adjacency matrix of the graph G.

Further centrality measurements are stated on network paths. They give information about the importance of certain paths by using information about path length. The first presented measurement is called eccentricity centrality. For every vertex it determines the maximum distance to all other vertices. The vertex with the shortest paths to all other vertices is the vertex with the highest eccentricity value. Formally, the **eccentricity centrality** is defined as [36]

$$\mathscr{C}_{\text{ecc}}(v_1) := \frac{1}{\max\{\text{dist}(v_1, v_2) : v_2 \in V\}} \tag{8.6}$$

The second important centrality measurement is the **closeness centrality**, which assigns a vertex v a high value if the shortest path distances for all other vertices to v is minimized. Formally, it is defined as [37]

$$\mathscr{C}_{\text{clo}}(v_1) := \frac{1}{\sum_{v_2 \in V} \text{dist}(v_1, v_2)} \tag{8.7}$$

The **shortest path betweenness centrality** measures the ability to monitor communication between other vertices. These vertices, which are on the shortest paths between all other vertices, are the most relevant ones. Let $\sigma_{v_1 v_2}$ be the number of shortest paths between v_1 and v_2, whereas more than one shortest path can exist. $\sigma_{v_1 v_2}(w)$ denotes the number of shortest paths, including w as an interior vertex which is neither start nor end vertex of the paths. The communication rate is given by

$$\delta_{v_1 v_2}(w) := \frac{\sigma_{v_1 v_2}(w)}{\sigma_{v_1 v_2}} \tag{8.8}$$

If no shortest path between v_1 and v_2 exists, then $\delta_{v_1 v_2}(w) := 0$. With these definitions the shortest path betweenness centrality can be defined as [38]

$$\mathscr{C}_{\text{spb}}(w) := \sum_{v_1 \in V \wedge v_1 \neq w} \sum_{v_2 \in V \wedge v_2 \neq w} \delta_{v_1 v_2}(w) \tag{8.9}$$

A further centrality measurement is based on the eigenvector. It is used on strongly connected graphs such as protein-protein interaction networks, to determine essential elements within a network. The **eigenvector centrality** is the eigenvector C_{eiv} of the largest eigenvalue $\lambda_{\max}$ in absolute value of the equation system $\lambda C_{\text{eiv}} = A C_{\text{eiv}}$, where A is the adjacency matrix of the graph G [39].

The **clustering coefficient**, a basic measurement for the local cohesiveness of a network, measures the probability that two vertices with a common neighbor are connected. In the case of undirected graphs, there exist $E_{\max} = k_i(k_i-1)/2$ possible edges between neighbors. The clustering coefficient C_i of the vertex n_i is then given as the number of edges E_i between the neighbors to the maximal number $E_{\max}$ with [9]:

$$C_i = \frac{2E_i}{k_i(k_i - 1)} \tag{8.10}$$

The **matching index** quantifies the similarity between two vertices on the number of common neighbors. The index is based on the following definition [9]:

$$M_{ij} = \frac{\sum \text{common neighbors}}{\sum \text{total number of neighbors}} = \frac{\sum_{k,l}^{N} A_{ik} A_{jl}}{k_i + k_j - \sum_{k,l}^{N} A_{ik} A_{jl}} \tag{8.11}$$

8 Biological Network Modeling and Analysis

Table 8.2 For a given network size, many different graphs can be reconstructed, where the difference between isomorphic and non-isomorphic graphs is significant

Nodes	Number of connected isomorphic graphs	Number of connected non-isomorphic graphs
3	8	2
4	64	6
5	1,024	21
6	32,768	112
7	2,097,152	853
8	268,435,456	11,117
9	68,719,476,736	261,080
10	35,184,372,088,832	11,716,571

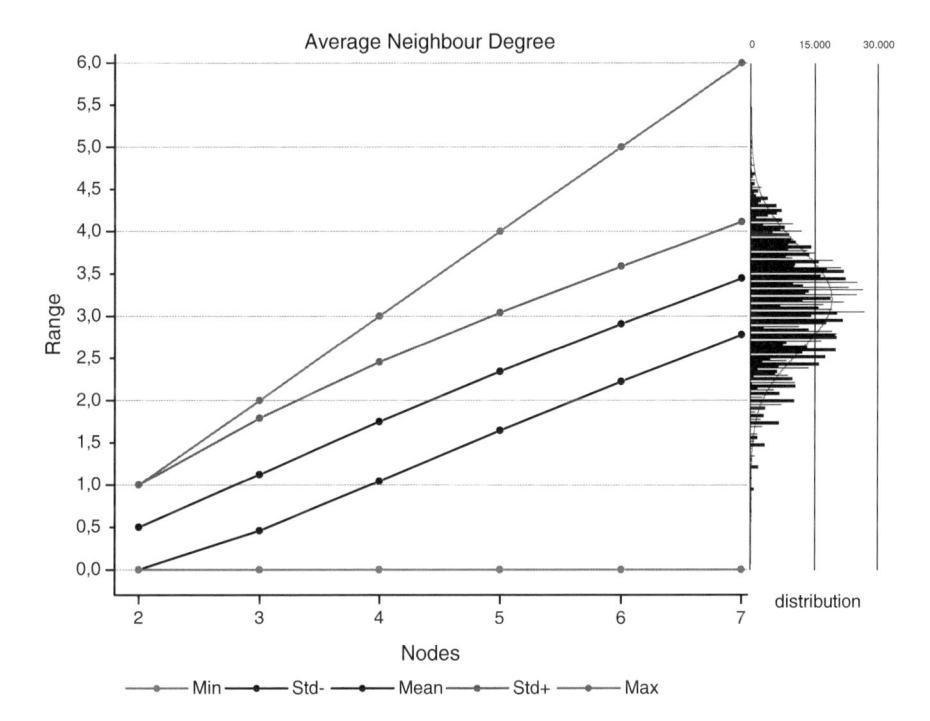

Fig. 8.2 The analysis of the distribution of graphs with the same average neighbor degree resembles a Gaussian curve, where thousands of different networks share the same average neighbor degree. The conclusion is that one specific average neighbor degree cannot characterize a unique network type [40]

In summary, all presented measurements are able to identify important elements within a graph. However, without a clear scientific question, the presented approaches can be misleading. Furthermore, scientists need to have in mind that a large set of graphs can share the same graph topological values [40]. In general, the number of possible graphs for a given node size is very large as presented in Table 8.2 [41]. Based on the non-isomorphic graphs, it was examined how many graphs share the same graph topology. Figure 8.2 presents the distribution of graphs with the same topological values.

Inferentially, thousands of different graphs share the same topological values. And having in mind that the discussed and examined graphs in biology have, in most cases, more than 30 nodes, the number of different graphs with the same topological values increases dramatically. Thus, graph theory has to be very carefully considered and only applied when it is linked to a specific scientific question. However, based on the presented definitions, a variety of analysis techniques are possible. The approaches enable structural as well as individual node analysis. Thus, it is not surprising, that applied to biological networks, it has become an important aspect in systems biology, bioinformatics, and theoretical biology [9].

8.4 How Biological Networks Can Be Modeled and Simulated

Modeling biological phenomena with the use of computer applications has become a common task. Therefore, different modeling techniques exist to study and analyze the dynamic details of biological systems. In general, biologists are more familiar with mathematical modeling, whereas computer scientists are accustomed to computational formalism. However, several approaches provide mathematical as well as computational capacities. In order to give an overview of existing modeling languages, the most important techniques in systems biology and biological network modeling are briefly described in the following subsections.

8.4.1 Ordinary Differential Equations

One of the most powerful techniques in modeling system dynamics is ordinary differential equations (ODEs), which provide a theoretical framework for discrete, continuous, deterministic, and stochastic models. In general, they describe the change rate of variables in the modeled system as a function of time. ODEs have been applied and used in many application cases and proved themselves very useful [8, 42, 43]. Furthermore, ODEs can be used to model entire systems with given kinetics [44, 45]. One common example for modeling gene activation or positive control is the Hill function in which the equilibrium binding of the transcription factor to its site on the promoter is modeled from zero to its maximal saturated level with Definition 2 (see Fig. 8.3 for a graphical representation).

Definition 2. A Hill function is defined by $F(X^*) = \dfrac{\beta X^{*n}}{K^n + X^{*n}}$, where:

- K is termed as the activation coefficient.
- β the maximal expression level of the promoter.
- n the steepness of the input function (the larger the n is, the more steplike the curve).

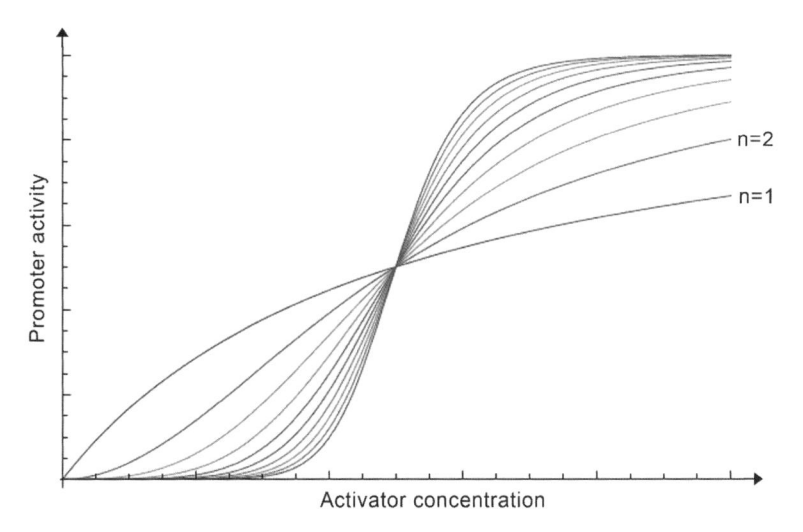

Fig. 8.3 Graphical plot of one Hill function with different steepness parameters (n) for the modeling of gene activation and positive control in biology

However, the model reconstruction with ODEs has some major drawbacks when the kinetic system parameters involved are unknown. With increasing network size and complexity, it becomes almost impossible to estimate all missing parameters. Due to high-throughput techniques, a huge amount of qualitative data is available, but the parameter estimation still remains challenging. Furthermore, precise quantitative measurements for parameter estimations are difficult to parametrically explore. A further disadvantage of ODE network modeling and analysis is that ODE-based models do not support any detailed insights into signal and information flow within biological networks. Thus, information flow, biological cascades, and system dependencies cannot be examined in detail.

8.4.2 Object-Oriented Modeling

Object-oriented modeling is a paradigm in which a system is primarily modeled with a set of related, interacting objects and the functions and services they provide [46]. These objects represent all entities relevant to the application (see Fig. 8.4 for an example). Nearly anything can be an object, which is defined as an assembly of classes. A class is a discrete reusable code block that has attributes, takes variables, performs functions, and returns values, among others. In general, objects do not exist in isolation from another. The relationships between the objects represent a wide set of different connections and interactions, for example, how one protein is related to a gene, or how one protein changes the state of another protein by phosphorylation. However, the modeling task is always specified for one specific

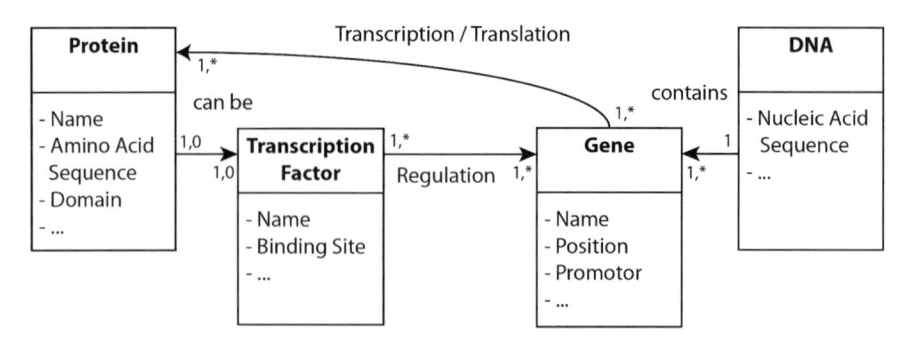

Fig. 8.4 An example of an object-oriented model in molecular biology. The model is focused on a mandatory set of properties, whereas a complete model is made up of more attributes and relationships. However, here, a protein can be a transcription factor regulating one or more specific genes. One gene can be even regulated by more than one transcription factor. The genes are derived from the class DNA, which contains a set of genes. Each gene alone or in combination with others can be transcribed and translated into one or more proteins. Each class is characterized by specific attributes, such as binding sites and nucleic acid sites, which are necessary for biological functions and molecular processing

context, where objects belong to each other and share a set of properties and methods to imitate the real-world system [47–49]. Using the standardized Unified Modeling Language (UML) [50], the object-oriented models can be made visually accessible through a set of graphic notation techniques.

8.4.3 Rule-Based Models

Rule-based specifications and formal grammars play an important role in the creation of photorealistic virtual organisms. Particularly plants and scientific models of vegetation structure are modeled with rule-based models [51]. One widely used formalism is the Lindenmayer system, a parallel rewriting system on strings. Based on an alphabet of symbols, a finite set of rules for string manipulations, a start string called axiom, and a mechanism to visualize data, it is possible to model the morphology of a variety of organisms. With an iterative process, which expands the model with new structures in each time step, growth processes can be modeled and simulated.

For example, having the axiom A and the rules $A \rightarrow B$ (letter A will be transformed into letter B) and the rule $B \rightarrow AB$ (letter B will be transformed into substring AB), a new string is generated in each time step by applying the aforementioned rules. Based on the system settings the development sequence for this model is described by $A \rightarrow B \rightarrow AB \rightarrow BAB \rightarrow ABBAB \rightarrow BABABBAB \rightarrow \ldots$. Finally, the expanded string only needs to be visualized to see developmental growth. In order to visualize this model, additional geometric rules have to be defined, which reconstruct geometric structures based on the appearance and order of the letters in the development sequence. One of the first examples of branching structures generated by an L-system was given by Prusinkiewicz and Lindenmayer in 1990 [52].

8.4.4 Constraint-Based Models

Constraint-based models are mainly used for cellular metabolism. The main idea of this approach is to describe detailed dynamic models with a set of constraints which characterize the models' possible behaviors. Therefore, stoichiometric, thermodynamic, and enzyme capacity constraints are defined. Instead of single solutions, a set of possible solutions represents different phenotypes which comply with the constraints. Thus, models can comprise thousands of reactions, such as the metabolic reconstruction of the bacterium *Escherichia coli*, where 2,583 constraint reactions were defined [53]. Furthermore, these models and constraints can be used for other metabolic engineering applications. However, the classical constraint-based models focus at flux balance analysis of metabolic networks [54, 55].

8.4.5 Interacting State Machines

Interacting state machines are mathematical models for the description of temporal behavior within a system. The model is based on the states of its parts and not on its components. Therefore, hierarchies are expressed by diagram-based formalisms. Each of the parts can be in one of a finite number of states, whereas the machine is in only one state at a given time. However, by initiating a trigger event, the machine can change its condition. The main advantage of interacting state machines is that they require little quantitative data, as they model biological behavior in a qualitative way [56, 57]. Usually, models described with interacting state machines are used for model checking and interactive execution.

8.4.6 Process Algebras

Process algebras are used for the modeling of concurrent systems. The language provides a framework for the high-level description of interactions, communications, and synchronizations using a set of process primitives. Operators are used to combine these primitives. Therefore, this approach provides algebraic laws for the manipulation and analysis of process expressions using equational reasoning. In most of the cases, process algebras are used in signal processing, as presented in the work of Danos and Laneve. The authors introduced a protein algebra to demonstrate how standard biological events can be expressed in simplified signaling pathways [58].

Initial state   Second generation   Third generation

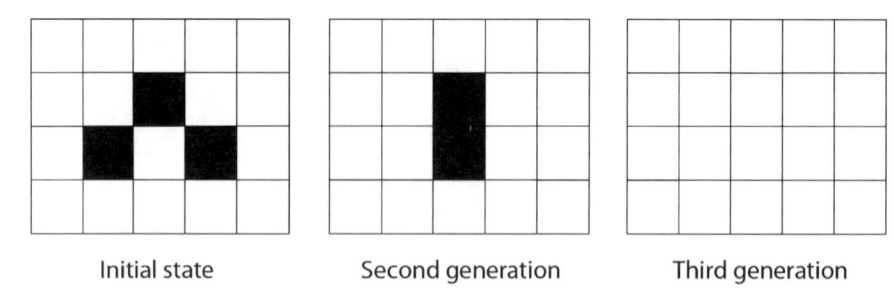

Fig. 8.5 An example of a simple cellular automaton with rules and settings of the "Game of Life" approach by John Horton Conway. From *left* to *right*: initial state and configuration (generation 1), second generation, and third generation

8.4.7 Cellular Automata

Cellular automata (CA) are used to model and simulate biological self-organization. They use a paradigm of fine-grained, uniform, parallel computation, which was used in many aspects of developmental biology [59–61]. With CA whole population dynamics can be simulated in which each individual's fate is dependent on its neighbor's behavior and existence. Therefore, a set of simple rules is defined that mimics the physical laws of the given system. The evolution of a CA is determined by its initial state, requiring no further input. The simulation is discrete in time, space, and state and, once running, evolves with its own given rules.

The most prominent example of a CA is the "Game of Life" devised by the British mathematician John Horton Conway in 1970 [62]. The example is based on a simple deterministic CA consisting of a regular two-dimensional grid of cells, in which each cell has a certain state: alive or dead. Every cell interacts with its neighbors based on the set of applied rules at each time step (see Fig. 8.5).

The following rules are applied to the "Game of Life" to calculate and simulate next generations:

- Any living cell with less than two living neighbors dies because of under population.
- Any living cell with two or three living neighbors does not change in the next generation.
- Any living cell with more than three living neighbors dies due to overcrowding.
- Any dead cell becomes alive by reproduction, when exactly three neighbors are alive.

Those rules are applied repeatedly to create further generation. Finally after n generations, a picture results that describes population structure, dynamics, population features, and system robustness, among others.

8 Biological Network Modeling and Analysis

a	Mutation probability	Mutation probability	Mutation probability
Grandfather		1	1
Father			1
Son	0.001	0.025	0.5

b

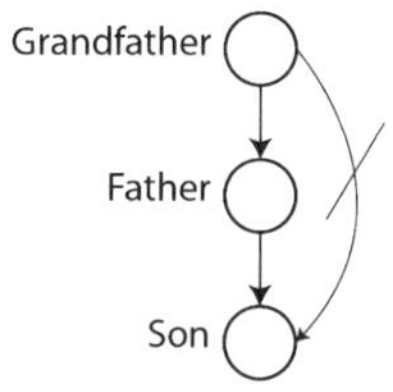

Grandfather ○
Father ○
Son ○

Fig. 8.6 A Bayesian network example from classical genetics studying mutations. (**a**) The probability that the son has a mutation is 0.001. If we know that his grandfather has the same mutation, the probability increases to 0.025. Thus, their genotypes are clearly dependent. But if we also know that his father has the mutation as well, the son's probability increases to 0.5. This additional information indicates that his father, independent of whether his grandfather has or does not have the mutation, only affects the son's probability. Therefore, only one conditionally network can be reconstructed (**b**), which matches the experimental data. All other possible networks are disregarded

8.4.8 Agent-Based Systems

Agent-based systems are similar to the concept of cellular automata, focusing on complex system behavior, structures, and phenomena in dynamics. This approach describes and simulates operations and interactions of autonomous agents in a given space. System operations and interactions are based on simple rules. However, in contrast to CAs, the agents are not placed on a grid or any similar environment. Moreover, the autonomous agents can freely move within the given 2D or 3D space. The most prominent examples are from multicellular studies, such as tumor growth studies [63], morphogenesis [64], and immune response [65].

8.4.9 Bayesian Networks

A technique for biological network modeling is the so-called "Bayesian networks" theory. Bayesian networks are used for the automatic reconstruction of causal signaling network models from experimentally derived data [66–68]. The core of this approach is the notion of conditional independency. This approach calculates probabilistic relationships to estimate which network structures, circuits, and motifs can be derived from the given biological data. This results in one or a set of possible directed acyclic graphs that match the experimental data conditions best. Nodes, which are not connected within the graph, represent variables which are conditionally independent. Nodes that are connected to each other represent strong probabilistic relationships based on experimental conditions. One example of such an approach is presented in Fig. 8.6.

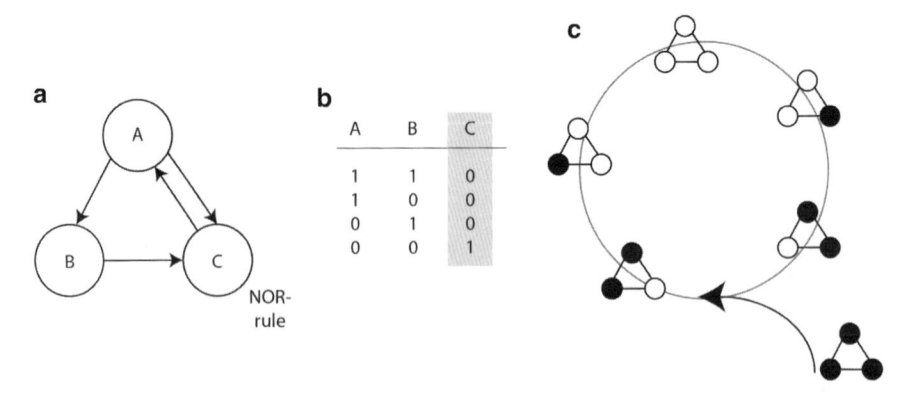

Fig. 8.7 A possible Boolean network based on three nodes (**a**), each having a state 0 (OFF) or 1 (ON). The states for each node are determined by the input of the other nodes. Nodes 1 and 2 copy their single input, while node 3 performs the Boolean function NOR on its inputs as described in the table (**b**). The dynamic system is described in (**c**), where filled nodes are on and lights are off

However, the reconstruction of such networks demands a large number of datasets. The greater the network, the larger the necessary experimental datasets must be. Otherwise, probabilistic relationships and independencies cannot be determined.

8.4.10 Boolean Networks

In 1969, Boolean networks were introduced by Kauffman to model gene regulatory networks [69]. Here, genes are modeled by Boolean variables which represent their active and inactive states within the model. A Boolean network is a directed graph, where all nodes are equivalent and receive information inputs from their neighbors. Every node can only take two binary values, 0 (OFF) and 1 (ON). These values represent the dynamic activity and behavior of the involved elements. Information flow and statement acting is determined by a logic rule. Therefore, the logical operators *and*, *or*, and *not* are used. If the statement is true, the logical operation results in an ON state; otherwise it remains in the OFF state (an example is given in Fig. 8.7).

The main advantage of this technique is the reduced number of parameters necessary while still capturing network dynamics and producing biologically predictions and insights [70]. However, quantitative measurements cannot be included for precise predictions and analysis.

8.4.11 Boolean Formalization

This approach formalizes in Boolean terms genetic situations for the description of complex circuits [71–73]. The main goal of this language is to formalize a complex model in a compact and unambiguous way by functions of binary variables. Therefore, three different types are defined and used. The genetic variable describes the gene state, being normal or mutated, and the recognition site, being a promoter, operator, terminator, or other. The environment describes temperature and the presence of different substances. Internal variables are used to memorize previous system states at a given time. Associated functions calculate the proceeding periods of the system with regard to the present variables. In order to reduce the algebraic expressions to its simplest form, tabulations of the logic equations as Veitch matrices are used. The Veitch matrices give a clear and exhaustive view of all calculated system states and show which states are stable and how the model proceeds from state to state.

8.4.12 Petri Net

A Petri net is a mathematical modeling language for the description and analysis of complex and distributed systems. Therefore, it provides an exact mathematical definition of its execution semantics. The language was introduced by Carl Adam Petri in 1962 [74] and constantly developed. Thus, this language comes with a well-developed mathematical theory for process analysis.

Reisig et al. presented the first basic definition in their article "A Primer in Petri Net Design" in 1982 [75]. This resulted in the general formalism presented in Definition 3.

Definition 3. A basic Petri net is defined by the tuple PN $= (P, T, F, W, m_0)$, where:

- $P = \{p_1, p_2, \ldots, p_n\}$ is a finite set of places.
- $T = \{t_1, t_2, \ldots, t_n\}$ is a finite set of transitions.
- P and T are pairwise disjoint.
- $F \subseteq (P \times T) \cup (T \times P)$ is a set of arcs from places to transitions and transitions to places, where $(p_i \to t_j)$ denotes the arc from place p_i to transition t_j and $(t_j \to p_i)$ the arc from transition t_j to place p_i,
- W is the weight function ($W : F \to \mathbb{R}$) which assigns every arc a non-negative integer, where $(f : p_i \to t_j)$ denotes the weight of the arc from place p_i to transition t_j.
- m_0 is the initial marking $\forall p_i \in P$.

A Petri net is based on a directed bipartite graph, in which the nodes represent transitions and places. Regarding the graphical representation, places are drawn as circles, transitions are drawn as rectangles, and arcs are drawn as directed arrows. The directed arcs describe which places are pre- and/or post-conditions for which transitions. Each place can contain tokens, which are drawn as black dots. The start configuration of a Petri net model is described by the state m_0, which assigns tokens to each place. With this graphical notation, processes such as choice, iteration, and concurrent execution can be modeled stepwise and analyzed (see Fig. 8.8).

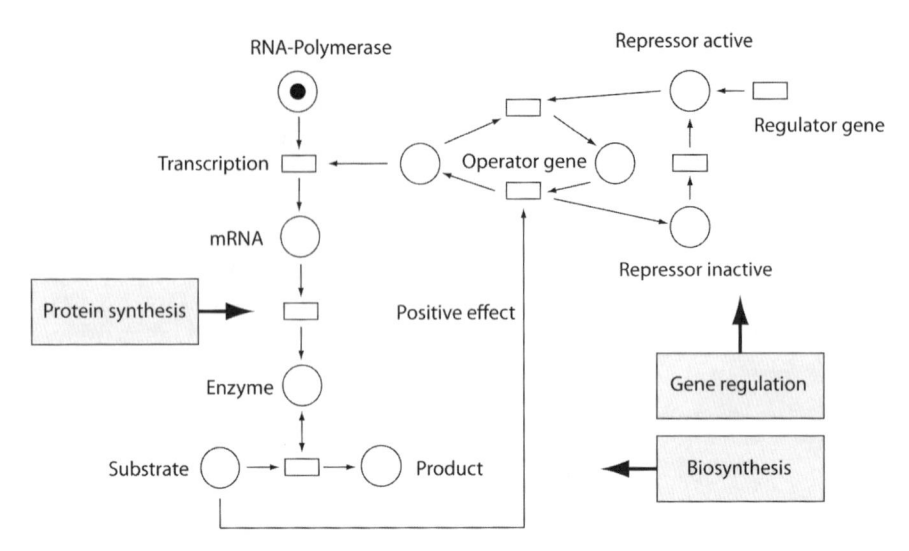

Fig. 8.8 The possibility of modeling abstract biological processes with Petri nets. The model is based on gene-controlled biochemical reactions, such as gene regulation and protein synthesis

Due to the presented formalism, Petri nets stand out by their balance between modeling power and analyzability in comparison to other modeling techniques. Furthermore, concurrent systems can be automatically determined, although some of the systems are difficult and expensive to determine [76]. Thus, the various modeling possibilities and analytic power of the proposed formalism offer a well-developed basis for the description of chemical processes and a mathematical theory for process analysis.

8.4.13 Visual Modeling

A further way to model a biological system is by using a standard graphical notation, such as the Systems Biology Graphical Notation (SBGN) [77]. SBGN is a visual language which focuses on the graphical notation of biological networks. It provides a common notation to represent interactions and regulations between molecular species, such as binding, complexation, and protein modification, among others. It consists of three complementary languages: process diagram, entity relationship diagram, and activity flow diagram. Together the different notations enable scientists to represent biological networks in a standard and unambiguous way (see Fig. 8.9 for an example).

In summary, each modeling technique comes with specific features and constraints. In order to model and analyze a biological system a powerful theoretical framework is necessary. Thus, visual languages such as SBGN are not suitable for

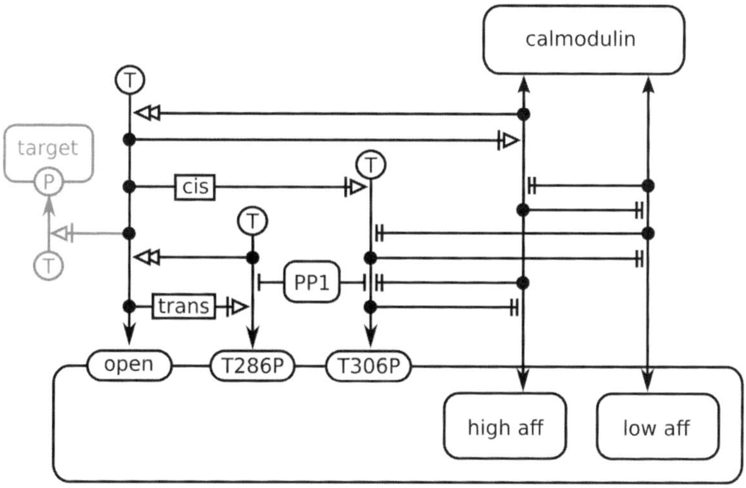

8 Biological Network Modeling and Analysis

Fig. 8.9 SBGN entity relationship diagram representing the effect of calmodulin binding on CaMKII activity, using the nested entities of ER L2 V1 [78]

systems biology analysis, as they do not provide any kind of analytical environment. Furthermore, these languages consider only a limited graphical representation of the biological components. Object-oriented models are software-intensive and complex systems. As systems evolve, classes and the function they perform need to be changed more often. This can result in a schema, where complexity continuously grows. Thus, a clean programming, organization, and notation are necessary during model design and software implementation. Furthermore, well-defined interfaces between objects are mandatory to keep the model maintainable. Otherwise, model parameters can become distorted or even incorrect. Ambiguities in data flow can also occur. Therefore, the following review only focuses on modeling techniques that provide sophisticated analysis power and are clean and well defined in their semantics. To show how often and in which application cases the aforementioned techniques are used, Machado et al. summarized literature references, classified by the type of biological process [79] (see Table 8.3). Boolean formalizations are not considered in this review as this approach is frequently used in systems biology and bioinformatics. Furthermore, the same or similar results can be produced with Boolean networks, ODEs, or Petri nets, among others.

The first thing to point out is that all formalisms have been applied to signaling networks. This is not surprising, as signaling networks have the largest number of features, such as spatial localization, multistate components, network information flow, and robustness, among others. Therefore, each of the presented formalisms contributes with powerful features. A smaller number of formalisms are applied to metabolic networks. However, this does not indicate that other formalisms are not able to model those systems. Moreover, it seems that Petri nets, process algebras,

Table 8.3 Overview of the amount of literature references using the presented formalism classified by the type of biological process [79]. Based on the evaluated information, signaling networks have been modeled and analyzed with all formalisms. Gene regulatory networks and metabolic networks have only been modeled with specific techniques due to their specific system dynamics and topology. However, differential equations, constraint-based models, and Petri nets have been used as universal techniques to examine all of the mentioned networks

	Signaling networks	Gene regulatory networks	Metabolic networks
Boolean networks	+	++	
Bayesian networks	+	++	
Petri nets	++	+	++
Process algebras	++		
Constraint-based models	+	+	++
Differential equations	++	++	++
Rule-based models	++		
Interacting state machines	++		
Cellular automata	+	+	
Agent-based models	++		+

constraint-based models, and differential equations seem to be powerful enough to consider all aspects of metabolic system dynamics. A further observation indicates that Petri nets, constraint-based models, differential equations, and cellular automata are applied to all kinds of biological networks. This makes them potential candidates for whole-cell modeling. The most powerful technique is still differential equations modeling, which is also reflected by the data provided in the table. However, Petri nets are among the formalisms that cover most of the features to model all kinds of biological networks as described in Table 8.4. It is a universal graphical modeling concept for representing processes from different application fields in nearly all degrees of abstraction. Petri nets provide the qualitative modeling approach as well as the quantitative one. Furthermore, qualitative and quantitative formalism can be combined to one paradigm. The formalism is easy to understand and use.

Once a basic qualitative model is established, it can be successively enriched with quantitative data. Thus, parameter estimations based on experimentally derived data are not implicitly necessary in the network reconstruction process. Furthermore, models can be modeled discretely as well as continuously. It is even possible to integrate ODEs for precise model description.

Besides, Petri nets allow hierarchical structuring of models and thus offer the possibility of different detailed views for every observer of the model. Petri net theory provides a variety of established analysis techniques that are well suited and applicable to biological network modeling. Moreover, database information, as described in the following section, can be used to automatically reconstruct sophisticated network and Petri net models.

8 Biological Network Modeling and Analysis 225

Table 8.4 Overview of implemented features for each modeling formalism based on [79]: (+) supported feature and (e) available through extension. Based on the provided data, the most powerful technique is the Petri net modeling as it includes the advantages and features of all other formalisms

	Visualization	Topology	Modularity	Hierarchy	Multistate	Compartments	Spatial	Qualitative	Synchronized	Stochastic	Continuous
Boolean networks	+	+						+	+	e	
Bayesian networks	+	+						+		+	
Petri nets	+	+	+	e	e			+	e	e	e
Process algebras			+	e		e		+		+	
Constraint-based models		+						+			
Differential equations							e			e	+
Rule-based models	+		+		+	+	e	+		+	+
Interacting state machines	+		+	+	+	+				+	
Cellular automata	+					+		+	+	+	
Agent-based models	+				+	+	+			+	

8.5 Network Reconstruction

A biological network, as described in Sect. 8.2, consists of a set of different biological elements being in interaction with each other. Such a network can be reconstructed by hand, with experimental data, information from literature, and/or database knowledge. In the first case, users need to put all involved elements into relation and draw the resulting models as a graph. They have several possibilities to model the system. They can use directed, undirected, mixed, or other graphs as presented in Sect. 8.3. Furthermore, they can use a standard graphical notation, such as SBGN for the visual modeling as presented in Sect. 8.4.13.

In terms of a network reconstruction with experimental data correlation, networks have to be reconstructed as described in Sect. 8.2. Therefore, a well-established modeling and analysis technique is necessary. One possible approach is the Bayesian networks as described in Sect. 8.4.9. Bayesian networks offer one way to automatically reconstruct signaling networks from experimentally derived data. The only disadvantage of this approach is the necessary input data. To be able to produce unambiguous results, a huge set of experimental data is mandatory.

A further way to reconstruct biological networks is by using text mining approaches [80, 81]. Text mining is equivalent to text analytics, with the goal of turning text into data for further analysis. This approach can be used, for example, to find interaction partners for a gene by analyzing a set of publications. The collected data is then modeled as a graph. In general, this technique is based on statistical pattern learning. The main disadvantage of this approach is still the interpretation of the input text. In many cases relations are identified which are positive false or false positive. Although the analysis and results are becoming better and better, the resulting networks need to be evaluated by an expert.

A more reliable way to reconstruct biological networks is by querying biological databases. Therefore, more than 1,300 different biological databases exist that can be accessed. Using complex queries, data transformations, and data integration techniques, rudimentary data such as genes and proteins can be linked with each other. Many databases provide links between the different biological compounds. If such a link does not exist, it is even possible to establish connections by mining genomic databases. Hence, several attempts have been made to reconstruct metabolic pathways via genome sequence comparison [82, 83]. Such attempts have a certain limit, as the results do not reflect all involved molecular functions. Due to cellular functions, such as translation, transcription, post-modification, and many more processes with genome sequence comparison and analysis, it is often not possible to predict direct correlations and further regulatory or metabolic processes.

However, several databases do exist, which contain more detailed information about metabolic pathways, such as the KEGG database [21]. The information about the networks can be accessed via the Internet or by parsing provided flat-files. The disadvantage with online access is that the elements cannot be analyzed and combined with other -omic level data and experimental datasets. Therefore, flat files have to be processed, filtered, normalized, and integrated into one model. Actually, the KEGG database consists of more than 121 tables, where at least 23 tables are necessary to reconstruct the backbone of a biological network. The other tables store further information, such as diseases, drugs, and taxonomies (see Fig. 8.10 for a simplified scheme of the KEGG database structure). With access to that data, it is possible to reconstruct metabolic networks as they are presented by KEGG and to analyze the biological elements in detail or overall context.

In terms of biological network reconstruction using database information, each scientist should follow some basic recommendations:

1. All databases should be free of charge and accessible by using a SOAP or an API.
2. All databases should use the same terms, identifiers, and publication structures as cited in literature.
3. Provided datasets must be up to date and should not overlap.
4. The selected databases should be well curated.
5. Only databases which can be used for the reconstruction of biological networks should be integrated.
6. The used databases should be focusing on the mechanisms which should be modeled, such as metabolic pathways, signaling pathways, and protein-protein interaction networks.
7. It should be possible to query each integrated database separately or in combination with each other.

8 Biological Network Modeling and Analysis 227

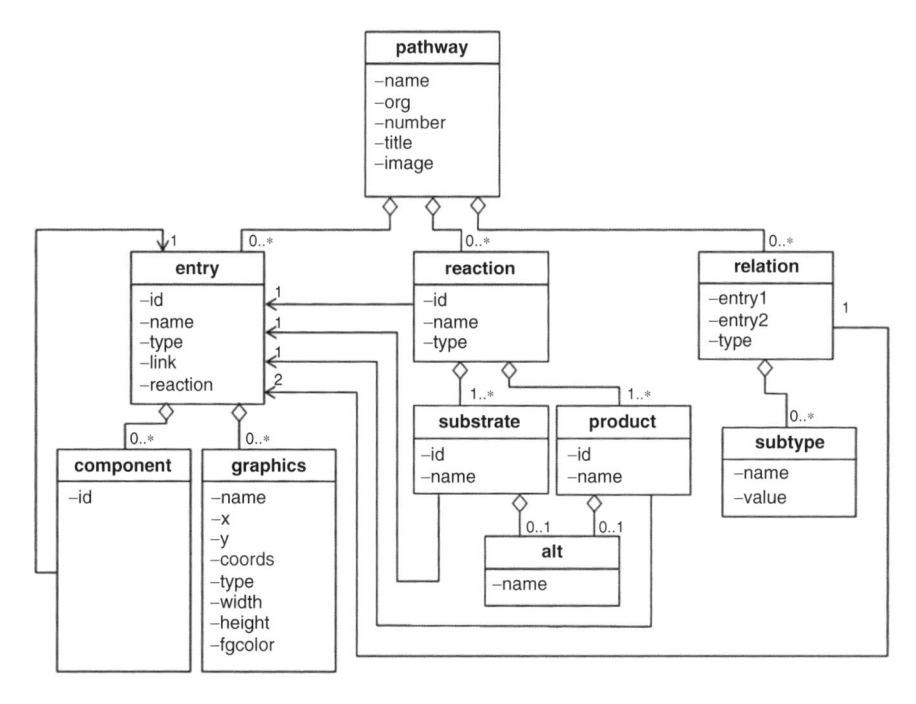

Fig. 8.10 Simplified scheme of the KEGG database structure [84]. The pathway element is the root element of the biological network, consisting of a list of entry, relation, and reaction elements. Theses entities specify the graph information. Additional elements specify more detailed information about the biological compounds, relations, and reactions within the model

8.6 Biological Network Exchange Formats

Molecular biotechnology, systems biology, bioinformatics, and many other disciplines in biology make it possible to reconstruct and analyze biological systems. More than 300 pathway or molecular interaction-related data resources, visualization, and analysis software tools have been developed.[1] However, the diversity of tools shows several problems in sharing and moving models between each other. An attempt to overcome this problem is the creation of standards [85–87].

In an online survey, Klipp et al. asked 125 researchers (75 % modelers, 4 % experimentalists, or 21 % both) covering various fields, such as modeling of individual pathways, investigation of complex processes, development and application of computational methods, and software development about their opinion on standards

---

[1]The number of software applications has been approximated by counting software tools that support SBML and CellML. Software tools are listed at http://www.sbml.org/ and http://www.cellml.org/

[88]. About 80 % of the scientists considered the creation of standards necessary or desirable. This is not surprising that science standards have many advantages as listed in the following:

- Model definitions and entities are based on ontologies, defined nomenclature, and restrictions. Thus, they become accessible and readable to a wide community.
- Standards improve communication between software tools, free exchange of information, and comparison between different studies, which results in more productive collaborations.
- Complementary resources from multiple simulation/analysis tools can work together, instead of redefining and reconstructing models in each tool.
- Reimplementation of models becomes easier or dispensable, which reduces duplication and redundancy.
- If tools are no longer supported, models developed within the tools can be still used if they are based on standards. Information, knowledge, and research progress is not lost and can be reused.
- Data curation teams can evaluate models without being restricted to a certain tool or formalism.
- In the publication process, any curator can process annotation and normalization before data is published and made available to the scientific community.

Scientists, simultaneously with both tool development and modeling projects, have developed standards to share, evaluate, and analyze knowledge and information. Standards are definitions in the form of common, inclusive, and computable languages. Here, only XML-based formats are considered, since it is used as universal language in data exchange. McEntire et al. [89] and Achard et al. [90] have shown in their studies that this language is very flexible and simple to use and, therefore, a powerful standard in bioinformatics and systems biology in comparison to Comma Separated Values (CSV), Excel, and other file formats. More than 85 standards can be found within systems biology [87].

For the modeling and sharing of biological models, main standards exist, such as the Systems Biology Ontology (SBO) [91], Systems Biology Markup Language (SBML) [92, 93], the CellML [94], and BioPAX [95]. For the graphical representation of biological pathways, languages such as the SBGN [77] have been introduced (see Sect. 8.4). Model description achieves human and computational usability, reusability, and interoperability when the encoded format is standardized. Models or software tools without standardization are only of limited use, as they do not provide the possibility to share, compare, and/or integrate large amount of systems. Thus, it is important to use common standards as described in the following section:

- **Systems Biology Ontology (SBO)**
  The SBO ontology [91] is a well-defined logic about biological terms, including single identifiers for each distinct entity, allowing clear reference and identification. Furthermore, it is augmented with terminological knowledge such as synonyms, abbreviations, and acronyms. The terminology is also used to

specify the type of the components being represented in a model and their role in systems biology descriptions. Thus, the ontology allows unambiguous and explicit understanding of the meaning of the involved components in a system and, moreover, enables mapping between elements of different models encoded in this format.

The ontology is a well-defined logic about biological terms, including a single identifier for each distinct entity, allowing clear reference and identification. It is composed of seven vocabulary branches: systems description parameter, participant role, modeling framework, mathematical expression, occurring entity representation, physical entity representation, and metadata representation. The terminology is also used to specify the type of components represented in a model and their role in systems biology descriptions. Thus, the ontology allows unambiguous and explicit understanding of the meaning of the involved components in a system and, moreover, enables mapping between elements of different models encoded in this format.

- **BioPAX**
  BioPAX is a standard language to represent biological pathways at the molecular and cellular level [95]. The main goal of BioPAX is the exchange of information between several pathway databases such as Reactome [96] and BioCyc [23]. It was introduced through a community process to make complete representation of basic cellular processes substantially easier to collect, to index, to interpret, and to share. BioPAX covers concepts such as metabolic and signaling pathways, gene regulatory networks, and genetic and molecular interactions. Therefore, it has a structure for substances, interactions, pathways, and links to organisms and experiments. The language is distributed as an ontology definition with associated documentation and a validator for checking. Therefore, the BioPAX community cooperates with the SBML and CellML mathematical modeling language communities. For better accessing and manipulating data in the BioPAX format, a house-implemented Java library called "Paxtool" is available. BioPAX Level 3 is currently available at http://www.biopax.org.

- **BioXSD**
  BioXSD is common exchange format for basic bioinformatics data [97]. Using this format, it should be possible to establish a common web service for the exchange of data for bioinformaticians in the World Wide Web. This format should fill gaps between specialized XML formats such as SBML [92, 93], MAGE-ML [98], GCDML [99], PDBML [100], MIF [101], and PhyloXML [102]. Therefore, BioXSD defines data formats such as biological sequences, sequence alignments, sequence annotation, and references to data, resources, and vocabularies in a variety of possibilities. BioXSD serves as a canonical data model and is available at http://bioxsd.org as version 1.1.

- **CellML**
  CellML [94, 103] is a language for representing mathematical models. Using differential algebraic equations, any cellular model can be represented in CellML. In addition, CellML represents entities using a component-based approach, where relationships between components are represented by connections. The

developers have implemented an API for working with CellML models and files. Thus, software developers do not need to reinvent the same functionality each time they develop a new tool. The API enables users to retrieve information, to manipulate, and to extend a model. The API interfaces are designed to be independent in any programming language, platform, or vendor. At the present time, CellML is available at http://www.cellml.org in version 1.1.

- **MathML**
  MathML is a low-level specification for describing mathematics [104, 105]. It is used wherever mathematics needs to be handled by software, such as mathematical expressions in web pages and workflows in science and technology. Actually, MathML is available at http://www.w3.org/Math/ as version 3.

- **PDBML**
  The PDB database is the single worldwide repository for macromolecular structure data [106]. For more than 30 years, the data resources have used a column-oriented format to store and share archival entries [100]. Facing more and more complex data for macromolecular structures, the used data format constrained several limitations such as internal structure and the organization of records. Therefore, a new XML-based data format, called PDBML, has been introduced [100]. It builds the content of the PDB exchange dictionary and can be used as a specific exchange medium for detailed molecular protein structures, such as data derived from experimental crystallography. PDBML is currently available at http://pdbml.pdb.org as version 3.3 to all users.

- **Systems Biology Markup Language (SBML)**
  SBML is an exchange format for representing biochemical reaction networks [92, 93]. Using SBML, users are able to describe models in many areas of computational biology, including cell signaling pathways, metabolic pathways, and gene regulation. Therefore, SBML has the structure, ontology, and links, for pathways and interactions. To enable mathematical descriptions, the SBML Level 2 uses MathML for more complex mathematical formulas. This extends the features of SBML and also results in a greater compatibility with CellML. Furthermore, it provides the possibility to specify delay functions and define discrete events that can occur at specified transitions in a certain state in biological models. In order to help users to read, write, manipulate, translate, and validate SBML files and data streams, the LibSBML API is available in different common programming languages, such as Java, C, and C++. Presently, SBML Level 2 is available at http://sbml.org/Software/libSBML and SBML Level 3 is being developed.

One of the main standards for the modeling of biological systems is the Systems Biology Ontology. Using this standard ensures the usability, reusability, and interoperability of biological models. Furthermore, data exchange standards can easily access models encoded in this format. For instance, SBML, MathML, and CellML support SBO definitions, which makes it easy to translate any kind of SBO model into such an exchange format. However, there is a significant difference in the scope of the mentioned standard exchange formats. By studying the most important formats and considering recommendations from literature [86,87], SBML and CellML are proposed as a means for the exchange of biochemical reaction networks and models between different software tools. They provide an ontology and structure that can even be used for simulations. They also provide constructs that are similar to the object models used in packages specialized for simulating and analyzing biochemical networks. CellML and SBML, embedding MathML, provide users with the possibility for the representation of whole models in differential algebraic expressions. Besides, SBML and CellML have an API, which allows reading, writing, and manipulating models in an easy manner. Furthermore, SBML and CellML have much in common, since the development of both standards takes place cooperatively. Formats such as PDBML only focus on particular substances. Thus, they are not appropriate for network models. This also applies to MathML, which only provides basic mathematics. Furthermore, BioXSD and BioPAX exist and can be used as data standards. However, BioXSD is focused on data that is not supported by the main formats and thus very specialized and not capable of representing the entire biological systems. BioPAX is only focused on pathway maps, which can be shared between databases and tools. SBML and CellML can support dynamic systems in ways not possible for BioPax.

8.7 Where to Find Biological Databases and Tools for Network Reconstruction and Modeling

The first biological database emerged in 1965 when Margaret Dayhoff published the Atlas of Protein Sequence and Structure [107]. In the 1970s the first protein structure database, called PDB was found [108–110]. A few years later in 1981, the first repository for nucleotide sequences was established called EMBL [111, 112] and 1 year later the GenBank [113,114]. Since then, more and more biological databases have developed. The 19th annual database issue of NAR now lists more than 1,380 databases in molecular biology [115]. The Pathguide [116], a meta-database with an overview of more than 325 biological pathway-related resources, with more than 100 databases focused on protein-protein interaction, is an additional important resource for biological databases. To make it easier for researchers to quickly find relevant information about useful molecular resources, tools, and databases, community-curated databases with content and links to other biological databases were established. Some of the most important are MetaBase [117], OBRC [118], BioDBCore [119], and the Bioinformatics Links Directory [120, 121]. Currently, more than 1,800 entries are listed in MetaBase, each describing different biological databases. BioDBCore gives a brief description of the core attributes of biological databases, whereas OBRC contains annotations and links for more than 1,700 bioinformatics databases and software tools. The Bioinformatics Links Directory curates links to software tools and databases. Using these resources, users have the possibility to contribute, update, and maintain database content.

Concerning software tools in bioinformatics, in 2011, the SBML website[2] listed more than 200 software tools which provide biological modeling based on the SBML [92, 93]. Going further into details, Copeland et al. highlighted a small, representative portion of available tools from each -omic area [122]. Still, this review lists more than 30 tools specialized in biological modeling. However, the state of the-art applications CellDesigner [123], Cell Illustrator [124], Cytoscape [125], E-Cell [126], Gepasi [127, 128], JDesigner [129], VANESA in combination with the PNlib [130, 131], and Snoopy [132, 133] are able to model, reconstruct, visualize, and simulate biological systems in one single comprehensive framework.

8.7.1 CellDesigner

CellDesigner is a structured diagram editor for drawing gene regulatory and biochemical networks. It was developed by the Systems Biology Institute (SBI) in Tokyo, Japan [123]. The core members of this software application are Akira Funahashi, Hiroaki Kitano, and Akiya Jouraku. The main goal of this application is to visually represent biochemical reactions in a comprehensive graphical notation such as SBGN (Systems Biology Graphical Notation) [77]. Besides, in the new version it enables users to connect from species name or ID to the databases Saccharomyces Genome Database [134], iHOP (Information Hyperlinked over Proteins) [135], and the Genome Network Platform (http://genomenetwork.nig.ac.jp). Furthermore, it is possible to get basic information about a biological element from PubMed [136] or Entrez Gene, the search engine from NCBI (http://www.ncbi.nlm.nih.gov). To assist users in the simulation, CellDesigner is able to connect to the SBML ODE Solver [137] and Copasi, a biochemical network simulator [138]. Simulations can be set up in a control panel, where users are able to adjust system amounts and parameters. CellDesigner is free of charge and available at http://www.celldesigner.org in version 4.2 running under Windows and Linux.

8.7.2 Cell Illustrator

The software application Cell Illustrator [124] is a software platform for systems biology that uses the concept of the Petri net language for the modeling and simulating of biological networks. The first version of Cell Illustrator was published as Genomic Object Net [139] in 2000 under Matsuno et al. at the Faculty of Science, Yamaguchi University, Japan. The software application employs the concept of a hybrid Petri net as the modeling and simulation method. To handle any type of objects, the existing paradigm has been extended to hybrid functional Petri nets

---

[2]http://sbml.org/

with extension (HFPNe). This paradigm is more suitable for biological network modeling and simulation, since HFPNe can handle discrete and continuous events simultaneously. Any kind of function can be assigned to delay, weight, and speed parameters of these elements. Additionally, ordinary differential equations can be modeled and integrated into a subset of HFPNe.

Furthermore, Cell Illustrator is able to import pathways or single reactions from the TRANSPATH database [140]. To import networks from other tools, SBML, CellML, and BioPAX data exchange formats are supported. In addition, Cell Illustrator has its own format called CSML. Simulation results can be visualized in either 2D or 3D plots in an all-in-one-window environment. To make the network visualization more legible, graph grid layout algorithms are implemented. The latest version of Cell Illustrator is version 5.0, which is commercially an online version available at http://www.cellillustrator.com.

8.7.3 Cytoscape

Cytoscape is an open-source bioinformatics software platform for data integration and visualization [125]. The first version of Cytoscape was published by Shannon et al. from the Institute for Systems Biology, Seattle, Washington [141]. Nowadays, it is supported and funded by many different institutions, particularly by Agilent Technologies, University of Toronto, Institute Pasteur, Memorial Sloan-Kettering Cancer Center, Institute for Systems Biology, and the University of California San Diego. Primarily, Cytoscape enables users to visualize molecular interaction networks and biological pathways and integrate these with any type of attribute data, such as gene expression profiles. Furthermore, Cytoscape supports standard network and annotation files such as BioPAX [95], and SBML. Additional features are available as plugins, which are developed by third parties focusing on network and molecular profiling analyses, new layouts, additional file format support, scripting, and connection with databases. For network reconstruction there is the plug-in BioNetBuilder [142], which uses the databases KEGG [21], HPRD [143], BioGrid [144], and GO [145], among others for its modeling. Furthermore, simulation plug-ins exist, such as the SimBoolNet [146], for the simulation of Boolean networks or FERN for the stochastic simulation and evaluation of reaction networks [147]. Most of the plug-ins are available free of charge. Cytoscape uses an open API based on Java technology and version 2.8.3 is available at http://www.cytoscape.org.

8.7.4 E-Cell

The E-Cell project [126] is an international research project aimed at modeling and reconstructing biological phenomena in silico. The main goal of this software application is to develop a dynamical cell with all its functions. It has been developed by Hashimoto et al. at the Institute for Advanced Biosciences, Keio University, Yokohama, Japan. The software platform allows precise whole-cell simulations with object-oriented modeling. Therefore, numerical integration methods are encapsulated into biologically related object classes. Virtually any integration algorithm can be used for simulation [148]. Thus, users have the possibility to define functions of proteins, protein-protein interactions, protein-DNA interactions, regulation of gene expressions, and other cellular cell processes with a set of functions rules. Therefore, hundreds of reaction rules are provided and available for simulation progress. E-Cell version 3 is freely available at http://www.e-cell.org and runs on several different platforms such as Microsoft Windows and Linux.

8.7.5 Gepasi

Gepasi is a software application for the modeling and simulating of biochemical systems [127, 128]. It has been developed by Pedro Mendes at the Department of Biological Sciences, University of Wales, Aberystwyth, UK. Gepasi uses mathematical formulas to transform biochemical properties into kinetic models. It provides a number of tools to fit data, to optimize any function of the model, and to perform metabolic control analysis and linear stability analysis. Sophisticated numerical algorithms realize simulation processes and analysis tasks. The simulation results can be plotted in 2D and 3D. Furthermore, the software application supports SBML 1.0 import and export. The latest version of Gepasi is 3.30 and freely available at http://www.gepasi.org. It only runs using Microsoft Windows.

8.7.6 JDesigner

JDesigner is a software application that enables users to draw a biochemical network, which can be exported to SBML for further processing [129]. The development of JDesigner was supported by the California Institute of Technology, Pasadena, California, and more recently by the KECK Institute of applied sciences, Claremont, California USA. JDesigner represents networks by using one notation for chemical species, which can be decorated with visual cues. This is also possible for reactions. Although it is a network design tool it also supports simulations. It has the ability to use JARNAC as a simulation server via the Systems Biology Workbench (SBW) [129] which is an open-source framework connecting heterogeneous software applications. JDesigner is an open-source project distributed under the LGPL license and available at http://sbw.kgi.edu/software/jdesigner.htm.

8.7.7 VANESA

VANESA is a modeling software for the automatic reconstruction and analysis of biological networks based on life-science database information [131, 149–153] and constantly developed at the Bielefeld University. VANESA is platform independent and available free of charge at www.vanesa.sf.net. Using VANESA, scientists are able to model any kind of biological processes and systems as biological networks. Scientists have the possibility to automatically reconstruct important biomedical systems with information from the databases KEGG, MINT, IntAct, HPRD, and BRENDA. Furthermore, users have the possibility to use graph theoretical approaches in VANESA to identify regulatory structures and significant actors within the modeled systems. These structures can then be further investigated in the Petri net environment PNlib for hypothesis generation and in silico experiments.

The PNlib is the powerful new state-of-the-art Petri net simulation library [130]. Proß et al. have developed the PNlib library using the Modelica language [154] at the Department of Engineering and Mathematics, University of Applied Sciences, Bielefeld, Germany. Modelica was developed and promoted by the Modelica Association since 1996 for modeling, simulation, and programming. Primarily it is focused on physical and technical systems and processes. Now, Modelica, embedding the PNlib, provides the possibility to simulate biological systems. VANESA and the PNlib are based on the xHPNbio formalism [131]. The mathematical modeling concept xHPNbio was specially developed for scientists, based on the demands of biological processes. The focus of this formalism is the processing of experimental data to gain usable new insights about biological systems.

8.7.8 Snoopy

Snoopy [132, 133] is a unifying Petri net framework to investigate biomolecular networks. It has been designed and implemented by Heiner et al. at the Brandenburg University of Technology at Cottbus, Germany. The simulation environment comprises a family of related Petri net classes, such as time Petri nets, stochastic Petri nets, continuous Petri nets, hybrid Petri nets, colored Petri nets, and extended Petri nets, among others. The mentioned classes enhance standard Petri nets in various ways to meet the demands of biological scientists. For example, the extended Petri nets are characterized by read arcs, inhibitor arcs, equal arcs, and reset arcs. Using these formalisms, scientists are able to reconstruct and simulate any kind of dynamic network. Larger networks can be hierarchically structured. If further demands on the supported Petri nets should arise, the software application can be extended by new properties and even by new Petri net classes. This is possible due to the generic data structure of the software application. Furthermore, users are able to move between the qualitative, stochastic, and continuous modeling paradigms. However, this transformation from one paradigm into another is not possible without information loss.

Simulation results are visualized within a built-in animation environment. To be able to share results with other scientists and software applications, Snoopy offers SBML support with both import and export functions. Snoopy is available for all major operating systems, such as Windows, Linux, and Mac OS-X. It is available free of charge at http://www-dssz.informatik.tu-cottbus.de/snoopy.html.

8.8 Discussion

Cellular life is very complex and governed by thousands of macroscopic functions being constantly carried out. To produce good theoretical models which can be used for hypothesis testing, the models need to be manageable. This can only be achieved by reducing a biological system to the known and essential parts, which are necessary to answer the underlying research questions. By trying to model a complete system, regardless of the lack of data and parameters, it is very likely that the modeled systems can be misleading. Therefore, any model needs to have a clear focus rather than model all levels of biological details.

One of the best ways to start modeling a biological system is by using biological networks. A small network consisting of known and already analyzed elements can be the initial point for the reconstruction of a more significant system. Therefore, there are different biological networks which can be used as powerful integrated frameworks to present, integrate, and visualize knowledge. As these networks are intuitive and easy to extend in knowledge, any scientist can work with them. With biological networks different -omic levels can be modeled, describing elements such as genes, RNAs, proteins, and metabolites being in interactions and relationships with each other. Moreover, biological databases can be used to reconstruct or enrich those networks with relevant information and new data. Kinetics and other information can be queried to model a system in a more precise way. With database integration modules, it is even possible to query multiple databases with one view instead of consulting each database separately. Besides, data integration tools filter, normalize, and link heterogeneous data from different distributed data sources.

A further advantage of biological networks is that a wide range of graphical theoretical analysis techniques can be applied on reconstructed models. Graph theory can give important clues about topological network properties, such as the identification of the most important nodes within a system, or average path lengths between different elements in a biological model. This is important in as much as biological networks can become large and complex. Scientists need a tool which assists them in identifying relevant information.

When it comes to simulating cell behavior, scientists often speak about ODE modeling. Indeed, it is one of the most powerful approaches, but needs prior knowledge in mathematics and a complete set of biological data and parameters. These are high requirements for a modeling approach when scientists try to reconstruct and understand system behavior or unknown regulatory processes. Thus, a more intuitive approach is necessary, which can be used in the beginning without biological data and is still able to imitate and predict cell behavior. Therefore, Petri nets can be used for the description, simulation, and analysis of complex and distributed systems. Petri nets cover most of the needed features for network modeling and provide qualitative as well as quantitative modeling features. Furthermore, it is possible to integrate ODEs for precise model descriptions. Another advantage of these modeling techniques is that each result can be shared within the scientific community using data exchange formats.

References

1. Alberts B, Johnson A, Lewis J, Raff M, Roberts K, Walter P (2007) Molecular biology of the cell, 5th edn. Garland Science, New York
2. Eigen M (1971) Self organization of matter and the evolution of biological macromolecules. Naturwissenschaften 58(10):465–523
3. Schrödinger E (1955) What is life? The physical aspect of the living cell. The University Press, Cambridge, pp 1–12
4. Dronamraju KR, Schrödinger E (1999) Erwin Schrödinger and the origins of molecular biology. Genetics 153(3):1071–1076
5. Holum JR (1998) Fundamentals of general, organic, and biological chemistry, 6th edn. Wiley, New York
6. Crick FHC (1958) Ideas on protein synthesis. Symp Soc Exp Biol XII:1–2
7. Crick FHC (1970) Central dogma of molecular biology. Nature 227(5258):561–563
8. Alon U (2006) An introduction to systems biology: design principles of biological circuits. Simplicity in biology, 1st edn. Chapman and Hall/CRC, Boca Raton
9. Junker BH, Schreiber F (2011) Analysis of biological networks, 1st edn. Wiley-Interscience, Hoboken
10. De Las Rivas J, Fontanillo C (2010) Protein-protein interactions essentials: key concepts to building and analyzing interactome networks. PLoS Comput Biol 6(6):1–8
11. Waksman G (2005) Proteomics and protein-protein interactions. Volume 3 of biology, chemistry, bioinformatics, and drug design. Springer, New York, pp 50–89
12. Panchenko A, Przytycka T (2008) Protein-protein interactions and networks. Identification, computer analysis, and prediction, 1st edn. Springer, New York Incorporated, pp 1–53
13. Suthram S (2008) Dissertation: understanding cellular function through the analysis of protein interaction networks. ProQuest, University of California, San Diego, 191p
14. Kepes F (2007) Biological networks, 1st edn. World Scientific, Singapore
15. Zhang A (2009) Protein interaction networks: computational analysis, 1st edn. Cambridge University Press, Cambridge/New York, pp. 1–62
16. Klussmann E, Scott J, Aandahl EM (2008) Protein-protein interactions as new drug targets, 1st edn. Springer, Berlin
17. Stelzl U, Worm U, Lalowski M, Haenig C, Brembeck FH, Goehler H, Stroedicke M, Zenkner M, Schoenherr A, Koeppen S, Timm J, Mintzlaff S, Abraham C, Bock N, Kietzmann S, Goedde A, Toksöz E, Droege A, Krobitsch S, Korn B, Birchmeier W, Lehrach H, Wanker EE (2005) A human protein-protein interaction network: a resource for annotating the proteome. Cell 122(6):957–968

18. Altman A (2002) Signal transduction pathways in autoimmunity, vol 5. Karger, Basel/New York
19. Cheong R, Hoffmann A, Levchenko A (2008) Understanding NF-$kappa$B signaling via mathematical modeling. Mol Syst Biol 4:192–196
20. Kearns JD (2006) IκB provides negative feedback to control NF-κB oscillations, signaling dynamics, and inflammatory gene expression. J Cell Biol 173(5):659–664
21. Kanehisa M, Goto S, Sato Y, Furumichi M, Tanabe M (2012) KEGG for integration and interpretation of large-scale molecular data sets. Nucleic Acids Res 40(Database Issue):109–114
22. Keseler IM, Bonavides-Martinez C, Collado-Vides J, Gama-Castro S, Gunsalus RP, Johnson DA, Krummenacker M, Nolan LM, Paley S, Paulsen IT, Peralta-Gil M, Santos-Zavaleta A, Shearer AG, Karp PD (2009) EcoCyc: a comprehensive view of Escherichia coli biology. Nucleic Acids Res 37(Database Issue):464–470
23. Karp PD, Ouzounis CA, Moore-Kochlacs C, Goldovsky L, Kaipa P, Ahrén D, Tsoka S, Darzentas N, Kunin V, López-Bigas N (2005) Expansion of the BioCyc collection of pathway/genome databases to 160 genomes. Nucleic Acids Res 33(19):6083–6089
24. Hopfield JJ (1982) Neural networks and physical systems with emergent collective computational abilities. Proc Natl Acad Sci 79(8):2554–2558
25. Van Pelt J, Kamermans U, Levelt CN (2004) Development, dynamics and pathology of neuronal networks: from molecules to functional circuits. Elsevier Science and Technology, Amsterdam/Boston
26. Kunin V, Goldovsky L, Darzentas N, Ouzounis CA (2005) The net of life: reconstructing the microbial phylogenetic network. Genome Res 15(7):954–959
27. Gray SH (2008) Food webs: interconnecting food chains. Exploring science. Compass Point Books, Minneapolis, pp 1–48
28. De Ruiter PC, Wolters V, Moore JC (2005) Dynamic food webs. Multispecies assemblages, ecosystem development, and environmental change. Academic, Amsterdam/Boston, pp 3–10
29. Pimm SL (2002) Food webs. University of Chicago Press, Chicago, pp 1–34
30. Hoppensteadt FC (1982) Mathematical methods of population biology. Cambridge University Press, Cambridge/New York, pp 1–28
31. Diestel R (2000) Graphentheorie, vol 2. Springer, Berlin, pp 1–25
32. Pavlopoulos GA, Secrier M, Moschopoulos CN, Soldatos TG, Kossida S, Aerts J, Schneider R, Bagos PG (2011) Using graph theory to analyze biological networks. BioData Min 4:10
33. Koschützki D, Lehmann KA, Peeters L, Richter S, Tenfelde-Podehl D, Zlotowski O (2005) Network analysis: centrality indices. Lecture notes in computer science, vol 3418. Springer, Berlin, Heidelberg
34. Jeong H, Mason SP, Barabási A, Oltvai ZN (2001) Lethality and centrality in protein networks. Nature 411(6833):41–42
35. Hahn MW, Kern AD (2005) Comparative genomics of centrality and essentiality in three eukaryotic protein-interaction networks. Mol Biol Evol 22(4):803–806
36. Hage P, Harary F (1995) Eccentricity and centrality in networks. Soc Netw 17:57–63
37. Sabidussi G (1966) The centrality index of a graph. Psychometrika 31(4):581–603
38. Freeman LC (1977) A set of measures of centrality based on betweenness. Sociometry 40:35–41
39. Bonacich P (1972) Factoring and weighting approaches to status scores and clique identification. J Math Sociol 2:113–120
40. Lewinski M (2012) Mathematischer Netzwerkvergleich anhand multipler Netzwerk Charakteristika. Master thesis at the Bielefeld University, pp 58–70
41. Stein ML, Stein PR (1967) Enumeration of linear graphs and connected linear graphs up to p = 18 Points. Report LA-3775, Los Alamos Scientific Laboratory of the University of California
42. van den Berg H (2011) Mathematical models of biological systems (Oxford biology), 1st edn. Oxford University Press, Oxford
43. Tyson J, Chen KC, Novak B (2003) Sniffers, buzzers, toggles and blinkers: dynamics of regulatory and signaling pathways in the cell. Curr Opin Cell Biol 15(2):221–231

44. Gizzatkulov M, Goryanin I, Metelkin EA, Mogilevskaya EA, Peskov KV, Demin OV (2010) DBSolve Optimum: a software package for kinetic modeling which allows dynamic visualization of simulation results. BMC Syst Biol 4:109
45. Demin O, Goryanin I (2008) Kinetic modelling in systems biology. Chapman & Hall/CRC, Boca Raton
46. Rumbaugh J, Eddy F (1991) Object-oriented modeling and design. Prentice-Hall, Englewood Cliffs, pp 15–57
47. Johnson CG, Goldman JP, William JG (2004) Simulating complex intracellular processes using object-oriented computational modelling. Prog Biophys Mol Biol 86(3):379–406
48. Francesca C, Daniele M, Marco G (2009) Modeling biological pathways: an object-oriented like methodology based on mean field analysis. Trans IRE 1:117–122
49. Dorkeld F, Perrière G, Gautier C (1993) Object-oriented modelling in molecular biology. In: Ganascia JG (ed) Proceedings of the artificial intelligence and genome workshop, IJCAI, pp 99–106
50. Lee M, Starr LLJ (1994) Object-oriented analysis in the real-world. Embed Syst Program 7(6):24–37
51. Kurth W (2007) Specification of morphological models with L-systems and relational growth grammars. J Interdiscip Image Sci 5:1–25
52. Prusinkiewicz P, Lindenmayer A (1990) The algorithmic beauty of plants. Springer, New York, pp 1–46
53. Orth JD, Conrad TM, Na J, Lerman JA, Nam H, Feist AM, Palsson BO (2011) A comprehensive genome-scale reconstruction of Escherichia coli metabolism. Mol Syst Biol 7:1–9
54. Kauffman KJ, Prakash P, Edwards JS (2003) Advances in flux balance analysis. Curr Opin Biotechnol 14(5):491–496
55. Wiechert W (2001) 13C metabolic flux analysis. Metab Eng 3(3):195–206
56. Efroni S, Harel D, Cohen IR (2003) Toward rigorous comprehension of biological complexity: modeling, execution, and visualization of thymic T-cell maturation. Genome Res 13(11):2485–2497
57. Kam N, Cohen IR, Harel D (2001) Proceedings IEEE symposia on human-centric computing languages and environments (Cat. No.01TH8587). In: HCC 2001. IEEE symposium on human-centric computing languages and environments, vol 1, Auckland. IEEE, pp 15–22
58. Danos V, Laneve C (2004) Formal molecular biology. Theor Comput Sci 325:69–110
59. Ermentrout GB, Edelsteinkeshet L (1993) Cellular automata approaches to biological modeling. J Theor Biol 160:97–133
60. Wishart DS, Yang R, Arndt D, Tang P, Cruz J (2005) Dynamic cellular automata: an alternative approach to cellular simulation. In Silico Biol 5(2):139–161
61. Wurthner JU, Mukhopadhyay AK, Peimann CJ (1999) A cellular automaton model of cellular signal transduction. Comput Biol Med 30:1–21
62. Gardner M (1970) Mathematical games: the fantastic combinations of John Conway's new solitaire game "life". Sci Am 1:120–123
63. Zhang L, Athale CA, Deisboeck TS (2007) Development of a three-dimensional multiscale agent-based tumor model: simulating gene-protein interaction profiles, cell phenotypes and multicellular patterns in brain cancer. J Theor Biol 244:96–107
64. Grant MR, Mostov KE, Tlsty TD, Hunt CA (2006) Simulating properties of in vitro epithelial cell morphogenesis. PLoS Comput Biol 2(10):129–134
65. Li NYK, Verdolini K, Clermont G, Mi Q, Rubinstein EN, Hebda PA, Vodovotz Y (2008) A Patient-Specific in silico model of Inflammation and Healing Tested in Acute Vocal Fold Injury. PLoS ONE 3(7):1–11
66. Lee S, Dudley AM, Drubin D, Silver PA, Krogan NJ, Pe'er D, Koller D (2009) Learning a Prior on Regulatory Potential from eQTL Data. PLoS Genet 5:1–24
67. Pe'er D (2005) Bayesian network analysis of signaling networks: a primer. Trans IRE Prof Group Audio 2005(281):14–24

68. Segal E, Shapira M, Regev A, Pe'er D, Botstein D, Koller D, Friedman N (2003) Module networks: identifying regulatory modules and their condition-specific regulators from gene expression data. Nat Genet 34(2):166–176
69. Kauffman SA (1969) Metabolic stability and epigenesis in randomly constructed genetic nets. J Theor Biol 22(3):437–467
70. Gershenson C (2004) Introduction to random boolean networks, pp 1–14. eprint arXiv nlin/0408006
71. Thomas R (1973) Boolean formalization of genetic control circuits. J Theor Biol 42(3):563–585
72. Thomas R, D'Ari RTR (1990) Biological feedback. CRC, Boca Raton, pp 1–316
73. Bernot G, Cassez F, Comet JP, Delaplace F, Müller C, Roux O (2007) Semantics of biological regulatory networks. Electron Notes Theor Comput Sci 180(3):3–14
74. Petri CA (1962) Dissertation: Kommunikation mit Automaten. Schriften des Rheinisch-Westfälischen Institutes für Instrumentelle Mathematik an der Universität Bonn
75. Reisig W (1992) A primer in Petri net design, 1st edn. Springer, Berlin/New York
76. Petri C, Reisig W (2008) Petri net. Scholarpedia 3(4):6477
77. Le Novère N, Hucka M, Mi H, Moodie S, Schreiber F, Sorokin A, Demir E, Wegner K, Aladjem MI, Wimalaratne SM, Bergman FT, Gauges R, Ghazal P, Kawaji H, Li L, Matsuoka Y, Villeger A, Boyd SE, Calzone L, Courtot M, Dogrusoz U, Freeman TC, Funahashi A, Ghosh S, Jouraku A, Kim S, Kolpakov F, Luna A, Sahle S, Schmidt E, Watterson S, Wu G, Goryanin I, Kell DB, Sander C, Sauro H, Snoep JL, Kohn K, Kitano H (2009) The systems biology graphical notation. Nat Biotechnol 27(8):735–741
78. SBGN Documentation: SBGN entity relationship diagram. http://www.sbgn.org/Documents/ER_L1_Examples. Online; accessed 17 Aug 2013
79. Machado D, Costa RS, Rocha M, Ferreira EC, Tidor B, Rocha I (2011) Modeling formalisms in Systems Biology. AMB Express 1:45–59
80. Krallinger M, Alonso-Allende Erhardt M, Valencia A (2005) Text-mining approaches in molecular biology and biomedicine. Drug Disc Today 10(6):439–445
81. Hoffmann R, Krallinger M, Andres E, Tamames J, Blaschke C, Valencia A (2005) Text mining for metabolic pathways, signaling cascades, and protein networks. Signal Transduct Knowl Environ 1(283):1–21
82. Mushegian AR, Koonin EV (1996) A minimal gene set for cellular life derived by comparison of complete bacterial genomes. Proc Natl Acad Sci 93(19):10268–10273
83. Bono H, Ogata H, Goto S, Kanehisa M (1998) Reconstruction of amino acid biosynthesis pathways from the complete genome sequence. Genome Res 8:203–210
84. KEGG Documentation: Overview of KGML. http://www.kegg.jp/kegg/xml/docs/. Online; accessed 17 Aug 2013
85. Milanesi L, Romano P, Castellani G, Remondini D, Lio P (2009) Trends in modeling Biomedical Complex Systems. BMC Bioinform 10(12):1–13
86. Sauro HM, Bergmann FT (2008) Standards and ontologies in computational systems biology. Essays Biochem 45:211–222
87. Strömbäck L, Hall D, Lambrix P (2007) A review of standards for data exchange within systems biology. Proteomics 7(6):857–867
88. Klipp E, Liebermeister W, Helbig A, Kowald A, Schaber J (2007) Systems biology standards–the community speaks. Nat Biotechnol 25(4):390–391
89. McEntire R, Karp P, Abernethy N, Benton D, Helt G, DeJongh M, Kent R, Kosky A, Lewis S, Hodnett D, Neumann E, Olken F, Pathak D, Tarczy-Hornoch P, Toldo L, Topaloglou T (2000) An evaluation of ontology exchange languages for bioinformatics. Int Conf Intell Syst Mol Biol 8:239–250
90. Achard F, Vaysseix G, Barillot E (2001) XML, bioinformatics and data integration. Bioinformatics 17(2):115–125
91. Courtot M, Juty N, Knüpfer C, Waltemath D, Zhukova A, Dräger A, Dumontier M, Finney A, Golebiewski M, Hastings J, Hoops S, Keating S, Kell DB, Kerrien S, Lawson J, Lister A, Lu J, Machne R, Mendes P, Pocock M, Rodriguez N, Villeger A, Wilkinson DJ, Wimalaratne

S, Laibe C, Hucka M, Le Novère N (2011) Controlled vocabularies and semantics in systems biology. Mol Syst Biol 7:1–12
92. Finney A, Hucka M (2003) Systems biology markup language: Level 2 and beyond. Biochem Soc Trans 31:1472–1473
93. Hucka M, Finney A, Sauro HM, Bolouri H, Doyle JC, Kitano H, Arkin AP, Bornstein BJ, Bray D, Cornish-Bowden A, Cuellar AA, Dronov S, Gilles ED, Ginkel M, Gor V, Goryanin II, Hedley WJ, Hodgman TC, Hofmeyr JH, Hunter PJ, Juty NS, Kasberger JL, Kremling A, Kummer U, Le Novere N, Loew LM, Lucio D, Mendes P, Minch E, Mjolsness ED, Nakayama Y, Nelson MR, Nielsen PF, Sakurada T, Schaff JC, Shapiro BE, Shimizu TS, Spence HD, Stelling J, Takahashi K, Tomita M, Wagner J, Wang J (2003) The systems biology markup language (SBML): a medium for representation and exchange of biochemical network models. Bioinformatics 19(4):524–531
94. Miller AK, Marsh J, Reeve A, Garny A, Britten R, Halstead M, Cooper J, Nickerson DP, Nielsen PF (2010) An overview of the CellML API and its implementation. BMC Bioinform 11:178–180
95. Demir E, Cary MP, Paley S, Fukuda K, Lemer C, Vastrik I, Wu G, D'Eustachio P, Schaefer C, Luciano J, Schacherer F, Martinez-Flores I, Hu Z, Jimenez-Jacinto V, Joshi-Tope G, Kandasamy K, Lopez-Fuentes AC, Mi H, Pichler E, Rodchenkov I, Splendiani A, Tkachev S, Zucker J, Gopinath G, Rajasimha H, Ramakrishnan R, Shah I, Syed M, Anwar N, Babur Ö, Blinov M, Brauner E, Corwin D, Donaldson S, Gibbons F, Goldberg R, Hornbeck P, Luna A, Murray-Rust P, Neumann E, Ruebenacker O, Reubenacker O, Samwald M, van Iersel M, Wimalaratne S, Allen K, Braun B, Whirl-Carrillo M, Cheung KH, Dahlquist K, Finney A, Gillespie M, Glass E, Gong L, Haw R, Honig M, Hubaut O, Kane D, Krupa S, Kutmon M, Leonard J, Marks D, Merberg D, Petri V, Pico A, Ravenscroft D, Ren L, Shah N, Sunshine M, Tang R, Whaley R, Letovksy S, Buetow KH, Rzhetsky A, Schachter V, Sobral BS, Dogrusoz U, McWeeney S, Aladjem M, Birney E, Collado-Vides J, Goto S, Hucka M, Le Novère N, Maltsev N, Pandey A, Thomas P, Wingender E, Karp PD, Sander C, Bader GD (2010) The BioPAX community standard for pathway data sharing. Nat Biotechnol 28(9):935–942
96. Matthews L, Gopinath G, Gillespie M, Caudy M, Croft D, de Bono B, Garapati P, Hemish J, Hermjakob H, Jassal B, Kanapin A, Lewis S, Mahajan S, May B, Schmidt E, Vastrik I, Wu G, Birney E, Stein L, D'Eustachio P (2009) Reactome knowledgebase of human biological pathways and processes. Nucleic Acids Res 37(Database Issue):619–622
97. Kalas M, Puntervoll P, Joseph A, Bartaseviciute E, Topfer A, Venkataraman P, Pettifer S, Bryne JC, Ison J, Blanchet C, Rapacki K, Jonassen I (2010) BioXSD: the common data-exchange format for everyday bioinformatics web services. Bioinformatics 26(18):540–546
98. Spellman PT, Miller M, Stewart J, Troup C, Sarkans U, Chervitz S, Bernhart D, Sherlock G, Ball C, Lepage M, Swiatek M, Marks WL, Goncalves J, Markel S, Iordan D, Shojatalab M, Pizarro A, White J, Hubley R, Deutsch E, Senger M, Aronow BJ, Robinson A, Bassett D, Stoeckert CJ, Brazma A (2002) Design and implementation of microarray gene expression markup language (MAGE-ML). Genome Biol 3(9):1–9
99. Kottmann R, Gray T, Murphy S, Kagan L, Kravitz S, Lombardot T, Field D, Glöckner FO (2008) A standard MIGS/MIMS compliant XML Schema: toward the development of the Genomic Contextual Data Markup Language (GCDML). OMICS 12(2):115–121
100. Westbrook J, Ito N, Nakamura H, Henrick K, Berman HM (2005) PDBML: the representation of archival macromolecular structure data in XML. Bioinformatics 21(7):988–992
101. Hermjakob H, Montecchi-Palazzi L, Bader G, Wojcik J, Salwinski L, Ceol A, Moore S, Orchard S, Sarkans U, von Mering C, Roechert B, Poux S, Jung E, Mersch H, Kersey P, Lappe M, Li Y, Zeng R, Rana D, Nikolski M, Husi H, Brun C, Shanker K, Grant SG, Sander C, Bork P, Zhu W, Pandey A, Brazma A, Jacq B, Vidal M, Sherman D, Legrain P, Cesareni G, Xenarios I, Eisenberg D, Steipe B, Hogue C, Apweiler R (2004) The HUPO PSI's molecular interaction format–a community standard for the representation of protein interaction data. Nat Biotechnol 22(2):177–183
102. Han MV, Zmasek CM (2009) phyloXML: XML for evolutionary biology and comparative genomics. BMC Bioinform 10:356

103. Cuellar AA, Lloyd CM, Nielsen PF, Bullivant DP, Nickerson DP, Hunter PJ (2003) An Overview of CellML 1.1, a biological model description language. Simulation 79(12):740–747
104. Sandhu P (2003) The MathML handbook. Charles River Media, Hingham, pp 1–10
105. Ausbrooks R, Buswell S, Carlisle D, Chavchanidze G, Dalmas S, Devitt S, Diaz A, Dooley S, Hunter R, Ion P, Kohlhase M, Lazrek A, Libbrecht P, Miller B, Miner R, Rowley C, Sargent M, Smith B, Soiffer N, Sutor R, Watt S (2010) Mathematical Markup Language (MathML) Version 3.0, pp 115–121. W3C recommendations-www.w3.org
106. Berman HM, Westbrook J, Feng Z, Gilliland G, Bhat TN, Weissig H, Shindyalov IN, Bourne PE (2000) The Protein Data Bank. Nucleic Acids Res 28:235–242
107. Strasser BJ, Dayhoff MO (2010) Collecting, comparing, and computing sequences: the making of Margaret O. Dayhoff's Atlas of Protein Sequence and Structure, 1954–1965. J Hist Biol 43(4):623–660
108. Bernstein FC, Koetzle TF, Williams GJ, Meyer EF, Brice MD, Rodgers JR, Kennard O, Shimanouchi T, Tasumi M (1977) The Protein Data Bank: a computer-based archival file for macromolecular structures. J Mol Biol 112(3):535–542
109. Meyer EF (1997) The first years of the Protein Data Bank. Protein Sci 6(7):1591–1597
110. Berman H, Henrick K, Nakamura H, Markley JL (2007) The worldwide Protein Data Bank (wwPDB): ensuring a single, uniform archive of PDB data. Nucleic Acids Res 35(Database Issue):301–303
111. Hamm GH, Cameron GN (1986) The EMBL data library. Nucleic Acids Res 14:5–9
112. Cochrane G, Akhtar R, Bonfield J, Bower L, Demiralp F, Faruque N, Gibson R, Hoad G, Hubbard T, Hunter C, Jang M, Juhos S, Leinonen R, Leonard S, Lin Q, Lopez R, Lorenc D, McWilliam H, Mukherjee G, Plaister S, Radhakrishnan R, Robinson S, Sobhany S, Hoopen PT, Vaughan R, Zalunin V, Birney E (2009) Petabyte-scale innovations at the European Nucleotide Archive. Nucleic Acids Res 37(Database Issue):19–25
113. Burks C, Fickett JW, Goad WB, Kanehisa M, Lewitter FI, Rindone WP, Swindell CD, Tung CS, Bilofsky HS (1985) The GenBank nucleic acid sequence database. Comput Appl Biosci 1(4):225–233
114. Benson DA, Karsch-Mizrachi I, Clark K, Lipman DJ, Ostell J, Sayers EW (2012) GenBank. Nucleic Acids Res 40(Database Issue):48–53
115. Galperin MY, Fernandez-Suarez XM (2012) The 2012 Nucleic Acids Research Database Issue and the online Molecular Biology Database Collection. Nucleic Acids Res 40(Database Issue):1–8
116. Bader GD, Cary MP, Sander C (2006) Pathguide: a pathway resource list. Nucleic Acids Res 34(Database Issue):504–506
117. Bolser DM, Chibon PY, Palopoli N, Gong S, Jacob D, Del Angel VD, Swan D, Bassi S, Gonzalez V, Suravajhala P, Hwang S, Romano P, Edwards R, Bishop B, Eargle J, Shtatland T, Provart NJ, Clements D, Renfro DP, Bhak D, Bhak J (2012) MetaBase–the wiki-database of biological databases. Nucleic Acids Res 40(Database Issue):1250–1254
118. Chen YB, Chattopadhyay A, Bergen P, Gadd C, Tannery N (2007) The Online Bioinformatics Resources Collection at the University of Pittsburgh Health Sciences Library System – a one-stop gateway to online bioinformatics databases and software tools. Nucleic Acids Res 35(Database Issue):780–785
119. Bateman A (2010) Curators of the world unite: the International Society of Biocuration. Bioinformatics 26(8):991
120. Fox JA, Butland SL, McMillan S, Campbell G, Ouellette BF (2005) The Bioinformatics Links Directory: a compilation of molecular biology web servers. Nucleic Acids Res 33(Web Server issue):3–24
121. Brazas MD, Yim DS, Yamada JT, Ouellette BF (2011) The 2011 Bioinformatics Links Directory update: more resources, tools and databases and features to empower the bioinformatics community. Nucleic Acids Res 39(Web Server issue):3–7
122. Copeland WB, Bartley BA, Chandran D, Galdzicki M, Kim KH, Sleight SC, Maranas CD, Sauro HM (2012) Computational tools for metabolic engineering. Metab Eng 14(3):270–280

123. Funahashi A, Morohashi M, Kitano H, Tanimura N (2003) CellDesigner: a process diagram editor for gene-regulatory and biochemical networks. Biosilico 1(5):159–162
124. Nagasaki M, Saito A, Jeong E, Li C, Kojima K, Ikeda E, Miyano S (2010) Cell Illustrator 4.0: a computational platform for systems biology. In Silico Biol 10:5–26
125. Smoot ME, Ono K, Ruscheinski J, Wang PL, Ideker T (2011) Cytoscape 2.8: new features for data integration and network visualization. Bioinformatics 27(3):431–432
126. Tomita M, Hashimoto K, Takahashi K, Shimizu TS, Matsuzaki Y, Miyoshi F, Saito K, Tanida S, Yugi K, Venter JC (1999) E-CELL: software environment for whole-cell simulation. Bioinformatics 15:72–84
127. Mendes P (1997) Biochemistry by numbers: simulation of biochemical pathways with Gepasi 3. Trends Biochem Sci 22(9):361–363
128. Mendes P (1993) GEPASI: a software package for modelling the dynamics, steady states and control of biochemical and other systems. Comput Appl Biosci 9(5):563–571
129. Sauro HM, Hucka M, Finney A, Wellock C, Bolouri H, Doyle J, Kitano H (2002) Next generation simulation tools: the Systems Biology Workbench and BioSPICE integration. OMICS: J Integr Biol 7(4):355–372
130. Proß S, Bachmann B (2011) An advanced environment for hybrid modeling of biological systems based on Modelica. J Integr Bioinform 8:152
131. Proß S, Janowski SJ, Bachmann B, Kaltschmidt C, Kaltschmidt B (2012) PNlib- a Modelica library for simulation of biological systems based on extended hybrid Petri nets. In: Proceedings of the 3rd international workshop on biological processes and Petri nets, vol 852, pp 1–16
132. Heiner M, Richter R, Schwarick M, Rohr C (2008) Snoopy-A tool to design and execute graph-based formalisms. Petri Net Newsl 74:8–22
133. Rohr C, Marwan W, Heiner M (2010) Snoopy- a unifying Petri net framework to investigate biomolecular networks. Bioinformatics 26(7):974–975
134. Cherry JM, Hong EL, Amundsen C, Balakrishnan R, Binkley G, Chan ET, Christie KR, Costanzo MC, Dwight SS, Engel SR, Fisk DG, Hirschman JE, Hitz BC, Karra K, Krieger CJ, Miyasato SR, Nash RS, Park J, Skrzypek MS, Simison M, Weng S, Wong ED (2012) Saccharomyces Genome Database: the genomics resource of budding yeast. Nucleic Acids Res 40(Database Issue):700–705
135. Hoffmann R, Valencia A (2004) A gene network for navigating the literature. Nat Genet 36(7):664–664
136. Wheeler DL, Church DM, Edgar R, Federhen S, Helmberg W, Madden TL, Pontius JU, Schuler GD, Schriml LM, Sequeira E, Suzek TO, Tatusova TA, Wagner L (2004) Database resources of the National Center for Biotechnology Information: update. Nucleic Acids Res 32(Database Issue):35–40
137. Machné R, Finney A, Müller S, Lu J, Widder S, Flamm C (2006) The SBML ODE Solver Library: a native API for symbolic and fast numerical analysis of reaction networks. Bioinformatics 22(11):1406–1407
138. Hoops S, Sahle S, Gauges R, Lee C, Pahle J, Simus N, Singhal M, Xu L, Mendes P, Kummer U (2006) COPASI – a COmplex PAthway SImulator. Bioinformatics 22(24):3067–3074
139. Matsuno H, Doi A, Drath R, Miyano S (2000) Genomic object net: object oriented representation of biological systems. Genome Inform Ser 11:229–230
140. Krull M, Pistor S, Voss N, Kel A, Reuter I, Kronenberg D, Michael H, Schwarzer K, Potapov A, Choi C, Kel-Margoulis O, Wingender E (2006) TRANSPATH: an information resource for storing and visualizing signaling pathways and their pathological aberrations. Nucleic Acids Res 34(Database Issue):546–551
141. Shannon P, Andrew M, Ozier O, Baliga NS, Wang JT, Ramage D, Amin N, Schwikowski B, Ideker T (2003) Cytoscape: a software environment for integrated models of biomolecular interaction networks. Genome Res 13(11):2498–2504
142. Avila-Campillo I, Drew K, Lin J, Reiss DJ, Bonneau R (2007) BioNetBuilder: automatic integration of biological networks. Bioinformatics 23(3):392–393

143. Keshava Prasad TS, Goel R, Kandasamy K, Keerthikumar S, Kumar S, Mathivanan S, Telikicherla D, Raju R, Shafreen B, Venugopal A, Balakrishnan L, Marimuthu A, Banerjee S, Somanathan DS, Sebastian A, Rani S, Ray S, Harrys Kishore CJ, Kanth S, Ahmed M, Kashyap MK, Mohmood R, Ramachandra YL, Krishna V, Rahiman BA, Mohan S, Ranganathan P, Ramabadran S, Chaerkady R, Pandey A (2009) Human Protein Reference Database – 2009 update. Nucleic Acids Res 37(Database Issue):767–772
144. Breitkreutz BJ, Stark C, Reguly T, Boucher L, Breitkreutz A, Livstone M, Oughtred R, Lackner DH, Bähler J, Wood V, Dolinski K, Tyers M (2008) The BioGRID Interaction Database: 2008 update. Nucleic Acids Res 36(Database Issue):637–640.
145. Blake JA, Dolan M, Drabkin H, Hill DP, Ni L, Sitnikov D, Burgess S, Buza T, Gresham C, McCarthy F, Pillai L, Wang H, Carbon S, Lewis SE, Mungall CJ, Gaudet P, Chisholm RL, Fey P, Kibbe WA, Basu S, Siegele DA, McIntosh BK, Renfro DP, Zweifel AE, Hu JC, Brown NH, Tweedie S, Alam-Faruque Y, Apweiler R, Auchincloss A, Axelsen K, Argoud-Puy G, Bely B, Blatter M, Bougueleret L, Boutet E, Branconi S, Breuza L, Bridge A, Browne P, Chan WM, Coudert E, Cusin I, Dimmer E, Duek-Roggli P, Eberhardt R, Estreicher A, Famiglietti L, Ferro-Rojas S, Feuermann M, Gardner M, Gos A, Gruaz-Gumowski N, Hinz U, Hulo C, Huntley R, James J, Jimenez S, Jungo F, Keller G, Laiho K, Legge D, Lemercier P, Lieberherr D, Magrane M, Martin MJ, Masson P, Moinat M, O'Donovan C, Pedruzzi I, Pichler K, Poggioli D, Porras Millan P, Poux S, Rivoire C, Roechert B, Sawford T, Schneider M, Sehra H, Stanley E, Stutz A, Sundaram S, Tognolli M, Xenarios I, Foulger R, Lomax J, Roncaglia P, Camon E, Khodiyar VK, Lovering RC, Talmud PJ, Chibucos M, Gwinn Giglio M, Dolinski K, Heinicke S, Livstone MS, Stephan R, Harris MA, Oliver SG, Rutherford K, Wood V, Bahler J, Lock A, Kersey PJ, McDowall MD, Staines DM, Dwinell M, Shimoyama M, Laulederkind S, Hayman T, Wang S, Petri V, Lowry T, D'Eustachio P, Matthews L, Amundsen CD, Balakrishnan R, Binkley G, Cherry JM, Christie KR, Costanzo MC, Dwight SS, Engel SR, Fisk DG, Hirschman JE, Hitz BC, Hong EL, Karra K, Krieger CJ, Miyasato SR, Nash RS, Park J, Skrzypek MS, Weng S, Wong ED, Berardini TZ, Li D, Huala E, Slonim D, Wick H, Thomas P, Chan J, Kishore R, Sternberg P, Van Auken K, Howe D, Westerfield M (2012) The Gene Ontology: enhancements for 2011. Nucleic Acids Res 40(Database Issue):559–564
146. Zheng J, Zhang D, Przytycki PF, Zielinski R, Capala J, Przytycka TM (2009) SimBoolNet – a Cytoscape plugin for dynamic simulation of signaling networks. Bioinformatics 26:141–142
147. Erhard F, Friedel CC, Zimmer R (2008) FERN – a Java framework for stochastic simulation and evaluation of reaction networks. BMC Bioinform 9:356
148. Takahashi K, Kaizu K, Hu B, Tomita M (2004) A multi-algorithm, multi-timescale method for cell simulation. Bioinformatics 20(4):538–546
149. Janowski S, Kormeier B, Töpel T, Hippe K, Hofestädt R, Willassen N, Friesen R, Rubert S, Borck D, Haugen P, Chen M (2010) Modeling of cell-to-cell communication processes with Petri Nets using the example of Quorum sensing. In Silico Biol 10:27–48
150. Kormeier B, Hippe K, Arrigo P, Töpel T, Janowski S, Hofestädt R (2010) Reconstruction of biological networks based on life science data integration. J Integr Bioinform 7(2):146–159
151. Sommer B, Tiys ES, Kormeier B, Hippe K, Janowski SJ, Ivanisenko TV, Bragin AO, Arrigo P, Demenkov PS, Kochetov AV, Ivanisenko VA, Kolchanov NA, Hofestädt R (2010) Visualization and analysis of a cardio vascular disease- and MUPP1-related biological network combining text mining and data warehouse approaches. J Integr Bioinform 7:148
152. Proß S, Janowski SJ, Hofestädt R, Bachmann B (2012) A new object-oriented Petri net simulation Environment Based On Modelica. In: Online proceedings of the 2012 Winter simulation conference, Berlin. IEEE, pp 1–13
153. Klenke C, Janowski S, Borck D, Widera D, Ebmeyer LA, Kalinowski J, Leichtle A, Hofestädt R, Upile T, Kaltschmidt C, Kaltschmidt B, Sudhoff H (2012) Identification of novel cholesteatoma-related gene expression signatures using full-genome microarrays. PloS One 7(12):1–14
154. Association M (2005) Modelica – a unified object-oriented language for physical systems modeling. Lang Specif 2:7–11

Chapter 9
Petri Nets for Modeling and Analyzing Biochemical Reaction Networks

Fei Liu and Monika Heiner

Abstract Petri nets have been widely used to model and analyze biochemical reaction networks. This chapter gives an overview of different types of Petri nets within a unifying Petri net framework that comprises the qualitative, stochastic, continuous, and hybrid paradigms at both uncolored and colored levels. The Petri net framework permits to investigate one and the same biological reaction network with different modeling abstractions in various complementary ways. We describe the use of the framework to investigate biochemical reaction networks with the help of the unifying Petri net tool, Snoopy, and its close friends Charlie and Marcie. The repressilator example serves as running case study.

Keywords Petri nets • Biochemical reaction networks • Unifying Petri net framework • Qualitative • Stochastic • Continuous and hybrid Petri nets • Colored Petri nets • Repressilator

9.1 Introduction

Modeling and analysis techniques have been widely used to study biochemical reaction networks. A large variety of modeling approaches, e.g., ordinary (partial) differential equations, Boolean networks, process algebras, and Petri nets, have been applied for modeling a wide range of biochemical reaction networks (for reviews,

---

F. Liu (✉)
Control and Simulation Center, Harbin Institute of Technology, Postbox 3006,
150080 Harbin, China
e-mail: liufei@hit.edu.cn

M. Heiner
Department of Computer Science, Brandenburg University of Technology,
Postbox 10 13 44, 03013 Cottbus, Germany
e-mail: monika.heiner@tu-cottbus.de

see, e.g., [22] and *Chap. 1 in this book*). Among them, Petri nets are particularly suitable for modeling the concurrent, asynchronous, and dynamic behavior of biological networks. Reddy et al. [48] and Hofestädt [36] were the first to pick up Carl Adam Petri's idea for a graphical representation of stoichiometric equations and applied qualitative Petri nets to model and analyze metabolic pathways. Since that time, a large variety of Petri net classes, e.g., stochastic Petri nets, continuous Petri nets, hybrid Petri nets, and colored Petri nets, have been developed for modeling and analyzing different types of biological networks; see, e.g., [2,9,24,38].

Petri nets offer a number of attractive advantages for investigating biological reaction networks [28]:

- Intuitive graphical and directly executable modeling formalisms
- Rich and mathematically founded analysis techniques
- Coverage of structural and behavioral properties as well as their relations,
- Integration of qualitative (i.e., time-free) and quantitative (i.e., time-dependent) analysis techniques and methods, including animation (the token flow)
- Coverage of discrete (stochastic), continuous (deterministic), and hybrid paradigms for quantitative analysis techniques and methods
- A wealth of computer tool support

This chapter gives an overview of different types of Petri nets within a unifying Petri net framework and describes how they can be used to model and analyze biochemical reaction networks with the help of the unifying Petri net tool, Snoopy [31, 49] and its close friends Charlie [14, 56] and Marcie [32].

This chapter has been deliberately written in an informal style; no formal definitions are given. We focus on an overview on the key concepts and their applications in our previous work. For formal definitions, see Heiner et al. [28], which also provides plenty of pointers where to continue reading.

This chapter is organized as follows. Section 9.2 gives an overview of our unifying Petri net framework, followed by a description of each net class contained in this framework from Sects. 9.3 to 9.7, respectively. After a brief description of the tools we use, this chapter is concluded.

9.2 A Unifying Petri Net Framework

Petri nets may easily serve as a convenient umbrella formalism integrating qualitative and quantitative (i.e., stochastic, continuous, or hybrid) modeling and analysis techniques. Thus, Petri nets are immediately ready to address distinctive modeling demands of systems and synthetic biology including those biochemical reaction networks that may need several modeling paradigms.

Motivated by this application scenario, a unifying Petri net framework (see Fig. 9.1) has been developed [25, 31], which can be divided into two levels: uncolored [28] and colored [38]. Each level comprises a family of related Petri net classes, sharing structure, but being specialized by their kinetic information.

9 Petri Nets for Modeling and Analyzing Biochemical Reaction Networks

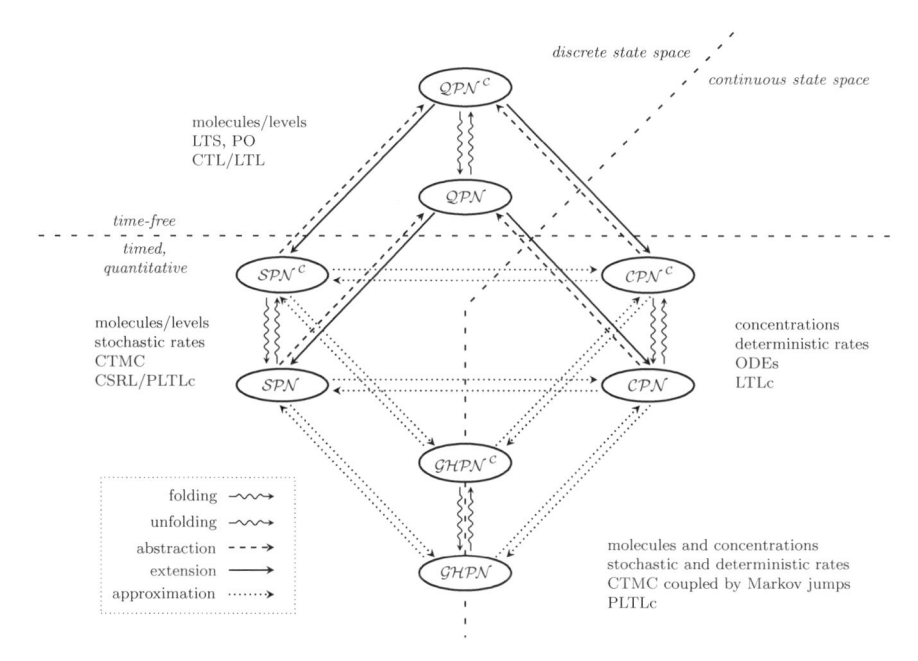

Fig. 9.1 A unifying Petri net framework, which has been implemented in the Petri net tool, Snoopy (Reprinted from Heiner et al. [31] with kind permission from Springer Science + Business Media B.V., Fig. 1, p. 399)

Specifically, the uncolored level contains qualitative (time-free) Petri nets ($\mathcal{QPN}$) as well as quantitative (time-dependent) Petri nets such as stochastic Petri nets ($\mathcal{SPN}$), continuous Petri nets ($\mathcal{CPN}$), and generalized hybrid Petri nets ($\mathcal{GHPN}$). The colored level consists of the colored counterparts of the uncolored level, thus containing colored qualitative Petri nets ($\mathcal{QPN}^\mathcal{C}$), colored stochastic Petri nets ($\mathcal{SPN}^\mathcal{C}$), colored continuous Petri nets ($\mathcal{CPN}^\mathcal{C}$), and colored generalized hybrid Petri nets ($\mathcal{GHPN}^\mathcal{C}$).

Petri nets of these net classes can be converted into each other; see arrows in Fig. 9.1. Obviously, there may be a loss of information in some directions (cf. arrows labeled with "abstraction" in Fig. 9.1). The conversion between colored and uncolored net classes is accomplished by means of user-guided folding or automatic unfolding (cf. arrows labeled with folding and unfolding in Fig. 9.1). Moving between the colored and uncolored level changes the style of representation but does not change the actual net structure of the underlying biochemical reaction network. Therefore, all analysis techniques available for uncolored Petri nets can be applied to colored Petri nets as well.

Snoopy supports the simultaneous use of different net classes, which provides the ground to investigate one and the same case study with different modeling abstractions in various complementary ways [24, 28, 38].

We will address each net class in the framework in the following sections by focusing on their application for investigating biochemical reaction networks.

9.3 Qualitative Petri Nets ($\mathcal{QPN}$)

9.3.1 Modeling

$\mathcal{QPN}$ comprise – first of all – the standard *place/transition nets* (*P/T nets*, *Petri nets* for short) which basically correspond to the original ideas introduced by Carl Adam Petri in 1962 [46]. Petri nets (see Fig. 9.2 for an introductory example) are bipartite-directed multigraphs with two types of nodes, called places and transitions, which are connected by arcs. Places (represented as circles) and transitions (represented as boxes) model in our context biochemical species and reactions, respectively. Arcs carry stoichiometric information, called weight or multiplicity. Tokens on places represent the (discrete) quantities of species, which may be understood as the number of molecules or the level of concentration of a species, or simply the presence of, e.g., a gene. A particular arrangement of tokens over all places of a Petri net specifies the current system state (*marking*). The initial state is called the initial marking. For example, the initial marking in Fig. 9.2a consists of five tokens on place H_2 and three tokens on place O_2.

The state of the system changes by the firing of transitions. A transition is enabled to fire if all its preconditions are fulfilled, i.e., each of its pre-places contains at least the number of tokens specified by the weight of the corresponding arc. Upon firing of a transition, tokens from all its pre-places are removed, and tokens are added to all its post-places, each according to the corresponding arc weights. See Fig. 9.2 for two state changes upon firing of transition t; that is, two tokens on pre-place H_2 and one token on pre-place O_2 are removed and two tokens are added to the post-place H_2O; i.e., we reach new markings. All markings, which can be reached

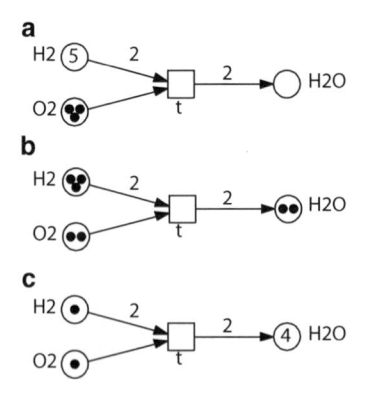

Fig. 9.2 A Petri net model of the chemical reaction $2H_2 + O_2 \rightarrow 2H_2O$. The places labeled with H_2 and O_2 are pre-places of the transition t, the place labeled with H_2O its post-place. (**a**) Initial marking before t fires, (**b**) marking reached by firing of t once, and (**c**) marking reached by a second firing of t. The transition is not enabled anymore in the marking reached after these two single firing steps

9 Petri Nets for Modeling and Analyzing Biochemical Reaction Networks

Fig. 9.3 A Petri net with marking-dependent arcs. Each arc may be marking-dependent, e.g., the multiplicity of the post-arc from transition t to place P3 is an addition expression, P1+P2. (**a**) Initial marking before t fires, (**b**) marking reached by firing of t

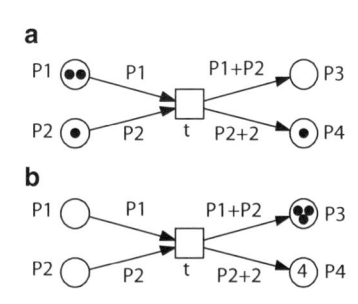

from a given marking by any firing sequence of arbitrary length, form the set of reachable markings. The set of markings reachable from the initial marking builds the state space of a given Petri net. The *reachability graph* of a Petri net comprises these reachable markings as nodes and the transitions between them as edges. The reachability graph is finite, iff (if and only if) the state space is finite.

$\mathcal{QPN}$ do not involve any timing aspects. The firing of a transition is atomic and does not consume any time. So they allow us a purely qualitative modeling of biochemical reaction networks.

The *P/T nets* have been enlarged to extended Petri nets ($\mathcal{XPN}$) by the provision of special arc types such as read arcs (often also called test arcs), inhibitor arcs, equal arcs, and reset arcs. All these special arcs are only allowed to go from places to transitions. Read, inhibitor, and equal arcs add constraints on the firing of a transition, but the connected places are not affected upon firing. A read arc (compare Fig. 9.16) allows to model that some resource is required, but not exclusively consumed upon firing. Hence, the same token can be used at the same time by many transitions. An inhibitor arc (compare Fig. 9.10) reverses the logic of the enabling condition of a place, i.e., it imposes a precondition that a transition may only fire if the place contains less tokens than the weight of the arc indicates. An equal arc imposes the precondition that a transition may only fire if the number of tokens on the place connected by the equal arc is equal to the arc weight. A reset arc empties the place connected by this arc once the transition fires; the number of tokens does not matter.

Finally, the $\mathcal{XPN}$ can be further enriched to include marking-dependent arcs, i.e., the arc multiplicities are allowed to be marking-dependent expressions of various types in terms of transitions' pre-places [10]. See Fig. 9.3 for a technical example.

Modeling repressilator. We now use the repressilator [5] as an example to illustrate a modular and stepwise construction of a Petri net model using Snoopy.

1. We start with designing a Petri net model of a gene, illustrated in Fig. 9.4a. The presence of one gene allows the generation of proteins without consuming the gene, while generated proteins can degrade. A possible run of this model is that the transition generate fires twice, adding two tokens to the place protein, and then transition degrade fires once, removing one token from place protein. We

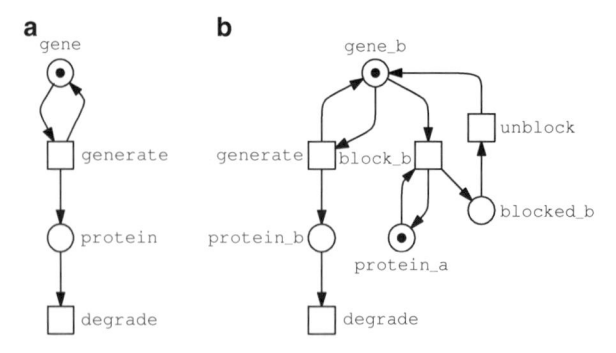

Fig. 9.4 (a) A Petri net model of a gene and (b) a Petri net model of a gene gate according to Blossey et al. [5], who also inspired the layout: gene b may be blocked by protein a (Reprinted from Heiner and Gilbert [25], Copyright 2013, with permission from Elsevier)

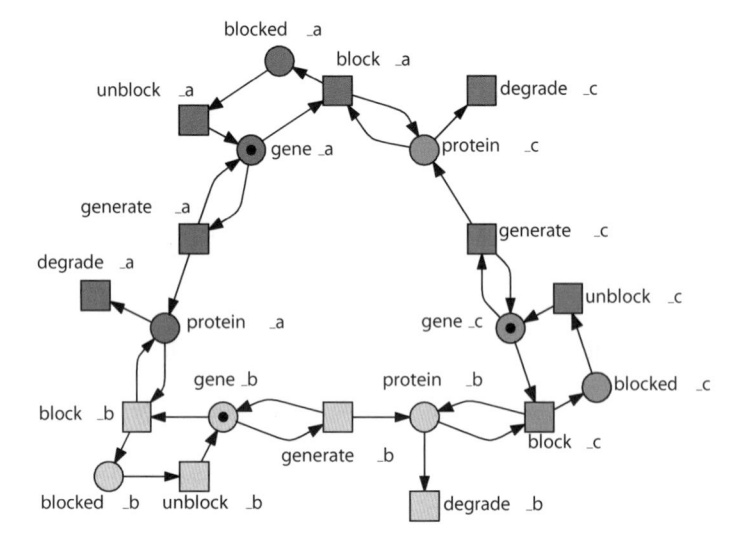

Fig. 9.5 The repressilator – Petri net for three genes in a regulatory cycle (Reprinted from Heiner and Gilbert [25], Copyright 2013, with permission from Elsevier)

obtain the marking where each place carries one token. It is easy to see that this Petri net has an infinite number of reachable markings.

2. Next, we extend the basic behavior in Fig. 9.4a by allowing the gene to be blocked by the protein produced by another gene, which makes a building block called gene gate; see Fig. 9.4b. The behavior of Fig. 9.4b is different from that of Fig. 9.4a, as a gene may be blocked or unblocked in Fig. 9.4b while it is always unblocked in Fig. 9.4a.
3. When genes repress each other in a circular manner, we obtain a gene regulatory cycle, the repressilator [5]; see Fig. 9.5, which is composed of three gene gates with identical structure.

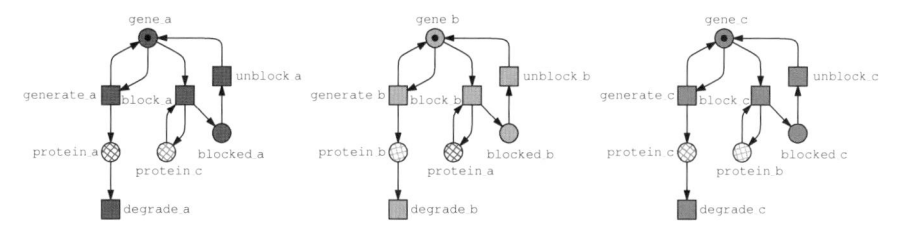

Fig. 9.6 The repressilator – Petri net for three genes in a regulatory cycle represented using logical nodes (here, places, cross-hatched) to preserve gene-centered modules. Logical nodes with identical names serve as connectors; they are multiple representations of the same node used for layout clarity. See also Fig. 9.7 (Reprinted from Heiner and Gilbert [25], Copyright 2013, with permission from Elsevier)

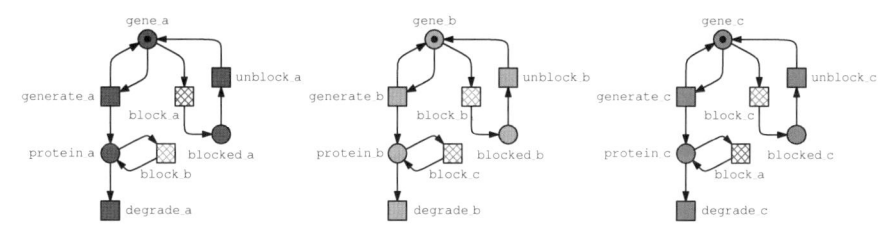

Fig. 9.7 The repressilator – Petri net for three genes in a regulatory cycle represented using logical transitions. See also Fig. 9.6

Snoopy supplies two features for the design and systematic construction of larger Petri nets – logical nodes and macro nodes. Logical nodes (i.e., logical places/transitions) serve as connectors to avoid lengthy arcs, and macro transitions (macro places) help to hide transition-bordered (place-bordered) subnets in order to design hierarchically structured Petri nets.

Using logical nodes, we are able to represent the repressilator model in alternative ways highlighting the modular structure of the Petri net, which are illustrated in Figs. 9.6 and 9.7, respectively, both of which are equivalent to Fig. 9.5.

Using macro transitions, we can hide all gene-related details while keeping the protein places as interface; see Fig. 9.8. We obtain a hierarchical Petri net; Fig. 9.9 gives its top level. This Petri net is also equivalent to Figs. 9.5–9.7, it just uses a different representation style.

9.3.2 Analysis

The $\mathcal{QPN}$ are time-free models; the qualitative analysis considers however all possible behavior of the system under any timing. Thus, the $\mathcal{QPN}$ model itself implicitly contains all possible time-dependent behaviors.

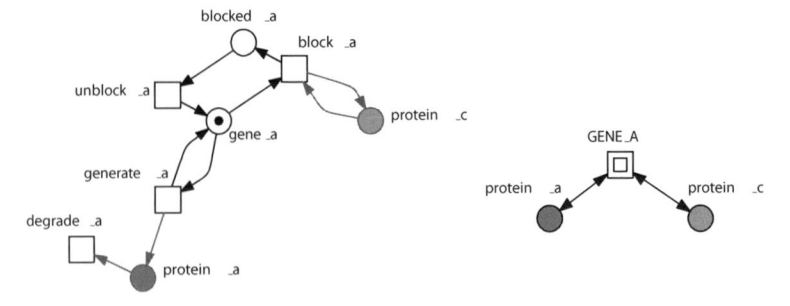

Fig. 9.8 Hierarchical structuring by the use of macro transitions. The uncolored nodes (*left*) make the contents of the macro transition GENE_A (*right*). The *blue arcs* highlight the connection to the interface places

Fig. 9.9 Hierarchical Petri net model of the repressilator using macro transitions; compare Fig. 9.8. Only the top level is shown

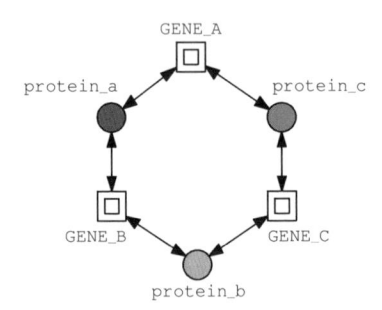

Behavioral properties. There are three orthogonal *general behavioral properties* which are usually explored first to gain some insights into the behavior of a Petri net.

- *Boundedness.* A place is said to be k-bounded (bounded for short) if the maximal number of tokens on this place is bounded by a constant k in all reachable markings. A Petri net is k-bounded (bounded for short) if all its places are k-bounded.
- *Liveness.* A transition is said to be live if it will always be possible to reach a state (marking) where this transition gets enabled, whatever happens. A Petri net is live if each transition is live.
- *Reversibility.* A Petri net is said to be reversible if the initial marking can be reached again from each reachable marking.

For example, by playing the token game for our repressilator model (take any Petri net in Figs. 9.5–9.7 and 9.9), we can easily figure out that the places gene_i and blocked_i, with $i = \{a, b, c\}$, are 1-bounded. But the net is unbounded as all places protein_i are unbounded. If the generation of a protein occurs faster than its degradation, infinite many tokens (molecules) will be accumulated. Furthermore, we can argue that this Petri net is likely to be live and reversible.

When models get more complicated, it might not be obvious anymore to decide behavioral properties by reasoning only. Then, we need mathematically sound analysis techniques. Petri net theory offers a rich body of such analysis techniques, most of them are implemented in our analysis tool Charlie. We sketch here only a few of them to give an impression of what kind of analysis techniques we have.

Structural properties. Structural properties [28, 41, 44] permit – if they hold – to deduce behavioral properties of Petri nets from their structure without constructing the complete or partial state space. If a property is proved structurally for a given Petri net, it holds for this Petri net in any initial marking. The most important structural properties can be classified as follows: elementary graph properties, siphons/traps, and place/transition invariants.

Elementary graph properties. The elementary graph properties relate to the following questions (see [28] for explanations of all the following terms):

- Is the Petri net pure (PUR), ordinary (ORD), homogeneous (HOM), conservative (CSV), static conflict-free (SCF), connected (CON), or strongly connected (SC)?
- Does the Petri net have boundary nodes; i.e., input transitions (FT0), output transitions (TF0), input places (FP0), or output places (PF0)?
- Does the Petri net structure obey the constraints of a state machine (SM), synchronization graph (SG), extended free choice net (EFC), or extended simple net (ES)?

Elementary graph properties occasionally permit on their own conclusions on behavioral properties. For example, a Petri net having input transitions, i.e., transitions without pre-places, is unbounded (as the firing of input transitions does not depend on any preconditions), or a Petri net having input places, i.e., places without pre-transitions, is not live (as the tokens on an input place are sooner or later used up). Our repressilator Petri net has output transitions, i.e., transitions without post-places, which tells us that the model is either not live or unbounded (at least the pre-place of the output transition had to be unbounded).

Siphons/traps. A nonempty set S of places of a Petri net is called a siphon if there is no transition which has post-places in S, but no pre-places in S. Consequently, every transition, which fires tokens onto a place in S, also has a pre-place in this set, i.e., the set of pre-transitions of S is contained in the set of post-transitions of S. Pre-transitions of a siphon cannot fire if the place set is clean, i.e., none of the places carries a token. Therefore, a siphon cannot get tokens again, as soon as it is clean, and then all its post-transitions are dead.

Contrary, a nonempty set Q of places of a Petri net is called a trap if there is no transition which has pre-places in Q, but no post-places in Q. Consequently, every transition, which subtracts tokens from a place of the trap set, also has a post-place in this set, i.e., the set of post-transitions of Q is contained in the set of pre-transitions of Q. Post-transitions of a trap always return tokens to the place set. Therefore, once a trap contains tokens, it cannot become clean again.

Siphon and trap are closely related but contrasting notions. When they come on their own, we usually get deficient behavior. However, both notions have the power to perfectly complement each other. A Petri net satisfies the siphon-trap property (STP) if every siphon includes an initially marked trap. For certain combinations of structural properties, we can derive behavioral properties. For example, if a net is ORD and ES, and the STP holds, then the net is live. The STP holds also for our repressilator model, but the net structure is beyond ES. Thus, we can only conclude the absence of dead states, i.e., states where no transition is enabled.

Place/transition invariants. Place and transition invariants (P- and T-invariants for short) play a crucial role in analyzing biological systems due to their biological interpretations. Both of them can be obtained by solving a linear equation system which describes the Petri net structure and which is independent of the initial marking. Any linear combination of P-invariants (T-invariants) yields again a P-invariant (T-invariant). Therefore, one is usually interested in minimal invariants, i.e., invariants which cannot be described by a linear combination.

A P-invariant represents a set of places over which the weighted token count keeps constant whatever happens in the Petri net. So a place belonging to a P-invariant is k-bounded. We get the upper bound k by multiplying the invariant with the initial marking. In metabolic networks, P-invariants often correspond to conservation laws in chemistry, reflecting substrate conservations, while in signal transduction networks, P-invariants often correspond to proteins and their possible states.

A T-invariant describes a multiset of transitions; it can be interpreted in two different ways. The multiset either specifies how often a transition has to fire to return to the original marking, or the multiset gives the relative firing rates required to keep the Petri net in the same state – the steady state.

Taking our repressilator model as an example, Charlie yields the following results. The Petri net has three minimal P-invariants, one for each gene gate: $x_i = $ (gene_i, blocked_i), where $i = a, b, c$. For each P-invariant, the constant token sum is 1, which confirms our expectations: a gene is either blocked or unblocked, it can neither disappear nor be multiplied.

The Petri net has also six minimal T-invariants, two for each gene gate: $y1_i = $ (block_i, unblock_i) and $y2_i = $ (generate_i, degrade_i), where $i = a, b, c$, which cover the whole Petri net, i.e., each transition belongs to a T-invariant. These T-invariants confirm our previous observations that a balanced firing of these transition sets reproduces the initial marking, and a balanced firing according to $y2_i$ makes the Petri net bounded.

Model checking. If the state space is finite and of manageable size, analytical model checking can be used to analyze $\mathcal{QPN}$; otherwise, simulative model checking may help to obtain an approximative answer. In any case, the behavioral properties of interest have to be expressed in temporal logics, e.g., in a branching time temporal logic, one instance of which is computational tree logic (CTL) [11] or in a linear-time logic (LTL) [47]. Both logics are supported by Marcie.

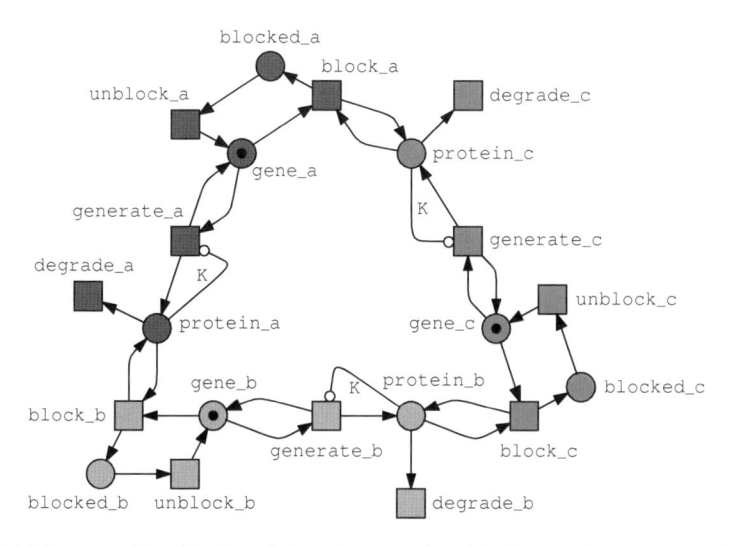

9 Petri Nets for Modeling and Analyzing Biochemical Reaction Networks 255

Fig. 9.10 The repressilator in a bounded version – maximal K tokens can be accumulated on each protein place. Inhibitor arcs (*hollow circle* as arc head) with arc weight K limit the generation of proteins. K is a constant which is used to conveniently parameterize the model; compare Table 9.1

Table 9.1 State space growth for increasing K (maximum number of each protein) (compare Fig. 9.10) computed with Marcie's symbolic state space representation. For the very specific case of our repressilator example, we are able to specify a general formula for the state space growth: $2^n * (K+1)^n$, with n being the number of genes in the regulatory circle (in our running example, we use $n = 3$)

K	Number of states	K	Number of states
1	64 (1)	1,000	8,024,024,008 (9)
50	1,061,208 (6)	5,000	1,000,600,120,008 (12)
100	8,242,408 (6)	10,000	8,002,400,240,008 (12)
150	27,543,608 (7)	50,000	1,000,060,001,200,008 (15)
500	1,006,012,008 (9)	100,000	8,000,240,002,400,008 (15)

To be able to deploy CTL model checking, we introduce a bounded version of our repressilator example; see Fig. 9.10. Its state space is finite but explosively grows as illustrated in Table 9.1.

Having a bounded repressilator model, we can check behavioral properties expressed as CTL properties. We give three examples for *special behavioral properties*:

- Forever it holds, gene b is either unblocked or blocked.
  AG [$(gene\_b = 1$ & $blocked\_b = 0) | (gene\_b = 0$ & $blocked\_b = 1)$]
- It is forever possible that there are at least k molecules of protein b; i.e., there will be new proteins b forever, which includes liveness of transition $degrade\_b$.
  AG EF [$protein\_b \geq k$]

- It is possible that there are at least k molecules of each protein at the same time.
  EF [$protein\_a \geq k$ & $protein\_b \geq k$ & $protein\_c \geq k$]

Here, **A** (for all paths) and **E** (there is one path) are path quantifiers, and **G** (globally) and **F** (finally) are temporal operators. The CTL can also be used to query the general behavioral properties. For more examples of temporal formulae, the reader may wish to check, e.g., Heiner et al. [28] where model checking has been explored for $\mathcal{QPN}$.

9.3.3 Applications

There are quite a number of applications of qualitative Petri nets for modeling of biochemical systems. In this chapter, we do not wish to give a review but just give some examples.

Model validation by means of P-/T-invariant analysis is discussed in Heiner and Koch [26] for three case studies: apoptosis, carbon metabolism in potato tuber, and the glycolysis and pentose phosphate metabolism. Structural analysis has also been used in Heiner [23] to derive coarse network structures highlighting the structural principles inherent in the functional modules identified by T-invariants and in Heiner and Sriram [27] to determine the core of a hypoxia response network and to identify its fragile node.

The $\mathcal{QPN}$ have been deployed in Heiner et al. [28, 30] to model signal transduction pathways, and their detailed analysis is exercised step by step.

In Blätke et al. [3], IL-6 signalling in the JAK/STAT signal transduction pathway serves as case study to illustrate a modular protein-centered modeling approach.

9.4 Stochastic Petri Nets ($\mathcal{SPN}$)

9.4.1 Modeling

The $\mathcal{SPN}$ extend $\mathcal{QPN}$ by assigning to transitions exponentially distributed waiting times, which are specified by firing rate functions (stochastic rates, compare Fig. 9.1). The underlying semantics of $\mathcal{SPN}$ is a continuous-time Markov chain (CTMC). The $\mathcal{SPN}$ have been previously extended to *generalized stochastic Petri nets* ($\mathcal{GSPN}$) [41] and later to *deterministic and stochastic Petri nets* ($\mathcal{DSPN}$) [18].

Our *extended stochastic Petri nets* ($\mathcal{XSPN}$) [29], which comprise the $\mathcal{GSPN}$ and $\mathcal{DSPN}$, provide the four special arc types and marking-dependent arcs as available for $\mathcal{XPN}$ and furthermore three special transition types: immediate transitions (zero waiting time), deterministic transitions (deterministic waiting time, relative to the time point where the transition gets enabled), and scheduled

Table 9.2 Rate functions for the $\mathcal{SPN}$ repressilator model. $MA(c)$ denotes the mass action function, where c is a kinetic parameter. See last column for the explicit rate functions for gene a

Transition class	Kinetic parameter c	Rate function pattern	Example: gene a
Generate	0.1	$MA(0.1)$	$0.1 * gene\_a$
Block	1.0	$MA(1.0)$	$1.0 * gene\_a * protein\_c$
Unblock	0.0001	$MA(0.0001)$	$0.0001 * blocked\_a$
Degrade	0.001	$MA(0.001)$	$0.001 * protein\_a$

transitions (scheduled to fire, if any, at single or equidistant, absolute points of the simulation time). In Snoopy, we do not distinguish between these three classes of stochastic Petri nets. Thus, we usually call our extended stochastic Petri nets simply $\mathcal{SPN}$ if confusion is precluded.

In biological reaction networks, rate functions are often marking-dependent. In Snoopy, popular kinetics like mass action semantics [40] and level semantics [28] are supported by predefined function patterns.

Modeling repressilator. Let us return to our repressilator model. If we associate a rate function with each transition, e.g., the rate functions given in Table 9.2, we can consider it as a stochastic repressilator model.

9.4.2 Analysis

The CTMC for a given $\mathcal{SPN}$ is isomorphic to the reachability graph of its corresponding $\mathcal{QPN}$, but edges are enriched by the transition rates. Thus, all $\mathcal{QPN}$ analysis techniques can still be applied, and all behavioral properties which hold for a $\mathcal{QPN}$ are still valid for the $\mathcal{SPN}$. Additionally, we have the following techniques to explore stochasticity.

Stochastic simulation. Stochastic simulation like the Gillespie stochastic simulation algorithm (SSA) [21] generates random walks through the CTMC. Approximated traces can be obtained by averaging a number of simulation runs. Besides, the unrestricted use of special (immediate, deterministic, scheduled) transitions destroys the Markov property. But the adaptation of the Gillespie stochastic simulation algorithm is rather straightforward and supported in Snoopy.

For example, assigning rates with the given kinetic parameters to any of our repressilator Petri nets generates sustained oscillation for all proteins, with each single run behaving differently. See Fig. 9.11 for a plot with the rates given in Table 9.2.

Simulative model checking. To systematically explore simulation traces, we use PLTLc [12], a probabilistic extension of LTL with constraints, to express our behavioral properties of interest. Simulative model checking considers a finite set of finite outputs from Gillespie's exact SSA, i.e., a finite subset of the state space. This permits to explore very big or even infinite state spaces in reasonable time or just to

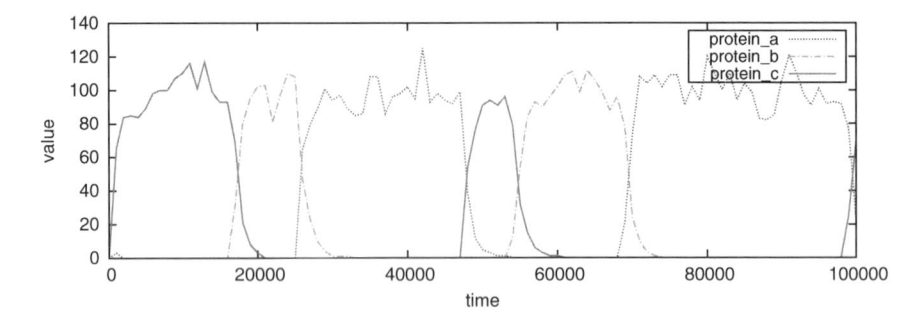

Fig. 9.11 Plot of one stochastic simulation run for the $\mathcal{SPN}$ repressilator model; for rate functions, see Table 9.2. Each single run looks differently in terms of oscillation, e.g., which gene starts rising

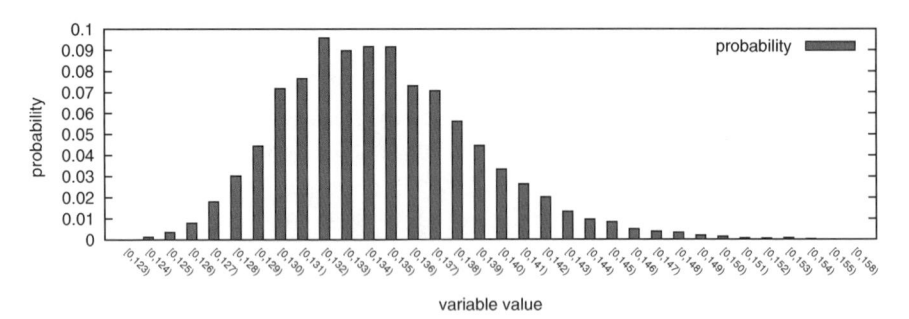

Fig. 9.12 Probability distribution of the value range for the protein places, determined by 10,000 stochastic simulation runs with $\tau = 200{,}000$ for the $\mathcal{SPN}$ repressilator model. Increasing the number of runs smoothes the bell shape but does not shift the value range. Values beyond 158 are most unlikely

obtain a first rough estimate. Each trace is evaluated to a Boolean truth value, and the probability of a behavioral property holding true is approximated by the number of traces with true values over the whole sample set. One has to consider a sufficient amount of simulation traces to obtain reliable approximations. The number of traces required increases with the expected confidence in the numerical results. Rare events may dramatically increase the required size of the sample set.

Let us return to our running example. We use a PLTLc-specific feature to explore the value range for the proteins – the *free variables* – which are specified by a leading $.

- What is the probability that up to time point τ one of the proteins rises above v? We do not know which protein will start rising, so we use the disjunction.
  $\mathbf{P}_{=?}[\ \mathbf{F}_{[0,\tau]}\ protein\_a > \$v\ |\ protein\_b > \$v\ |\ protein\_c > \$v\ ]$

Simulative model checking yields the domain of the free variable v and the probability of each interval; see Fig. 9.12. We observe that values beyond 150 are increasingly unlikely. Thus, we take $K = 160$ as upper bound (see Fig. 9.10), which cuts the infinite state space down to 33,386,248 states.

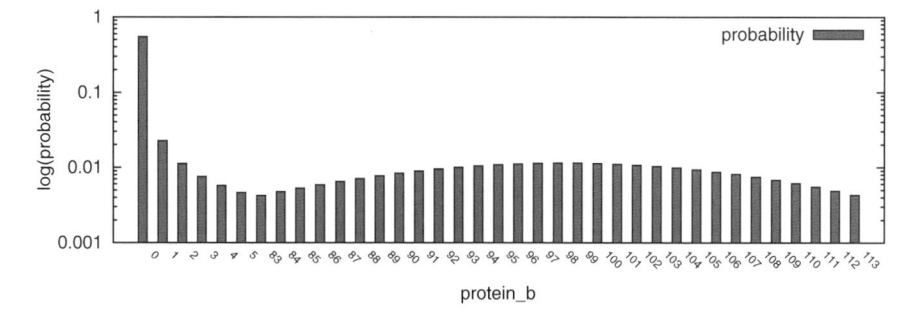

Fig. 9.13 Probability distribution for having k molecules of protein b in the steady state; x-axis: values with a probability below the (arbitrarily chosen) threshold (0.004) have been omitted, y-axis: given in log scale

Analytical model checking. As long as the underlying semantics of a stochastic Petri net is described by a finite CTMC of manageable size, it can be analyzed using such standard stochastic analysis techniques as transient analysis, steady-state analysis, or analytical model checking [50, 51].

Transient analysis means to compute the transient probabilities to be in a certain state at a specific time point using, e.g., the uniformization or Jensen method [54]. Steady-state analysis computes the steady-state probabilities using, e.g., Gaussian elimination or Jacobi iteration [45]. In analytical model checking, special behavioral properties can be checked, which have been expressed in, e.g., Continuous Stochastic Logic (CSL), a stochastic counterpart of CTL which was originally introduced in Aziz et al. [1].

For illustration, we compute for the bounded version of our repressilator model ($K = 160$) the probability that in the steady state, there are k molecules of protein b; see Fig. 9.13. The obtained probability distribution tells us that we have oscillations with very sharp rise and fall, with peaks around 100. However, most of the time there are only a few proteins. The expectation value $E[protein\_b] = 33.18$ corresponds to the steady-state value which we observe when averaging over a sufficient amount of simulation traces; see Fig. 9.14.

Likewise, we could use transient analysis to evaluate the following CSL formulae.

- What is the probability that at time point τ there are at least k molecules of protein b?
  $\mathbf{P}_{=?} \, [\, \mathbf{F}_{[\tau,\tau]} \, (\, protein\_b \geq k \,) \,]$
  Increasing τ will finally approach the steady-state values for any $k \leq K$.
- What is the probability that up to time point τ there are at least k molecules of each protein at the same time?
  $\mathbf{P}_{=?} \, [\, \mathbf{F}_{[0,\tau]} \, (\, protein\_a \geq k \, \& \, protein\_b \geq k \, \& \, protein\_c \geq k \,) \,]$
  The probability is technically larger than 0 for any $k \leq K$ (as the property holds for the $\mathcal{QPN}$), but drops dramatically with increasing k, and reaches very fast insignificant values which are practically 0.

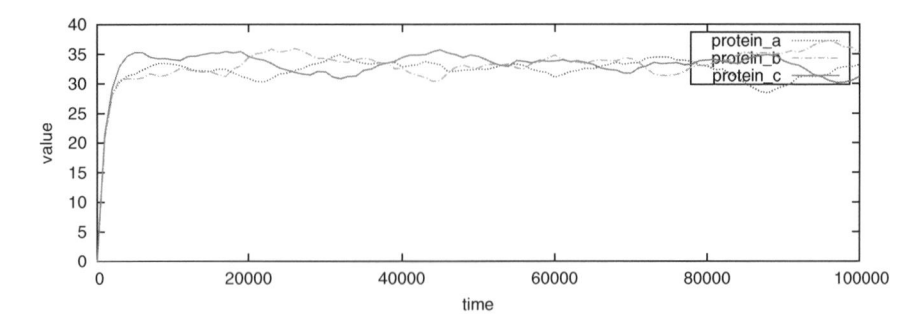

Fig. 9.14 Average of 1,000 simulation traces for the $\mathcal{SPN}$ repressilator model. Increasing the number of averaged traces smoothes the curves

Marcie supports standard stochastic analysis techniques and analytical model checking for (Markovian) $\mathcal{SPN}$ and simulative model checking for $\mathcal{XSPN}$. We recommend Heiner et al. [29] and Schwarick et al. [51] for more illustrative examples.

9.4.3 Applications

The $\mathcal{SPN}$ have been used in Heiner et al. [28, 30] to model and analyze signal transduction pathways; the detailed analysis exploits analytical and simulative model checking.

A classical example of prokaryotic gene regulation, the lac operon, is taken in Heiner et al. [29] to demonstrate the power of $\mathcal{XSPN}$ for model-based design of wet lab experiments. This paper may also serve as a gentle introduction into the use of simulative model checking.

In Marwan et al. [42], the $\mathcal{XSPN}$ have been applied to investigate phosphate regulation in enteric bacteria; modeling details and stochastic simulation runs are given. There one also finds more information on how to control stochastic simulation experiments in Snoopy.

9.5 Continuous Petri Nets ($\mathcal{CPN}$)

9.5.1 Modeling

Continuous Petri nets offer a graphical way to specify systems of ordinary differential equations (ODEs) [28]. The discrete tokens, which we had so far in $\mathcal{QPN}$ and $\mathcal{SPN}$, are exchanged in $\mathcal{CPN}$ by real-valued tokens, one token for each place. The instantaneous firing of a transition is carried out like a continuous flow. Its

Table 9.3 The unreduced ODEs induced by the $\mathcal{CPN}$ repressilator model, as generated by Snoopy. We deliberately give the unreduced ODEs to highlight the relation to the generating $\mathcal{CPN}$

$$\frac{d\ gene\_a}{dt} = (0.1 * gene\_a) + (0.0001 * blocked\_a)$$
$$-(0.1 * gene\_a) - (1 * protein\_c * gene\_a)$$

$$\frac{d\ protein\_a}{dt} = (0.1 * gene\_a) + (1 * gene\_b * protein\_a)$$
$$-(0.001 * protein\_a) - (1 * gene\_b * protein\_a)$$

$$\frac{d\ blocked\_a}{dt} = (1 * protein\_c * gene\_a) - (0.0001 * blocked\_a)$$

$$\frac{d\ gene\_b}{dt} = (0.1 * gene\_b) + (0.0001 * blocked\_b)$$
$$-(0.1 * gene\_b) - (1 * protein\_a * gene\_b)$$

$$\frac{d\ protein\_b}{dt} = (0.1 * gene\_b) + (1 * gene\_c * protein\_b)$$
$$-(0.001 * protein\_b) - (1 * gene\_c * protein\_b)$$

$$\frac{d\ blocked\_b}{dt} = (1 * protein\_a * gene\_b) - (0.0001 * blocked\_b)$$

$$\frac{d\ gene\_c}{dt} = (0.1 * gene\_c) + (0.0001 * blocked\_c)$$
$$-(0.1 * gene\_c) - (1 * protein\_b * gene\_c)$$

$$\frac{d\ protein\_c}{dt} = (0.1 * gene\_c) + (1 * gene\_a * protein\_c)$$
$$-(0.001 * protein\_c) - (1 * gene\_a * protein\_b)$$

$$\frac{d\ blocked\_c}{dt} = (1 * protein\_b * gene\_c) - (0.0001 * blocked\_c)$$

strength is determined by the continuous rate functions (deterministic rates, compare Fig. 9.1), which are assigned to each transition. The $\mathcal{CPN}$ and $\mathcal{SPN}$ have the power to approximate each other as it is depicted in Fig. 9.1.

Modeling repressilator. Let us return to the repressilator model in Fig. 9.5. If we read tokens on each place as (real-valued) concentrations of species and associate a deterministic rate function with each transition, e.g., the same rate functions as given in Table 9.2, we can consider it as a continuous repressilator model. The underlying ODEs of the continuous model as generated by Snoopy are given in Table 9.3.

From these ODEs, we can see that each place in the Petri net model is mapped to a variable in the ODEs, and each variable gets its own equation. A place's pre-transitions increase the token value; thus, their rate functions appear as plus terms in the equation. Contrary, post-transitions decrease the token value; thus, their rate

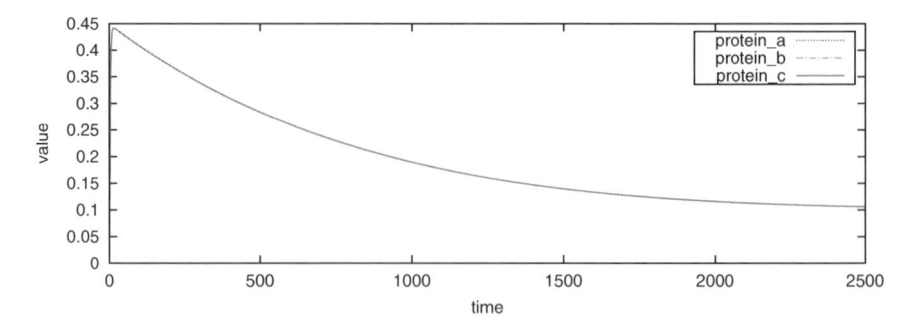

Fig. 9.15 Plot of a continuous simulation run for the $\mathcal{CPN}$ repressilator model. For rate functions, see Table 9.2. This plot suggests that the repressilator quickly reaches a steady state. The three curves for the three proteins coincide, so we see only one of them. Contrary, Fig. 9.11 suggests that the stochastic repressilator fluctuates around a steady-state value

functions appear as minus terms. A transition which is pre- and post-transition yields two terms, which can be reduced by algebraically transforming the right-hand side of the equation.

9.5.2 Analysis

Continuous simulation. The ODEs induced by a given $\mathcal{CPN}$ are usually not linear, which calls for numerical integration algorithms. Snoopy supports 14 different stiff/unstiff ODE integrators to numerically solve the ODEs. These ODE solvers range from simple fixed-step-size solvers (e.g., Euler), which are suitable for unstiff $\mathcal{CPN}$ models, to more sophisticated variable-order, variable-step, multistep solvers (e.g., backward differentiation formulas (BDFs)), which have to be used for stiff $\mathcal{CPN}$ models. Snoopy's implementation of the latter solvers builds on the library SUNDIALS CVODE [35].

Running continuous simulation for the $\mathcal{CPN}$ repressilator model yields plots as illustrated in Fig. 9.15.

Continuous model checking. The behavior of a $\mathcal{CPN}$ model is deterministic, i.e., each run with the same parameters yields the same results. Thus, the state space can be considered as being continuous and linear. It can be explored by using, for example, continuous linear temporal logic with constraints (LTLc) [7] or PLTLc [12] in a deterministic setting. Both are interpreted over the continuous simulation trace generated by numerically integrating ODEs. Please refer to Donaldson and Gilbert [12] and Heiner et al. [28] for details about how to use MC2 tools to do simulative model checking and a couple of biological examples.

For illustration, we specify the following PLTLc property for the $\mathcal{CPN}$ repressilator model.

- Does finally the value of protein *b* first rise and then fall; with other words: does there exist a peak in the trace?
  P$_{=?}$ [**F** [$(d(protein\_b) > 0)$ & **F** [$(d(protein\_b) < 0)$]]]
  The function $d(species)$ returns the derivative of the concentration of the species at each time point. The probability is 1, i.e., there is a peak; see Fig. 9.15.

Other analysis techniques. Besides, all standard ODEs analysis techniques, e.g., bifurcation analysis, sensitivity analysis, and parameter scanning [52], are applicable when the ODEs are exported to suitable tools, e.g., Matlab [43].

9.5.3 Applications

Gilbert and Heiner [19] present results of an investigation to integrate Petri nets and ODEs for the modeling and analysis of biochemical networks and apply their approach to a model of the influence of the Raf kinase inhibitor protein (RKIP) on the extracellular signal-regulated kinase (ERK) signalling pathway.

A novel methodology for the engineering of biochemical network models is proposed in Breitling et al. [6] and illustrated for signalling pathways. It includes the structured design of ODEs using $\mathcal{CPN}$ and their systematic composition.

Soliman and Heiner [53] discuss sufficient conditions for the unique construction of $\mathcal{CPN}$ models from ODEs. The challenge is to reveal the network structure which is hidden in a given ODE. Generally, this reverse problem does not have a unique solution, while the ODEs induced by a given $\mathcal{CPN}$ are uniquely defined.

9.6 Generalized Hybrid Petri Nets ($\mathcal{GHPN}$)

9.6.1 Modeling

Snoopy integrates all functionalities of its stochastic and continuous Petri nets ($\mathcal{SPN}$ and $\mathcal{CPN}$) into one net class, yielding generalized hybrid Petri nets ($\mathcal{GHPN}$) [33, 34]. $\mathcal{GHPN}$ are specifically tailored (but not limited) to models that require an interplay between stochastic and continuous behavior. They provide a trade-off between accuracy and runtime of model simulation by adjusting the number of stochastic transitions appropriately, which can be done either statically (by the user) or dynamically (by the simulation algorithms). A typical application of $\mathcal{GHPN}$ is the hybrid representation of biochemical reactions at different scales (also called stiff systems), where slow reactions are represented by stochastic transitions and fast reactions by continuous transitions.

Modeling repressilator. For illustration, we now interpret our repressilator Petri net, reusing the rate functions given in Table 9.2, as a $\mathcal{GHPN}$ model. The 1-bounded

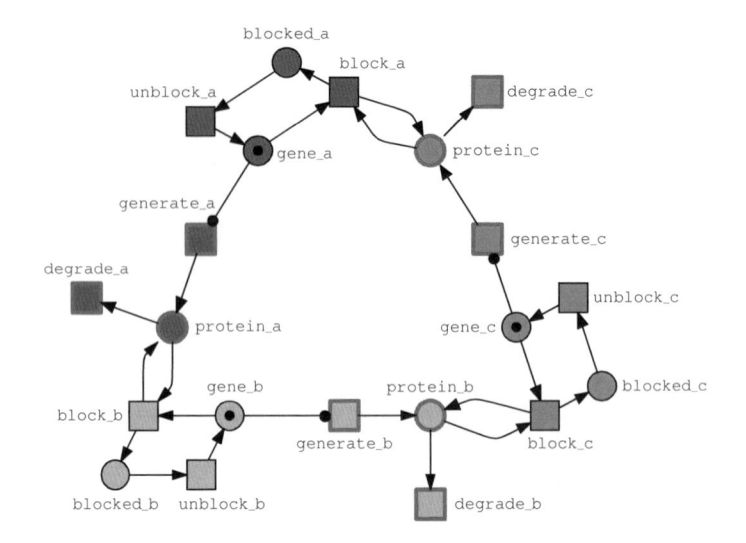

Fig. 9.16 The $\mathcal{GHPN}$ model of the repressilator, where the continuous places (transitions) are represented by *shaded line circles* (*squares*). Besides, the bidirectional arcs between places *gene_i* and transitions *generate_i* are replaced by read arcs (*black dots* as arc heads) in order to comply with the connection rules (a continuous transition is not allowed to remove from or write to a discrete place)

places as determined by P-invariant analysis and the related transitions as determined by T-invariant analysis are kept discrete. The unbounded places and related transitions are approximated by continuous places and transitions, respectively. That is, places *gene_i* and *blocked_i*, and transitions *block_i* and *unblock_i* are treated as discrete and all other nodes as continuous. To distinguish between discrete and continuous nodes, we choose different graphical representations; see Fig. 9.16.

9.6.2 Analysis

Hybrid simulation. Snoopy's hybrid simulation builds on Gillespie's direct method [21] to simulate stochastic transitions and on continuous simulation to integrate the ODEs induced by the continuous transitions using SUNDIALS CVODE [35].

For example, if we still assign the rates in Table 9.2 to the hybrid model and consider them as stochastic or deterministic rates, depending on the transition type, hybrid simulation yields plots as illustrated in Fig. 9.17.

Simulative model checking. Likewise, we can use PLTLc [12] to do simulative model checking of a $\mathcal{GHPN}$ model, which also handles a subset of the state space, e.g., a set of finite outputs from hybrid simulation in Snoopy.

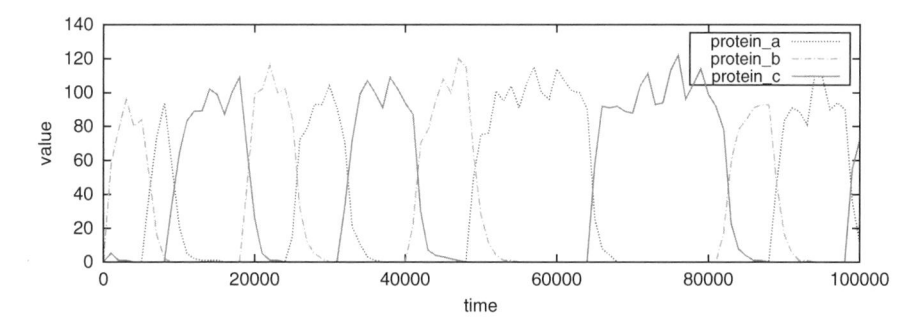

Fig. 9.17 Plot of one hybrid simulation run for the repressilator. For rate functions, see Table 9.2. This plot suggests that $\mathcal{GHPN}$ are able to capture the oscillation. Repeated runs look differently; thus, stochasticity is captured as well

For the hybrid repressilator model, we specify the following property using PLTLc:

- What is the probability of $protein\_i$ ($i = a, b, c$) to be sometimes (finally) greater than 120?
  $\mathbf{P}_{=?} [\mathbf{F} [protein\_i > 120]]$
  With the same data as used for Fig. 9.17, it is evaluated for this single trace to 1 for $protein\_c$ and 0 for the other proteins.
- What is the probability that there are k molecules of two proteins (here, $protein\_a$ and $protein\_b$) at the same time?
  $\mathbf{P}_{=?} [\mathbf{F} [(protein\_a \geq k) \, \& \, (protein\_b \geq k)]]$
  With the same data as used for Fig. 9.17, it is evaluated for this single run to 1 for $k = 1, 10, \ldots, 40$, and 0 for $k \geq 50$.

9.6.3 Applications

In Herajy [33], $\mathcal{GHPN}$ are used to model and analyze three case studies: the intracellular growth of bacteriophage T7, the eukaryotic cell cycle, and the circadian rhythm. In all three cases, chemical reactions are divided into two groups: fast and slow.

9.7 Colored Extensions

9.7.1 Modeling

Colored Petri nets [17, 37] are a high-level extension of standard Petri nets, where a group of similar model components are represented by one component, each of which is defined as and thus distinguished by a color.

Fig. 9.18 A colored Petri net model for the repressilator. The declarations: colorset GeneSet = enum with a, b, c, and variable x: GeneSet. With this color set, this model corresponds exactly to the Petri nets in Figs. 9.5–9.7 and 9.9 (Reprinted from Heiner and Gilbert [25], Copyright 2013, with permission from Elsevier)

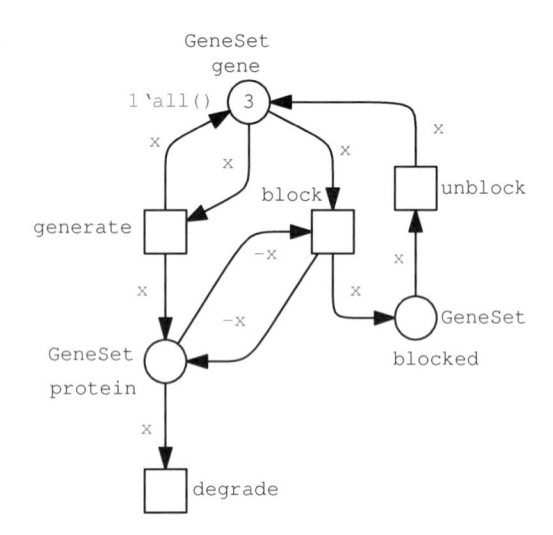

Colored Petri nets consist, as standard Petri nets, of places, transitions, and arcs. Additionally, a colored Petri net model is characterized by a set of data types [8], called color sets. Each place gets assigned a color set and may contain distinguishable tokens colored with a color of this color set.

Modeling repressilator. Figure 9.18 gives a colored Petri net model for the repressilator model in Fig. 9.6. A color set GeneSet is defined with three colors, a, b, and c to distinguish three genes. Each place gets assigned this color set GeneSet. By this way, we use one place to represent three similar objects, e.g., representing three protein objects as one colored place protein.

As there can be several tokens of the same color on a given place, the tokens on a place define a multiset over the place's color set. For example, in Fig. 9.18, we denote the initial marking for the place protein by a multiset expression, 1'a++1'b++1'c, which means one token of each color of GeneSet.

Each transition gets a guard, which is a Boolean expression over variables, constants, etc. The guard must be evaluated to true for the enabling of the transition. The trivial guard "true" is usually not explicitly given. For example, in Fig. 9.18, all colored transitions have the trivial guard "true."

Each arc gets assigned an expression; the result type of this expression is a multiset over the color set of the connected place. For example, in Fig. 9.18, we define a variable x of GeneSet, which is used in arc expressions. The predecessor operator "−" in the arc expression $-x$ returns the predecessor of x in an ordered finite color set. For example, if $x = b$, then $-x$ returns a. If x is the first color, then it returns the last color. For example, if $x = a$, then $-x$ returns c. The result type of each arc expression is a multiset over GeneSet.

Each uncolored net class has a colored counterpart [38], which inherits all features of its corresponding uncolored net class, e.g., $\mathcal{SPN}^C$ enjoy all special arc types and transition types of $\mathcal{SPN}$.

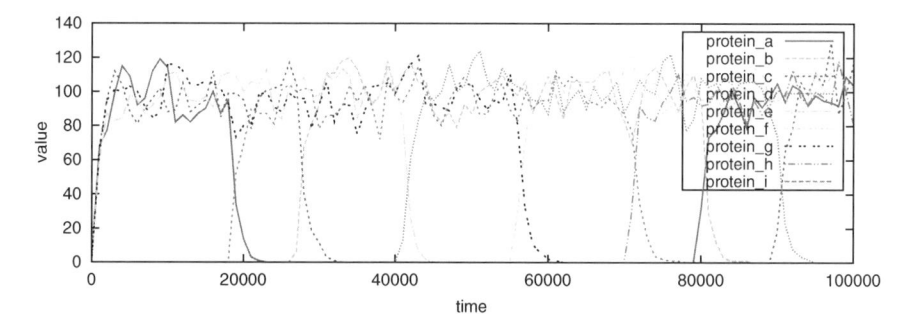

9 Petri Nets for Modeling and Analyzing Biochemical Reaction Networks 267

Fig. 9.19 Plot of a stochastic simulation run for the repressilator with nine genes. For rate functions, see Table 9.2

Snoopy provides various flexible ways to define declarations to be used in the annotations of colored Petri nets. For example, rich data types for color set definitions are supported: (1) simple types (dot, integer, string, Boolean, enumeration, and index) and (2) compound types (product and union). Concise initial marking specifications for larger color sets and individual rate function definitions for each transition instance are supported. Syntax checking ensures the syntactical correctness of constructed models.

A Petri net can be folded into a colored Petri net if the partition of place and transition nodes is given. After that, colored Petri nets can show their attractive advantage, scalability, e.g., changing the number of genes involved in the regulatory cycle just requires to adapt the color set GeneSet appropriately. For example, if we set GeneSet to, let's say, nine colors, a–i, a stochastic simulation plot for nine genes can be produced, illustrated in Fig. 9.19.

Vice versa, colored Petri nets with finite color sets can be automatically unfolded into uncolored Petri nets, which then allows the application of all of the existing powerful standard Petri net analysis techniques. For example, unfolding the colored Petri net in Fig. 9.18 generates the Petri net given in Fig. 9.6.

Modeling procedure. Usually, modeling with colored Petri nets follows the following procedure:

1. Convert an uncolored Petri net into a colored Petri net by using the predefined color set *Dot*, which contains only one color called *dot*.
2. Identify similar subnets in the uncolored Petri net model. For example, we can identify three similar subnets in Fig. 9.6, the three gene gates.
3. Define declarations, e.g., color sets, variables, or constants. For example, we define a color set with three colors a, b, c to distinguish three similar subnets in Fig. 9.6.
4. Assign color sets to places and define the initial marking.
5. Write and assign arc expressions.
6. Define guards for transitions.
7. Check the syntax of inscriptions of the net, e.g., arc expressions and guards.

9.7.2 Analysis

In Snoopy, we support automatic unfolding of colored Petri nets into uncolored Petri nets, so all the analysis techniques mentioned above for each uncolored Petri net class can be equally applied to the corresponding colored Petri net class.

The key challenge when unfolding colored Petri nets is the computation of all transition instances, which in fact is a combinatorial problem, suffering from combinatorial explosion. For overcoming this, a constraint satisfaction approach [55] has been employed. Specifically, the efficient search strategies of Gecode [16] have been used to greatly improve the unfolding efficiency of colored Petri nets; see Liu [38] for details.

9.7.3 Applications

In Liu et al. [39], colored Petri nets have been used to model cooperative ligand binding and the repressilator, with the aim to illustrate how to use colored Petri nets to model biological systems.

In Liu [38], three case studies have been given. The first case study illustrates how to model a multicellular *C. elegans* vulval development system, where each cell is encoded as color. The second describes the modeling of coupled Ca^{2+} channels that are arranged in two-dimensional space, where each channel is encoded as a color. The third explores the use of colored Petri nets to model membrane systems, where a membrane system composed of compartments is modeled as a colored Petri net model by encoding each compartment as a color.

In Gao et al. [15], colored Petri nets are applied to model a tissue comprising multiple cells hexagonally packed in a honeycomb formation in order to describe the phenomenon of planar cell polarity (PCP) signalling in Drosophila wing, which illustrates how to use colored Petri nets to address the multiscaleness problem in systems biology.

In Gilbert et al. [20], phase variation in bacterial colony growth has been studied using colored stochastic Petri nets and continuous diffusion using colored continuous Petri nets.

9.8 Tools

BioModel Engineering of nontrivial case studies requires adequate tool support. We deploy a sophisticated toolkit covering the whole framework:

- *Snoopy* [31, 49] is a platform to support the construction and animation/simulation of all the types of Petri nets discussed in this chapter, with an automatic conversion between them. Obviously, there may be a loss of information in some directions (cf. arrows labeled with abstraction in Fig. 9.1). The conversion between colored and uncolored net classes involves user-guided folding or automatic unfolding. Snoopy supports several data exchange formats, among them the following analysis tools in this list, as well as SBML import/export, which opens the door to a bunch of tools popular in systems and synthetic biology.

- *Charlie* [14, 56] permits the analysis of standard properties and applies standard techniques of Petri net theory, expanded by explicit CTL and LTL model checking.
- *Marcie* [32, 51] is a symbolic analysis tool of standard Petri net properties and CTL model checking for $\mathcal{QPN}$ and CSL model checking for $\mathcal{SPN}$. Exact analyses are complemented by approximative PLTLc model checking built on fast adaptive uniformization and distributed Gillespie simulation.
- *MC2(PLTLc)* [12] is a Monte Carlo Model Checker for properties written in (PLTLc). MC2(PLTLc) can operate with stochastic/deterministic simulation output, deterministic parameter scan output, or even wet lab data.

The Petri net tools are publicly available at http://www-dssz.informatik.tu-cottbus.de, and MC2(PLTLc) at http://www.brc.dcs.gla.ac.uk/software/mc2/. All the Petri net models of the repressilator used in this chapter can be downloaded at http://www-dssz.informatik.tu-cottbus.de/DSSZ/Software/Examples.

9.9 Conclusions

Petri nets have been widely used to model and simulate biochemical reaction networks, and a variety of extensions of standard Petri nets have been developed. In this chapter, we have first described a unifying Petri net framework that comprises a structured family of Petri net classes and then given a brief introduction of each net class in this framework.

Our running case study illustrates how to easily move between these Petri net classes by the help of our Petri net tool Snoopy. An elaborated treatment of another version of the repressilator [13] in the various paradigms can be found in Blätke et al. [4].

This chapter is meant to give a general idea of how to use Petri nets for modeling and analyzing biochemical reaction networks of various types with different modeling paradigms. We gave plenty of pointers to related literature where the interested reader may continue reading.

Acknowledgements This work has been supported by Germany Federal Ministry of Education and Research (0315449H), Natural Scientific Research Innovation Foundation in Harbin Institute of Technology (HIT.NSRIF.2009005), and National Natural Science Foundation of China (61273226). We would like to thank David Gilbert and Wolfgang Marwan for many fruitful discussions and Mary Ann Blätke, Mostafa Herajy, Christian Rohr, and Martin Schwarick for their assistance in model construction, software development, and model checking.

WWW-List of Tools

Snoopy http://www-dssz.informatik.tu-cottbus.de/DSSZ/Software/Snoopy
Charlie http://www-dssz.informatik.tu-cottbus.de/DSSZ/Software/Charlie
Marcie http://www-dssz.informatik.tu-cottbus.de/DSSZ/Software/Marcie
MC2 tool http://www.brc.dcs.gla.ac.uk/software/mc2/
Examples http://www-dssz.informatik.tu-cottbus.de/DSSZ/Software/Examples

References

1. Aziz A, Sanwal K, Singhal V, Brayton R (2000) Model checking continuous-time Markov chains. ACM Trans Comput Log 1(1):162–170
2. Baldan P, Cocco N, Marin A, Simeoni M (2010) Petri nets for modelling metabolic pathways: a survey. Nat Comput 9(4):955–989
3. Blätke M, Dittrich A, Rohr C, Heiner M, Schaper F, Marwan W (2013a) JAK/STAT signalling – an executable model assembled from molecule-centred modules demonstrating a module-oriented database concept for systems and synthetic biology. Mol Biosyst 9:1290–1307
4. Blätke M, Rohr C, Heiner M, Marwan W (2013b, in preparation) A Petri net based framework for biomodel engineering. In: Benner P, Findeisen R, Flockerzi D, Reichl U, Sundmacher K (eds) Large scale networks in engineering and life sciences. Lecture notes of a Summer School held in Magdeburg, 26–30 Sept 2011
5. Blossey R, Cardelli L, Phillips A (2008) Compositionality, stochasticity and cooperativity in dynamic models of gene regulation. HFSP J 2(1):17–28
6. Breitling R, Gilbert D, Heiner M, Orton R (2008) A structured approach for the engineering of biochemical network models, illustrated for signalling pathways. Brief Bioinform 9(5):404–421
7. Calzone L, Chabrier-Rivier N, Fages F, Soliman S (2006) Machine learning biochemical networks from temporal logic properties. In: Priami C et al (eds) Transactions on computational systems biology. LNCS 4220. Springer, Berlin/Heidelberg, pp 68–94
8. Cardelli L, Wegner P (1985) On understanding types, data abstraction, and polymorphism. Comput Surv 17(4):471–522
9. Chaouiya C (2007) Petri net modelling of biological networks. Brief Bioinform 8(4):210–219
10. Ciardo G (1994) Petri nets with marking-dependent arc cardinality. In: Valette R (ed) Properties and analysis, advances in petri nets. LNCS 815. Springer, Berlin/Heidelberg, pp 179–198
11. Clarke EM, Grumberg O, Peled DA (2001) Model checking. MIT, Cambridge
12. Donaldson R, Gilbert D (2008) A model checking approach to the parameter estimation of biochemical pathways. In: Proceedings of the 6th international conference on computational methods in systems biology, Rostock. LNCS 5307. Springer, pp 269–287
13. Elowitz MB, Leibler S (2000) A synthetic oscillatory network of transcriptional regulators. Nature 403(6767):335–338
14. Franzke A (2009) Charlie 2.0 – a multi-threaded Petri net analyzer. Master's thesis, Computer Science Department, Brandenburg University of Technology Cottbus
15. Gao Q, Gilbert D, Heiner M, Liu F, Maccagnola D, Tree D (2012) Multiscale modelling and analysis of planar cell polarity in the Drosophila wing. IEEE/ACM Trans Comput Biol Bioinform 10(2):337–351
16. Gecode (2013) Gecode: an open constraint solving library. http://www.gecode.org/. Accessed on 2 Nov 2013
17. Genrich HJ, Lautenbach K (1979) The analysis of distributed systems by means of predicate/transition-nets. In: Proceedings of the international symposium on semantics of

concurrent computation, Evian. LNCS 70. Springer, pp 123–146
18. German R (2001) Performance analysis of communication systems with non-Markovian stochastic Petri nets. Wiley, New York
19. Gilbert D, Heiner M (2006) From Petri nets to differential equations – an integrative approach for biochemical network analysis. In: Proceedings of the 27th international conference on applications and theory of Petri nets and other models of concurrency, Turku. LNCS 4024. Springer, pp 181–200
20. Gilbert D, Heiner M, Liu F, Saunders N (2013) Colouring space – a coloured framework for spatial modelling in systems biology. In: Proceedings of the PETRI NETS 2013, Milan. LNCS 7927. Springer, pp 230–249
21. Gillespie DT (1977) Exact stochastic simulation of coupled chemical reactions. J Phys Chem 81(25):2340–2361
22. Heath AP, Kavraki LE (2009) Computational challenges in systems biology. Comput Sci Rev 3(1):1–17
23. Heiner M (2009) Understanding network behaviour by structured representations of transition invariants – a Petri net perspective on systems and synthetic biology. Natural computing series. Springer, Berlin/Heidelberg, pp 367–389
24. Heiner M, Gilbert D (2011) How might Petri nets enhance your systems biology toolkit. In: Proceedings of the PETRI NETS 2011, Newcastle. LNCS 6709. Springer, pp 17–37
25. Heiner M, Gilbert D (2012) Biomodel engineering for multiscale systems biology. Prog Biophys Mol Biol 111(2–3)
26. Heiner M, Koch I (2004) Petri net based system validation in systems biology. In: Proceedings of the ICATPN 2004, Bologna, June 2004. LNCS 3099. Springer, pp 216–237
27. Heiner M, Sriram K (2010) Structural analysis to determine the core of hypoxia response network. PLoS ONE 5(1):e8600
28. Heiner M, Gilbert D, Donaldson R (2008) Petri nets for systems and synthetic biology. In: Proceedings of the 8th international conference on formal methods for computational systems biology (CMSB 2008), Rostock. LNCS 5016. Springer, pp 215–264
29. Heiner M, Lehrack S, Gilbert D, Marwan W (2009) Extended stochastic Petri nets for model-based design of wetlab experiments. In: Priami C et al (eds) Transactions on computational systems biology XI. LNBI 5750. Springer, Berlin/Heidelberg, pp 138–163
30. Heiner M, Donaldson R, Gilbert D (2010) Petri nets for systems biology, chapter 3. In: Iyengar MS (ed) Symbolic systems biology: theory and methods. Jones and Bartlett Publishers, LCC, pp 61–97
31. Heiner M, Herajy M, Liu F, Rohr C, Schwarick M (2012) Snoopy – a unifying Petri net tool. In: Proceedings of the PETRI NETS 2012, Hamburg. LNCS 7347. Springer, pp 398–407
32. Heiner M, Rohr C, Schwarick M (2013) MARCIE – model checking and reachability analysis done efficiently. In: Proceedings of the PETRI NETS 2013, Milan. LNCS 7927. Springer, pp 389–399
33. Herajy M (2013) Distributed collaborative and interactive simulation of large scale biochemical networks. PhD thesis, Brandenburg University of Technology Cottbus
34. Herajy M, Heiner M (2012) Hybrid representation and simulation of stiff biochemical networks. J Nonlinear Anal Hybrid Syst 6(4):942–959
35. Hindmarsh A, Brown P, Grant K, Lee S, Serban R, Shumaker D, Woodward C (2005) Sundials: suite of nonlinear and differential/algebraic equation solvers. ACM Trans Math Softw 31(3):363–396
36. Hofestädt R (1994) A Petri net application of metabolic processes. J Syst Anal Model Simul 16:113–122
37. Jensen K (1981) Coloured Petri nets and the invariant-method. Theor Comput Sci 14(3): 317–336
38. Liu F (2012) Colored Petri nets for systems biology. PhD thesis, Brandenburg University of Technology Cottbus
39. Liu F, Heiner M, Rohr C (2012) Manual for colored Petri nets in Snoopy. http://www-dssz.informatik.tu-cottbus.de/publications/btu-reports/Manual_for_colored_Petri_nets_2012_03.pdf

40. Lund EW (1965) Guldberg and Waage and the law of mass action. J Chem Educ 42(10):548
41. Marsan MA, Balbo G, Conte G, Donatelli S, Franceschinis G (1995) Modelling with generalized stochastic Petri nets. Wiley series in parallel computing. Wiley, New York
42. Marwan W, Rohr C, Heiner M (2012) Petri nets in Snoopy: a unifying framework for the graphical display, computational modelling, and simulation of bacterial regulatory networks. Volume 804 of Methods in molecular biology, chapter 21. Humana Press, pp 409–437
43. Matlab (2013) http://www.mathworks.cn
44. Murata T (1989) Petri nets: properties, analysis and applications. Proc IEEE 77(4):541–580
45. Parker D (2002) Implementation of symbolic model checking for probabilistic systems. PhD thesis, University of Birmingham
46. Petri CA (1962) Kommunikation mit Automaten. PhD thesis, Institut für Instrumentelle Mathematik, Schriften des IIM Nr. 2, Bonn
47. Pnueli A (1981) The temporal semantics of concurrent programs. Theor Comput Sci 13(1): 45–60
48. Reddy VN, Mavrovouniotis ML, Liebman MN (1993) Petri net representations in metabolic pathways. In: Proceedings of the 1st international conference on intelligent systems for molecular biology, Bethesda. AAAI, pp 328–336
49. Rohr C, Marwan W, Heiner M (2010) Snoopy – a unifying Petri net framework to investigate biomolecular networks. Bioinformatics 26(7):974–975
50. Schwarick M, Tovchigrechko A (2010) IDD-based model validation of biochemical networks. Theor Comput Sci 412(26)
51. Schwarick M, Rohr C, Heiner M (2011) MARCIE – model checking and reachability analysis done efficiently. In: Proceedings of the 8th international conference on quantitative evaluation of systems, Aachen. IEEE, pp 91–100
52. Segel LA (1980) Mathematical models in molecular cellular biology. Cambridge University Press, Cambridge
53. Soliman S, Heiner M (2010) A unique transformation from ordinary differential equations to reaction networks. PLoS ONE 5(12):e14284
54. Stewart W (1994) Introduction to the numerical solution of Markov chains. Princeton University Press, Princeton
55. Tsang EPK (1993) Foundations of constraint satisfaction. Academic, London/San Diego
56. Wegener J, Schwarick M, Heiner M (2011) A plugin system for Charlie. In: Proceedings of the international workshop on concurrency, specification, and programming (CS&P 2011), Pułtusk. Białystok University of Technology, pp 531–554. ISBN:978-83-62582-06-8

Part IV
BioData Mapping

Chapter 10
Network Analysis and Integration in a Virtual Cell Environment

Björn Sommer

Abstract Integrative Bioinformatics combines a number of different disciplines related to biology as well as informatics. One major target of this research area is the creation of a virtual cell. Naturally, this topic is accompanied by a vast amount of problems which arise due to the fact that a large number of highly specific disciplines have to be addressed. In this publication a subproblem will be discussed, functional cell modeling. Beginning with a virtual cell environment which provides cell components featuring different subcompartmental layers, protein-related networks will be localized and visualized. For the localization, a data warehouse will be accessed. Special interactive techniques will be applied to the semiautomatic analysis of localization entries found in databases. Finally, different visualization approaches will be shown, and 2D and 3D network visualization will be discussed, as well as quantitative illustrations using charts.

10.1 Introduction

One of the most demanding tasks of today's scientific community is revealing the secrets of the biological cell. Three aspects make it nearly impossible to accomplish this task: (1) the unimaginable complexity of the cell, (2) its extremely small scale, and (3) the amount of differing disciplines to unite for this task.

The modeling and simulation of a virtual cell is one important approach to understanding the functioning of the cell. In the past, approaches like VCell and E-Cell were developed which mathematically simulated cells by using differential

---

B. Sommer (✉)
Bielefeld University, Universitätsstraße 25, 33615 Bielefeld, Germany
e-mail: bjoern@CELLmicrocosmos.org

equations [15, 24]. Naturally, this mathematical simulations address only quantitative simulations ignoring molecular interactions. Atomistic simulations using, e.g., molecular dynamic techniques are able to simulate small fragments of a cell like a membrane patch or vesicles [4, 5, 10]. However, it is not possible to simulate an entire cell taking molecular interactions into account. Alternatively, many different cell visualization approaches were developed in recent years, such as The Interactorium, MetNetVR, or Meta!Blast [26–28]. They visualize cells at the mesoscopic scale, where cell components can be differentiated but not molecular structures.

10.1.1 *CELLmicrocosmos PathwayIntegration*

This book is devoted to Integrative Bioinformatics which – as was discussed in the preceding chapters – comprises a number of different disciplines uniting biology and informatics. In this chapter a set of these disciplines will be applied to correlate a virtual cell environment with metabolic networks. In this way the aforementioned cellular mesoscopic level is combined with the functional level. For this purpose, the CELLmicrocosmos 4.2 PathwayIntegration (CmPI) will be used, a cell modeling and visualization environment [20, 21].

The mesoscopic level is represented by a set of cell components, such as the cell membrane, the cytosol, and the extracellular matrix, the mitochondrion, and the nucleus. These cell components are represented by three-dimensional models as well as by color codes as shown in Fig. 10.5. These contrasting color codes are based on the color alphabet developed by Green-Armytage, enabling a good visual differentiation [9, 20].

The functional level is covered by two well-known metabolic pathways: the citrate cycle and the glycolysis. Both pathways are interrelated, because the glycolysis generates pyruvate which is needed to initiate the citrate cycle. An important fact for subsequent analysis is that the localization of both pathways is well known. The cytosol – the intracellular fluid surrounding all membrane-based organelles – is the reaction chamber of the glycolysis. After a number of reactions, the final product of the glycolysis, the pyruvate, is transported by a specific protein through the inner mitochondrial membrane to the mitochondrial matrix. There, the pyruvate is oxidatively decarboxylized by the pyruvate dehydrogenase complex, resulting in the product acetyl CoA. This compound enters the citrate cycle which is also located in the mitochondrial matrix. The final result of the citrate cycle is the citrate [2, 6].

The following sections will discuss a workflow combining the mesoscopic and the functional level by using Integrative Bioinformatics techniques.

10.2 Relevant Databases

10.2.1 Metabolic Pathways from KEGG

First of all, the metabolic pathways have to be acquired from an electronical source. One of the most acknowledged biochemical databases and one of the most frequently used sources in Bioinformatics is the Kyoto Encyclopedia of Genes and Genomes (KEGG) [14]. It includes genomic, chemical, as wells as systemic functional information, and it is available at http://www.kegg.jp/kegg/.

KEGG is partially freely, partly commercially available and has been developed in the Kanehisa Laboratories of Kyoto University. It is well known for its two-dimensional interlinked pathway maps. Figure 10.1 shows the human citrate cycle. The KEGG identifier is "hsa00020," where "hsa" is an abbreviation for the organism (Homo sapiens) and 00020 is the KEGG-internal number of the pathway map.

KEGG contains a large number of different eukaryotic and prokaryotic organisms which are all linked to different versions of the same pathway. The enzymes

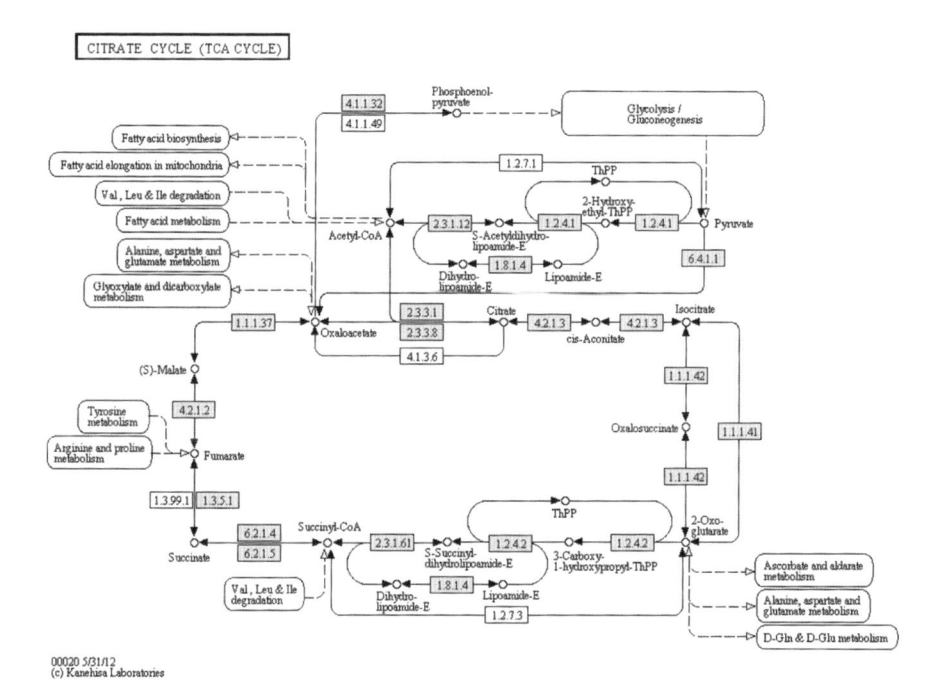

Fig. 10.1 KEGG: the citrate cycle pathway of Homo sapiens, hsa00020 (Courtesy of/Copyright 2013 by Kanehisa Laboratories, source: http://www.kegg.jp. Reprinted with permission from [14])

in Fig. 10.1 are coded following the EC standard discussed in Sect. 10.2.2.2. The enzymes are connected to their products and substrates which are named by their commonly known identifiers. The connections symbolize the reactions and the reactions' directions are indicated by arrows. All other elements are interrelated metabolic pathways. Here, the connection to the glycolysis can also be found.

Alternatively, it is possible to load externally created pathways into CmPI. For this purpose, a SBML import has been integrated [12]. Biological network reconstruction tools like VANESA (see Chap. 8: Biological Network Modeling and Analysis) can be used to model networks which are then imported to CmPI and localized in a virtual cell environment [13].

10.2.2 Protein Localization Databases

The digital data describing metabolic pathways is now available, and now the question arises, how can this structure – visualized as a two-dimensional image as was seen in Fig. 10.1 – be combined with the spatial structure of a cell? Because KEGG does not provide information about the localization of the networks, alternative sources have to be accessed. These sources will be discussed in this section. Four databases will be introduced which can be applied to this problem.

10.2.2.1 Reactome

First of all, the Reactome databases should be introduced. It is developed by the European Bioinformatics Institute (EBI) and different American institutes. It is a freely available, curated Open Source project. Similar to KEGG, it contains a large variety of different pathways and of course also a number of metabolic pathways. The major focus lies on the human organism. Expert users may integrate experimental data into Reactome. For each protein complex part, specific pathway localization information is available [7]. The database is found at http://www.reactome.org.

10.2.2.2 BRENDA and the Enzyme Classification

In contrast to the previous databases, the one following does not contain pathway maps. BRENDA (BRaunschweig ENzyme Database) is developed and curated at the TU Braunschweig. It can be freely accessed at: http://www.brenda-enzymes.org

A commercial version of BRENDA also exists which contains additional current information. In BRENDA, the user finds functional structural and property-related data which is mainly based on manually annotated references from primary literature [19].

The classification of the different enzyme types – which was already shown in Sect. 10.2.1 – follows the Enzyme Commission number (EC) classification, which basically consists of four numbers subdivided by a period [25]. The first number codes, e.g., (1) oxidoreductases, (2) transferases, (3) hydrolases, (4) lyases, (5) isomerases, and (6) ligases.

It is important to note that the EC numbers do not usually describe one particular protein. Instead, it applies to a number of different proteins which meet the criteria of the specific EC definition. Therefore, a single EC number may describe a large set of different proteins. The BRENDA databases link many EC numbers to specific databases based on manually curated literature. Of course, the aforementioned problem applies also to the localization, because it is not linked to a specific protein, but a protein family.

10.2.2.3 UniProt

In contrast to BRENDA, UniProt contains information linked to specific proteins. It is the freely accessible and regularly updated universal protein database for curated as well as automatic acquired data. UniProt is a collaboration between the European Bioinformatics Institute (EBI), the Protein Information Resource (PIR), and the Swiss Institute of Bioinformatics (SIB). It is linked to various external databases, such as BRENDA, Gene Ontology, and Reactome [8]. It is available at http://www.UniProt.org.

UniProt contains different sub-databases, but for the localization of proteins, this work will focus at the UniProt Knowledgebase (UniProtKB). The website and the database contain a number of categories holding localization information, for example:

- General annotation (comments)
  - Subcellular location
- Ontologies
  - Keywords
    · Cellular component
- Gene Ontology
  - Cellular component

The terms found in these categories may be a concrete cell component, an intracompartmental location, or a sentence describing location-related facts.

10.2.2.4 The Gene Ontology and the Redundancy of Terms

A problem which arises when dealing with the large variety of different components of a cell is redundancy. Using the different databases previously introduced, a large number of different terms may describe the same entity. Just to give an example, an excerpt of different terms provided by the databases BRENDA, UniProt, and Reactome will now be listed for the term "plasma membrane."

- BRENDA
    - Cellular component
    - Cell membrane
    - Plasma membrane
    - Cytoplasmic membrane
    - Cell outer membrane

- UniProt
    - Associated with the synaptic plasma membrane (by similarity)
    - Integral to plasma membrane
    - Intrinsic to internal side of plasma membrane
    - Localized on the cell surface

- Reactome
    - Integrin cell surface interactions

Obviously, each of these terms is associated with the plasma membrane. But it also can be seen that each term contains additional information which might be relevant in different contexts.

For a long time, the Gene Ontology (GO) has been addressing this problem [1, 3]. This database contains gene-related protein information in conjunction with structured, controlled vocabularies. The ontologies contain all these terms and link them to the so-called GO-terms which are a quasi standard in the Bioinformatics community. GO is located at http://www.geneontology.org.

Now, the GO vocabularies should be examined by looking again at the major term "plasma membrane" which has the GO identifier "GO:005886." Directly correlated to this term is each of the three terms "cell membrane," "plasma membrane," and "cytoplasmic membrane."

But the other previously listed terms are more specific and are linked to the following GO-terms:

- GO:0005887: integral to plasma membrane
- GO:0009279: cell outer membrane
- GO:0031235: intrinsic to internal side of plasma membrane

Now the question arises, how are these terms connected to the previously mentioned term GO:005886 representing the plasma membrane? These hierarchical dependencies are also addressed by GO, as can be seen in Fig. 10.2 showing the GO Graph View for GO:0005887. The terms GO:005886 and GO:0031235 are also contained and show spatial interdependencies: "integral to plasma membrane" → "intrinsic to plasma membrane" (→ "plasma membrane part") → "plasma membrane."

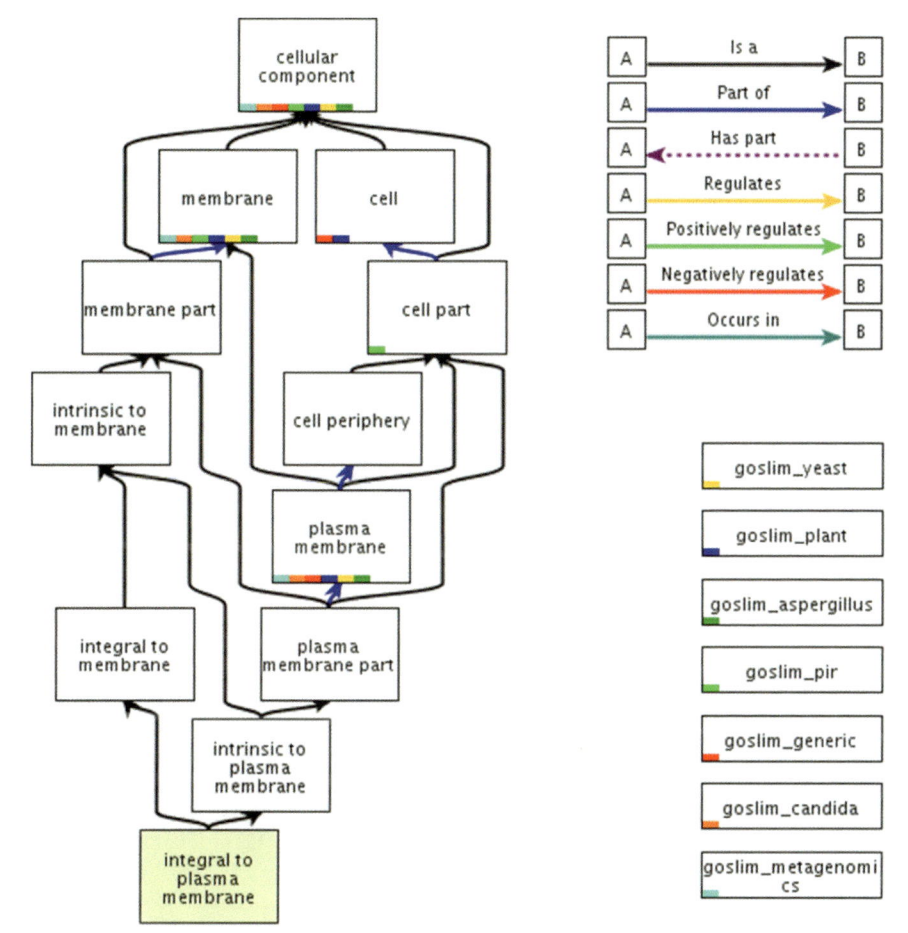

Fig. 10.2 Gene Ontology: graph view for the GO-term "integral to plasma membrane" (Courtesy of/Copyright 2013 by The Gene Ontology, AmiGO version 1.8, http://amigo.geneontology.org. Reprinted with permission from [1, 3])

But it also has to be mentioned that currently no GO identifiers were found for the following entries:

- Localized on the cell surface
- Associated with the synaptic plasma membrane (by similarity)
- Integrin cell surface interactions

The problem which applies to all databases discussed here is that they have to be continuously curated and extended.

10.2.3 DAWIS-M.D.

CmPI uses all previously discussed databases to solve the following application case. To enable the fast access of this large amount of diverse data, all databases were integrated in a data warehouse. The used data warehouse called DAWIS-M.D. was already extensively discussed in Chap. 4: Data Warehouses in Bioinformatics [11].

10.2.4 ANDCell

The previously mentioned databases contain a large amount of curated data. In many cases there is no data found for a specific localization in these databases. But if the user searches PubMed via its web interface, a publication might be found which was published in the most recent year. Of course, this information will usually find the way into the previously mentioned databases within the next new releases. But sometimes this will take some time.

Therefore, an alternative way is needed to acquire this information directly from PubMed. For this purpose, text mining is an appropriate approach. CmPI uses the ANDCell database for this purpose [18]. This database was already discussed in Chap. 6: Text Mining on PubMed.

10.3 Localizing Metabolic Pathways Using Integrative Bioinformatics Techniques

Now that the basics has been discussed, the correlation of the mesoscopic and the functional level has to be addressed by using CmPI. First of all, the metabolic pathways will be downloaded, and then these pathways will be correlated with a cellular environment by using the localization databases.

10.3.1 Downloading the Citrate Cycle and the Glycolysis

CmPI connects to DAWIS-M.D. to download the two pathways hsa00010, the glycolysis, and hsa00020, the citrate cycle. Figure 10.3 shows the resulting two-dimensional visualization of hsa00010 in the 2D viewer of CmPI and Fig. 10.4 hsa00020. Both layouts are based on the original KEGG layouts, the so-called KGML (KEGG Markup Language) pathway maps. These pathway maps are well known from the KEGG website which provides images for each pathway. If comparing the original KGML layouts (Fig. 10.1) from the websites with the 2D visualization of CmPI (Fig. 10.4), an important difference will be noted. The KGML layouts often contain multiple instances of a protein. For example, the original

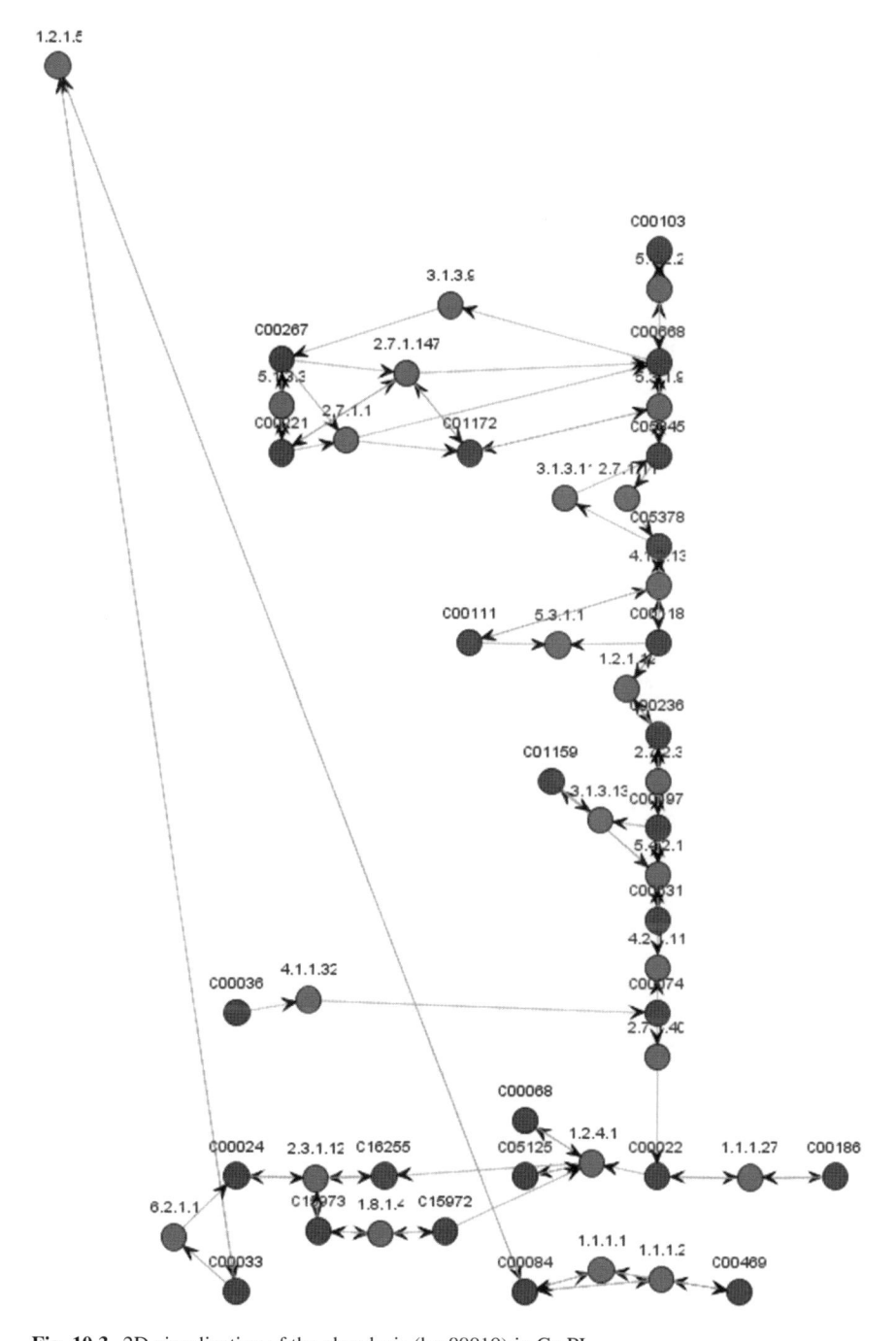

10 Network Analysis and Integration in a Virtual Cell Environment

Fig. 10.3 2D visualization of the glycolysis (hsa00010) in CmPI

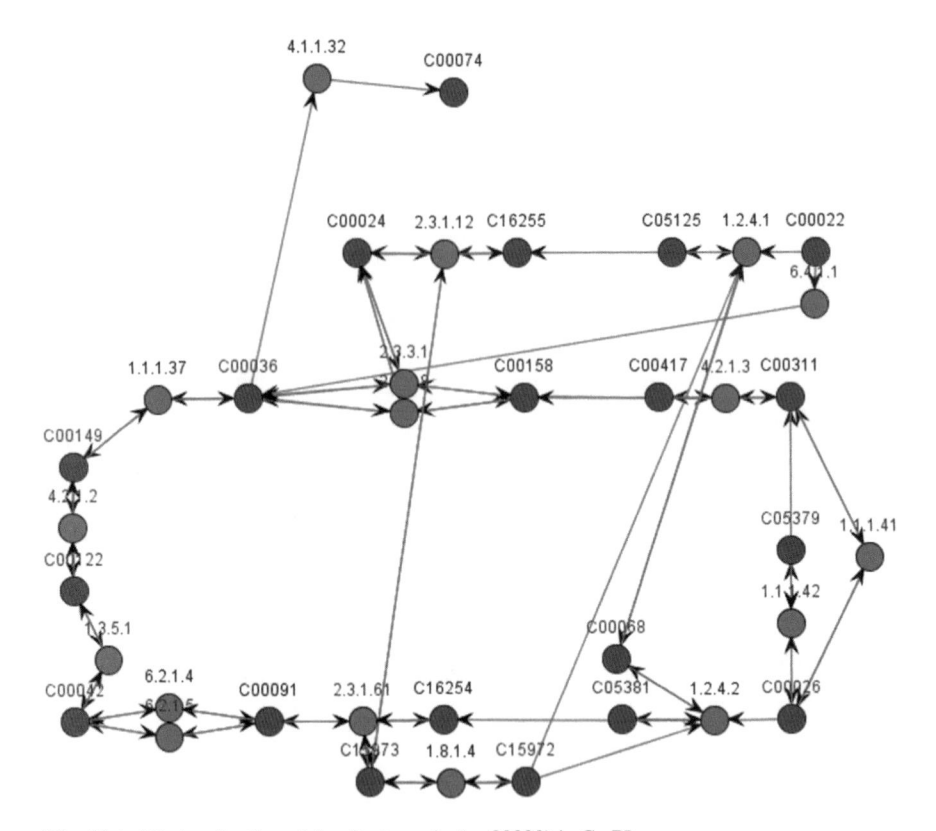

Fig. 10.4 2D visualization of the citrate cycle (hsa00020) in CmPI

hsa00020 map contains the compound C15973 twice. For CmPI, this is not possible, because it always contains a single distinct instance of a distinct protein. The reason is the localization-focused view of CmPI: each instance in a pathway has a unique position. C15973 is connected to the enzyme 2.3.1.61 on the bottom and 2.3.1.12 on the top. In Fig. 10.3, representing the 2D visualization of CmPI, there is only a single instance of compound C15973. Of course, all connections are still known.

10.3.2 First Localization Results

In Sect. 10.2.1 the localization of the citrate cycle and the glycolysis have been already discussed; basically, the glycolysis is located at the cytosol and the citrate cycle at the mitochondrion. Now, the question should be evaluated, if it is possible to reproduce the localization of these two pathways by using the results from the databases and by ignoring the previously mentioned advance information. If this is possible, it can be stated that CmPI may be used to analyze the localization of protein-related data sets where the localization is not known.

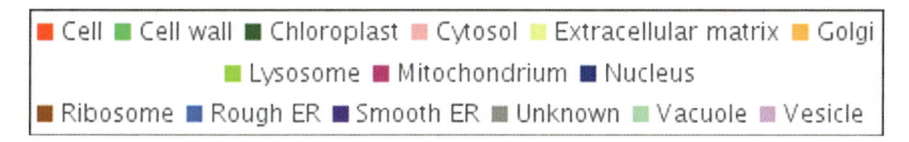

10 Network Analysis and Integration in a Virtual Cell Environment

Fig. 10.5 Color codes for all cell components coded in the following figures

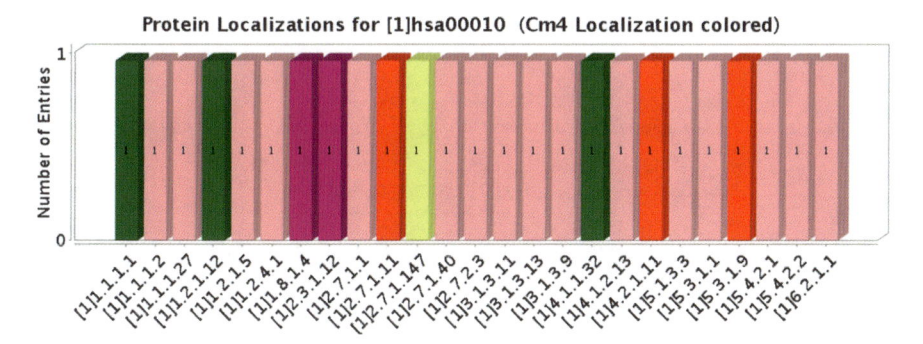

Fig. 10.6 Initial Localization Chart, category "Localizations/Cm4" for hsa00010

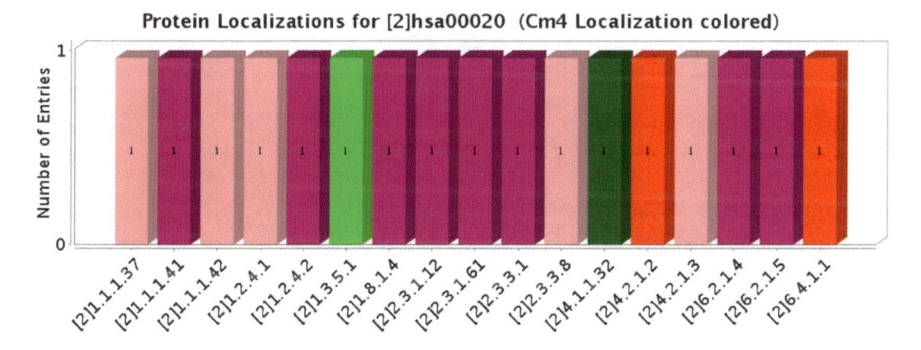

Fig. 10.7 Initial Localization Chart, category "Localizations/Cm4" for hsa00020

First, the downloaded metabolic pathways are localized by using the connection between CmPI and DAWIS-M.D. For this purpose, the EC identifier of each protein is send to the localization databases combined with the information that the organism of interest is "Homo sapiens." Otherwise, the localization results will contain a lot of information from other organisms. Now, a number of diverse localizations are retrieved, as shown in Fig. 10.5 which also lists the color codes used during the following analysis.

To analyze and filter this data, CmPI provides a special visualization, the Subcellular Localization Charts.

Figures 10.6 and 10.7 show the initially assigned localizations by CmPI which automatically selects the first entry in the alphabetically ordered list downloaded

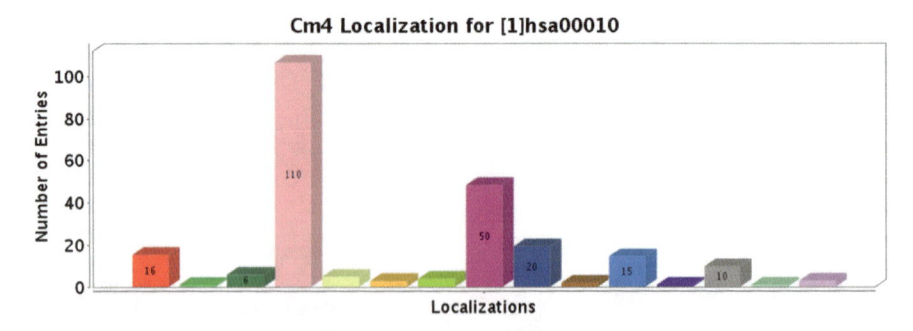

Fig. 10.8 Initial Localization Chart, category "Protein Localizations/Cm4" for hsa00010

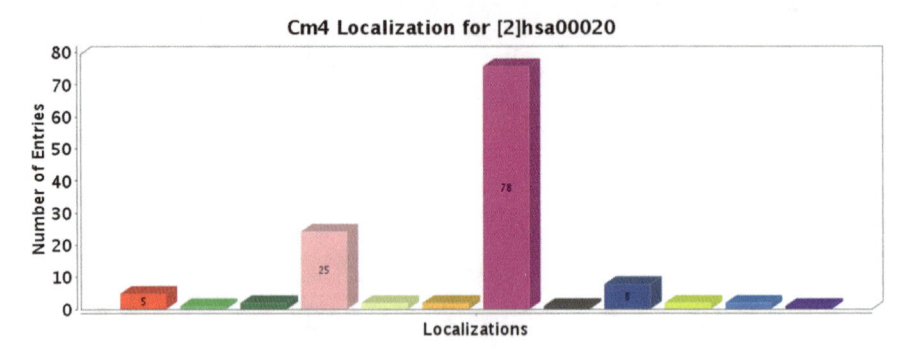

Fig. 10.9 Initial Localization Chart, category "Protein Localizations/Cm4/" for hsa00020

from the databases as the potential localizations for each protein. Of course, the result does not meet the expectations given by the literature. The glycolysis and the citrate cycle each show five different localizations. For example, the result "chloroplast" does not meet the expectations of a Homo sapiens-related cell. It will be shown that only a few clicks are needed to assign the correct localizations.

10.3.3 Investigating the Preliminary Localizations

Now the Localization Charts will be used to investigate all localization entries found in the databases. Each entry is one single result from one of the five localization databases pointing to one single localization. Therefore, a single database like UniProt usually provides multiple localization entries for a single protein. All localization entries for glycolysis is shown in Fig. 10.8 and the citrate cycle in Fig. 10.9.

And obviously, recalling the initial expectations for both pathways seem to be confirmed by a first glance at the images. While the enzymes of hsa00010 are mainly localized at the cytosol, those of the hsa00020 are concentrated at the mitochondrion. Moreover, it is interesting to note that the localization with the

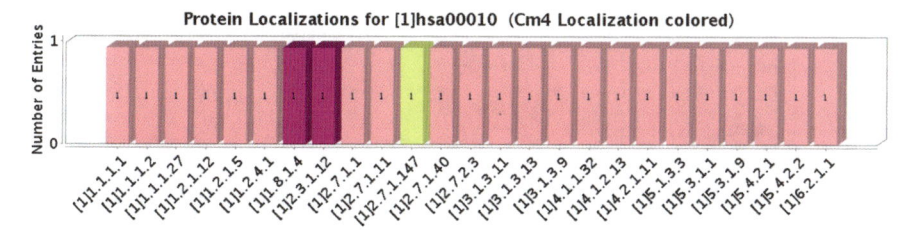

10 Network Analysis and Integration in a Virtual Cell Environment

Fig. 10.10 Localization Chart, category "Protein Localizations/Cm4" after assigning the localization "cytosol" to hsa00010

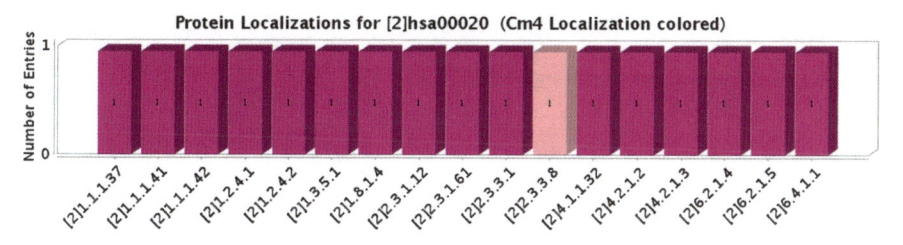

Fig. 10.11 Localization Chart, category "Protein Localizations/Cm4" after assigning the localization "mitochondria" to hsa00020

second highest incidence is – in both cases – the localization with the highest priority of the pathway vis-a-vis.

These images give a first idea of a localization, but it is very important to be aware of the fact that the shown localizations apply to the complete pathway. This visualization does not show any information about the single proteins. Therefore, it might be that 50 % of the proteins cannot be localized to the localization confirmed by the literature.

Using the Localization Charts, there are different ways to find a solution. Here, a method will be chosen which quickly leads to the final localization.

A double-click is performed on each of the bars representing the dominant localizations in Fig. 10.8, the cytosol, and Fig. 10.9, the mitochondrion. This action is already sufficient to assign most of the proteins to the correct localizations for both pathways, as can be seen by the resulting Localization Charts in Figs. 10.10 and 10.11.

Twenty two of twenty five enzymes were localized at the cytosol for the glycolysis and 16 of 17 enzymes at the mitochondrion for the citrate cycle. Therefore, the initial assumption was verified, but in addition to this, it is also possible to interpret more. Glycolysis and citrate cycle interact, but by checking the textbook, it is not known which concrete proteins are involved in the transition between cytosol and the mitochondrion. By looking at Fig. 10.10, it is possible to directly identify these enzymes; 1.8.1.4 and 2.3.1.12 are part of both pathways, and the only localization in the context of these two pathways is the mitochondrion. In addition, the enzyme 2.3.3.8, which here is only part of the citrate cycle, seems to be an enzyme not localized in the mitochondrion but in the cytosol.

2	1.8.1.4	mitochondrial matrix | ...	mitochondrial matrix	1/5 Matrix	REACTOME: by G.O. (1/5)
2	1.2.4.2	mitochondrial inner me...	mitochondrial inner membrane	2/5 Inner Mem...	BRENDA: Reviewed (1/7)
2	2.3.1.61	mitochondrial matrix | ...	mitochondrial matrix	1/5 Matrix	REACTOME: by G.O. (1/5)
2	6.2.1.5	mitochondrial chromos...	mitochondrial chromosome	1/5 Matrix	ANDCell: PubMed (1/5)
2	6.2.1.4	mitochondrial chromos...	mitochondrial chromosome	1/5 Matrix	ANDCell: PubMed (1/4)
2	1.1.1.42	mitochondrial matrix | ...	mitochondrial matrix	1/5 Matrix	REACTOME: by G.O. (4/9)
2	1.1.1.41	mitochondrial matrix | ...	mitochondrial matrix	1/5 Matrix	REACTOME: by G.O. (1/8)
2	2.3.3.8	citrate lyase complex | H...	citrate lyase complex	1/1 Cytosol	GO: TAS:ProtInc (1/7)
2	1.3.5.1	mitochondrial respirator...	mitochondrial respiratory cha...	2/5 Inner Mem...	GO: ISS:UniProtKB (3/9)
2	4.2.1.2	mitochondrial matrix | ...	mitochondrial matrix	1/5 Matrix	REACTOME: by G.O. (4/5)
2	4.2.1.3	mitochondrial chromos...	mitochondrial chromosome	1/5 Matrix	ANDCell: PubMed (10/20)
2	2.3.3.1	mitochondrial matrix | ...	mitochondrial matrix	1/5 Matrix	GO: IDA:UniProtKB (1/3)
2	1.1.1.37	mitochondrial matrix | ...	mitochondrial matrix	1/5 Matrix	REACTOME: by G.O. (4/7)
2	6.4.1.1	mitochondrial matrix | ...	mitochondrial matrix	1/5 Matrix	REACTOME: by G.O. (3/6)
2	1.2.4.1	mitochondrial matrix | ...	mitochondrial matrix	1/5 Matrix	GO: IEA:UniProtKB-SubCell (2/12)
2	2.3.1.12	mitochondrial pyruvate ...	mitochondrial pyruvate dehy...	1/5 Matrix	GO: NAS:UniProtKB (1/7)
2	4.1.1.32	mitochondrial matrix | ...	mitochondrial matrix	1/5 Matrix	REACTOME: by G.O. (8/10)

Fig. 10.12 An excerpt of the Localization Table showing only hsa00020

Now, a more accurate examination of the different localizations should be done by looking at the Localization Terms. The Localization Terms describe the concrete entries downloaded from the databases which were used to map them to the localizations inside CmPI. For example, the terms listed in Sect. 10.2.2.4 – "associated with the synaptic plasma membrane (by similarity)," "integral to plasma membrane," and "intrinsic to internal side of plasma membrane" – each are mapped to the Cm4 Localization, the cell membrane. The Localization Term used to map 2.3.3.8 to the cytosol is the "citrate lyase complex." By examining Fig. 10.4, it can be seen that 2.3.3.8 is directly involved in the generation of the citrate (C00158) which is processed in the citrate lyase complex converting citrate to oxaloacetate [2, 6].

Next, the accuracy of the Membrane Localization of the citrate cycle should be verified by looking at the Localization Table part of CmPI in Fig. 10.12. This figure shows an excerpt of the Localization Table showing all citrate cycle-associated enzymes. At a first glance it can be seen that most of the enzymes were correctly localized to the mitochondrial matrix. But for the localization of the enzymes 1.2.4.2 and 1.3.5.1, the mitochondrial inner membrane was selected. The Localization Table shown in Fig. 10.12 is interactive. Therefore, all found localizations can also be found here by clicking the corresponding entry. For 1.3.5.1, an alternative option to the mitochondrial matrix is only the outer membrane of the mitochondrion. Therefore, the selection of the inner membrane seems to be correct. In contrast to this, the enzyme 1.2.4.2 shows also an entry for the matrix which can be directly selected in the Localization Table.

10.3.4 Examining an Outsider by Direct Access to External Sources

Because the glycolysis is localized at the cytosol, Membrane Localizations as those discussed for the citrate cycle (e.g., mitochondrial matrix or mitochondrial inner

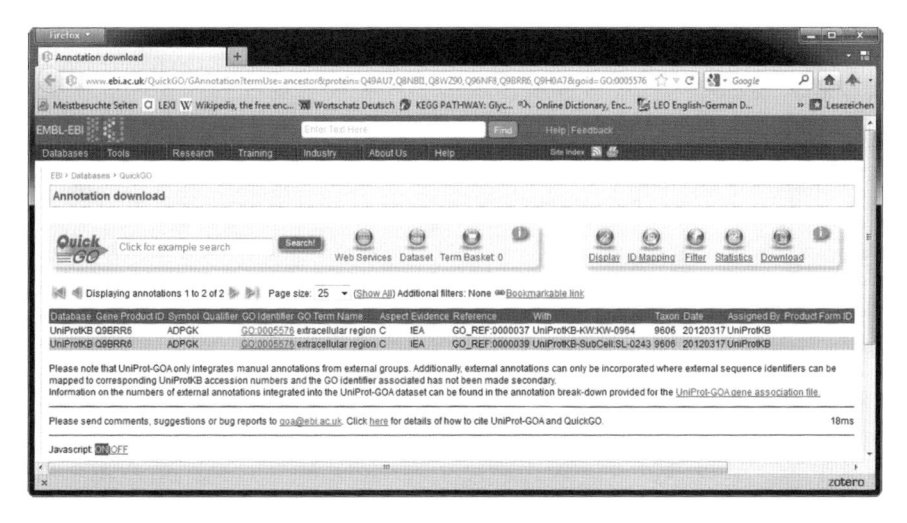

10 Network Analysis and Integration in a Virtual Cell Environment

Fig. 10.13 The link to 2.7.1.147 from the Localization Table in CmPI; it shows additional localization information

membrane) are not of relevance. But still there is one outsider enzyme – 2.7.1.147 – because this enzyme was localized at the extracellular matrix by the Localization Term "extracellular region." By using the Localization Table, the localization can be examined with respect to its source. The database which provided this result was GO based on the Localization Term "Inferred from electronic annotation. Source: UniProtKB-SubCell." The Localization Table can also be used to click on the provided link which directly opens a web browser with the address "http://www.ebi.ac.uk/QuickGO/." Here, the link to the UniProt entry Q9BRR6 is shown, based on two references (Fig. 10.13). Examining the UniProt entry shows that this enzyme is also involved in the glycolysis. Concluding these observations there are two potential reasons why the enzyme 2.7.1.147 was localized at extracellular matrix:

1. The enzyme 2.7.1.147 is in fact located in the extracellular matrix during the involvement in the glycolysis.
2. Due to the fact that the distance between the extracellular matrix and the cytosol is too large, the second option is more probable. There is currently no experimental proof available in the databases proving that this enzyme is located in the cytosol.

In both cases it can be predicted that 2.7.1.147 will most probably be located at the cytosol during the glycolysis, because all interacting enzymes are also found in this cell component.

10.3.5 Localization Result

Finally, it can be stated that the preliminary Localization Charts discussed in Sect. 10.3.2 are visually equal to the final result. But two changes according to the Membrane Localizations of two enzymes have to be done. The resulting localization priority list would look like this:

1. Mitochondrial matrix
2. Mitochondrial inner membrane
3. Cytosol
4. Extracellular space

Therefore, it was shown that nearly no foreknowledge would have been needed to predict the localization of these two pathways by using CmPI.

10.3.6 3D Visualization

Of course, the Subcellular Localization Charts of CmPI can be used to analyze protein-related data sets without the intention to generate a virtual cell environment. However, for a number of application cases, it will be relevant to visualize the networks in correlation with a cell model.

In the previous sections, it was discussed how the databases are queried to get (1) the metabolic pathways (the glycolysis and the citrate cycle) and (2) to localize these pathways. Now, the networks have to be correlated with the cell components inside the cell model. For this purpose, a geodesic layout is combined with an Inverted Self-Organizing Map (ISOM) layout [17] and then mapped onto the surface of the cell components [20, 21].

This process is subdivided into three steps. First, the nodes – representing the enzymes, substrates, and products – are distributed onto the surface of a unit sphere by using the geodesic layout [21]. By doing so, the layout tries to achieve a node distribution where the distance between a given node and its neighboring nodes is equal. In the second step, this initial layout is used to apply the ISOM layout [17]. This layout moves interconnected nodes – these are nodes connected by a reaction – closer to each other, whereas those nodes without interconnections try to move away from each other.

After the first two steps have been accomplished, the resulting layout is shown in Fig. 10.14. This image shows an abstract SphereCell containing only cell components relevant for the final localizations (from inside to outside): mitochondrion, cytosol, cell membrane, and extracellular matrix [20]. The applied contrast color-coding was also used for the enzyme localization and is described in Fig. 10.5.

The third step of the layout process is the mapping of the nodes onto the surface of the cell components. This process is not needed in the case of the SphereCell, because the layout is directly applied to its spherical cell components. But of course, this extremely simplified representation of a cell is usually not sufficient, because

10   Network Analysis and Integration in a Virtual Cell Environment   291

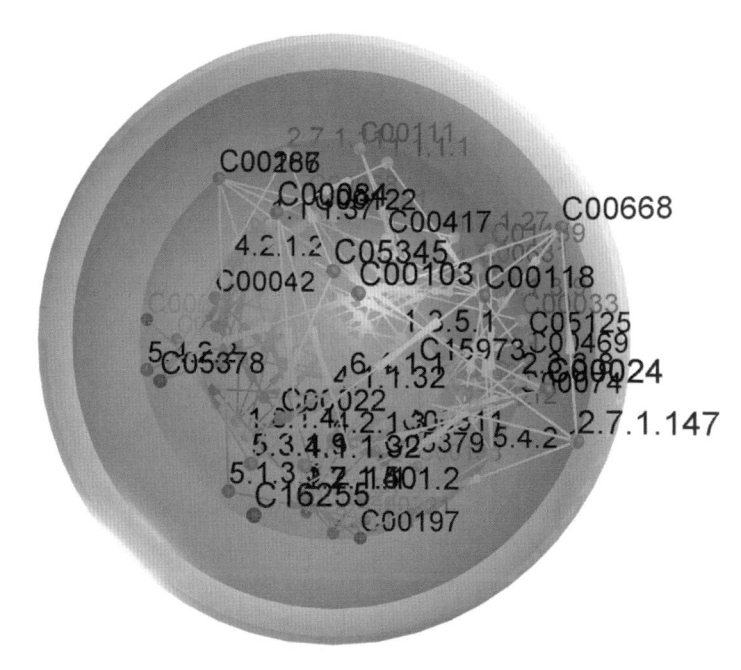

Fig. 10.14 An abstract SphereCell correlated with the citrate cycle and glycolysis (color figure online)

the cell components' structure is not shown as well as the hierarchical structure of the cell, which also is not represented correctly. In Fig. 10.15 the same layouts are used with an animal cell model. And here, the third step has to be applied. The layouted metabolic pathway has to be mapped onto the three-dimensional shapes of the cell components [20]. The nodes which are visually not located at a specific cell component are associated to the cytosol which is represented in the animal cell model as an invisible structure.

Figure 10.16 shows an excerpt of the animal cell model, the mitochondrion, including the largest part of the citrate cycle. The mitochondrion model is based on a tomographic data set derived from the Cell Centered Database [16, 29]. Here, the correlation of functional data with microscopic data is shown. It can be seen that mostly all enzymes are located in the matrix region. Moreover, examining the position of the 1.3.5.1 in Fig. 10.15 shows that it is correctly placed at the shape of the inner mitochondrial membrane (see also Fig. 10.12).

Of course it has to be mentioned that the two-dimensional projections shown here are not able to compete with an interactive three-dimensional visualization where the user is able to use the different CELLmicrocosmos navigation methods. For example, it is possible to navigate through the cell directly in the 3D viewer, but the 3D view can also be moved to the corresponding enzymes by clicking on their entries in the Localization Table (Fig. 10.12) or onto the nodes in the 2D viewer (Figs. 10.3 and 10.4).

Finally, three alternative ways to visualize the localizations will be shown. Figures 10.17 and 10.18 show the final localizations of the metabolic pathways

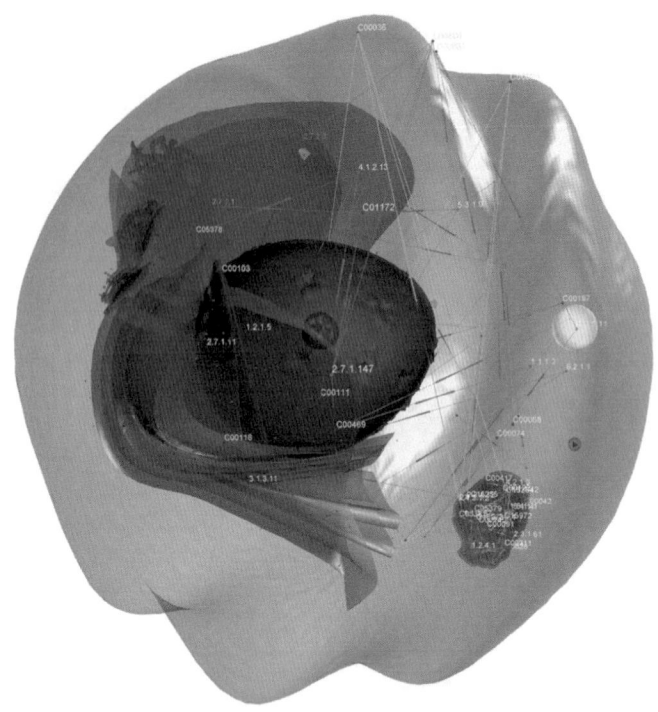

Fig. 10.15 An animal cell model correlated with the citrate cycle and glycolysis (color figure online)

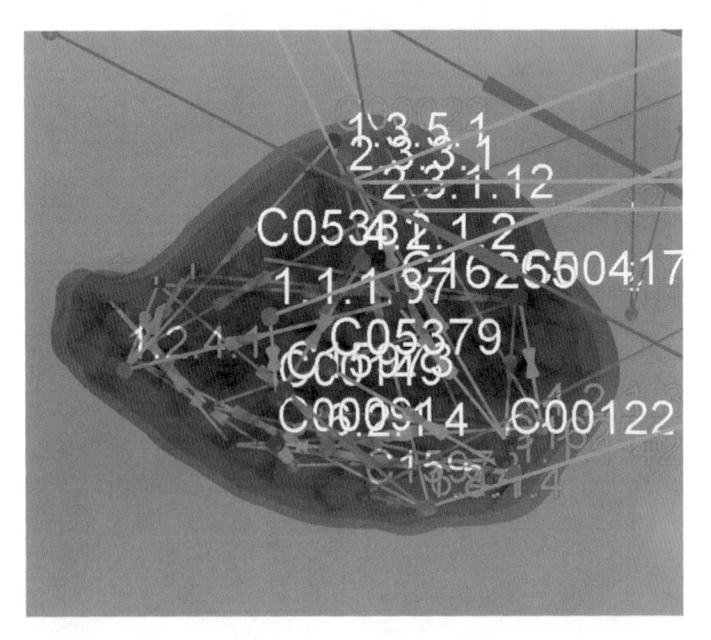

Fig. 10.16 A mitochondrion model based on microscopic data correlated with the citrate cycle and glycolysis (color figure online)

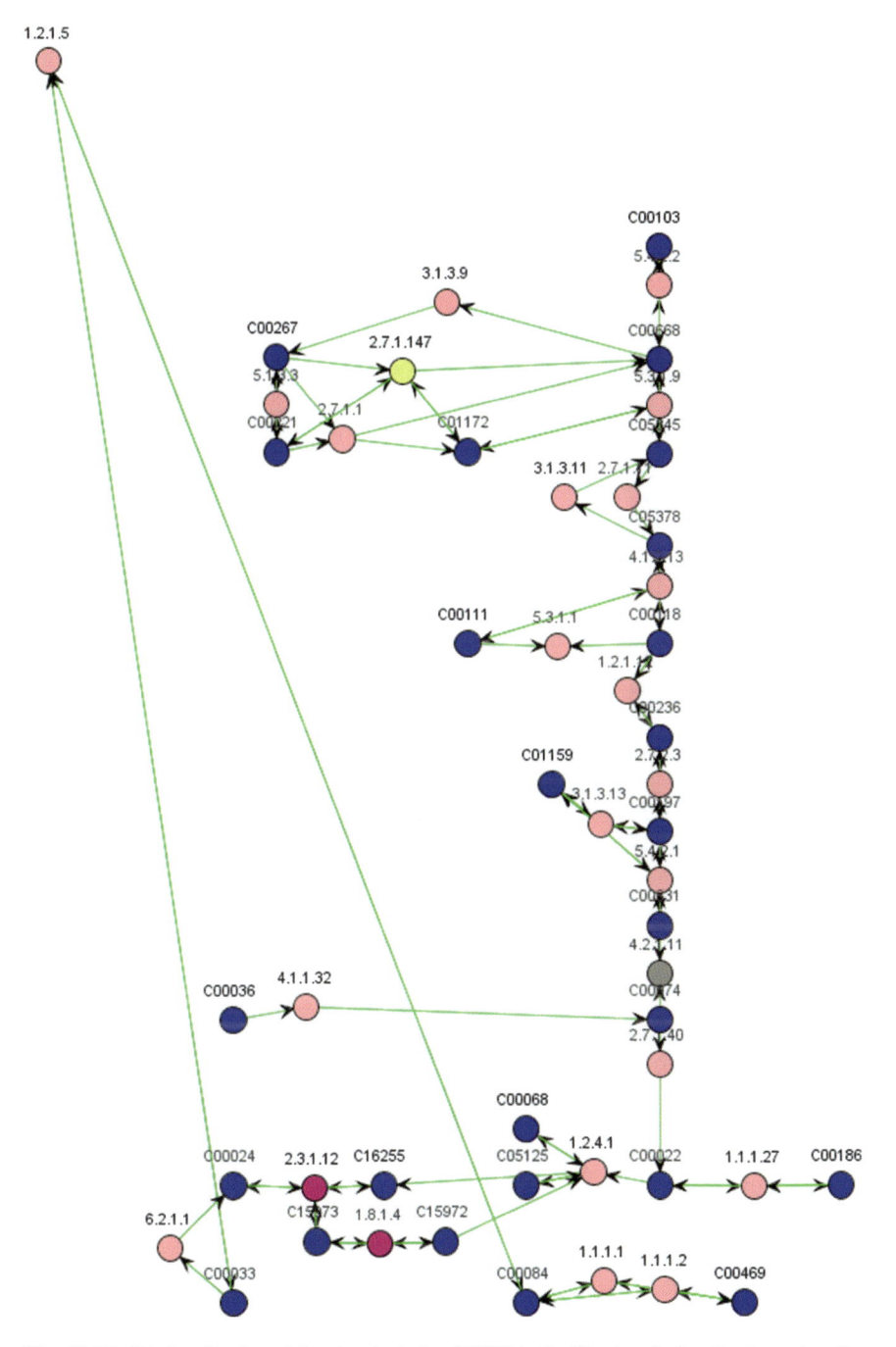

Fig. 10.17 2D visualization of the glycolysis (hsa00010) in CmPI using the localization colors for the enzymes

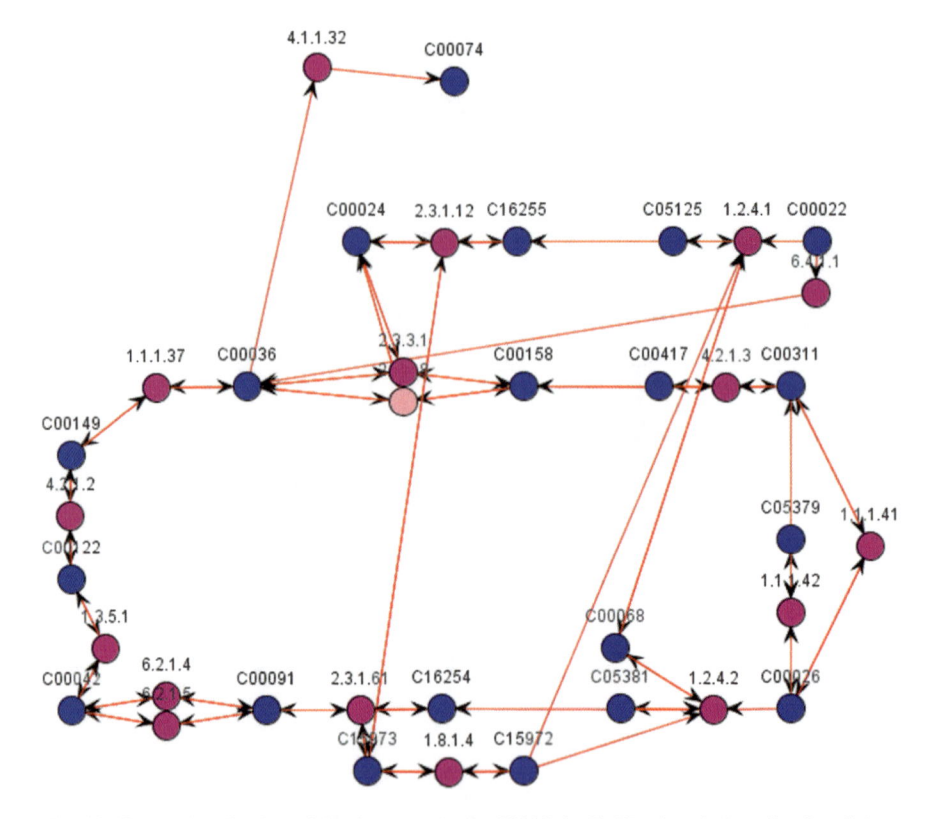

Fig. 10.18 2D visualization of the citrate cycle (hsa00020) in CmPI using the Localization Colors for the enzymes

by coloring the enzymes according the color codes known from Fig. 10.5. And Fig. 10.19 shows all potential enzyme localizations found in the databases, providing a good overview of all available localization information.

10.4 Conclusions

In conclusion, it can be stated that Integrative Bioinformatics techniques can be used to generate and analyze biological networks, to predict the localization of its components, and to use different visualization techniques to enable the discussion of potential results. It should be mentioned again that the information provided by KEGG concerning the involved enzymes is quiet vague, because the EC classification does not describe specific proteins, but protein families. Therefore, many different localizations for each enzyme were available. Of course, CmPI was

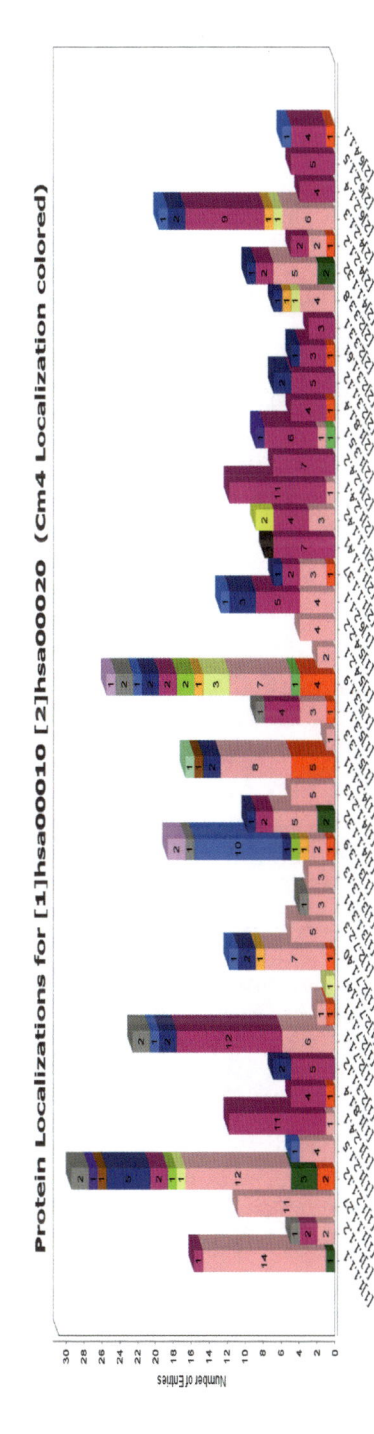

10 Network Analysis and Integration in a Virtual Cell Environment 295

Fig. 10.19 Localization Chart, category "Protein Localizations/Cm4" showing all potential enzyme localizations found in the databases

already used to localize specific proteins in a cellular environment, for example, to examine the localization of a cardiovascular disease-related protein set [22, 23].

Despite this fact, it was shown that the CmPI can be used (1) for the prediction and analysis of the localization of protein-related networks or sets by using the Subcellular Localization Charts and (2) to visualize and explore the results in two and three dimensions. Therefore, the functional and mesoscopic level of cytology can be combined; microscopic data sets – as known, e.g., from the Cell Centered Database [16] – can be used as a base to combine spatial cellular structures with metabolic pathways.

References

1. Ashburner M, Ball CA, Blake JA et al (2000) Gene Ontology: tool for the unification of biology (The Gene Ontology Consortium). Nat Gen 25:25–29
2. Berg JM, Tymoczko JL, Stryer L (2006) Biochemistry, 6th edn. W.H. Freeman, New York
3. Blake JA, Chan J, Kishore R (2013) Gene Ontology annotations and resources. Nucl Acids Res 41:D530–D535
4. Brooks BR, Brooks CL, Mackerell Jr AD et al (2009) CHARMM: the biomolecular simulation program. J Comput Chem 30:1545–1614
5. Case DA, Cheatham TE, Darden T et al (2005) The Amber biomolecular simulation programs. J Comput Chem 26:1668–1688
6. Cooper GM, Hausman RE (2007) The cell: a molecular approach, 4th edn. ASM, Washington
7. Croft D, O'Kelly G, Wu G (2011) Reactome: a database of reactions, pathways and biological processes. Nucl Acids Res 39:D691–D697
8. Dimmer EC, Huntley RP, Alam-Faruque Y (2012) The UniProt-GO annotation database in 2011. Nucl Acids Res 40:D565–D570
9. Green-Armytage P (2010) A colour alphabet and the limits of colour coding. Colour Des Creat 5:1–23
10. Hess B, Kutzner C, van der Spoel D et al (2008) Gromacs 4: algorithms for highly efficient, load-balanced, and scalable molecular simulation. J Chem Theory Comput 4:435–447
11. Hippe K, Kormeier B, Töpel T et al (2010) DAWIS-MD-a data warehouse system for metabolic data. GI Jahrestag 2:720–725
12. Hucka M, Finney A, Sauro HM et al (2003) The systems biology markup language (SBML): a medium for representation and exchange of biochemical network models. Bioinformatics 19:524–531
13. Janowski S, Kormeier B, Töpel T et al (2010) Modeling of cell-to-cell communication processes with Petri nets using the example of quorum sensing. In Silico Biol 10:27–48
14. Kanehisa M, Goto S, Sato Y, Furumichi M, Tanabe M (2012) KEGG for integration and interpretation of large-scale molecular data sets. Nucl Acids Res 40:D109–D114
15. Loew LM, Schaff JC (2001) The Virtual Cell: a software environment for computational cell biology. TRENDS Biotechnol 19:401–406
16. Martone ME, Gupta A, Wong M et al (2002) A cell-centered database for electron tomographic data. J Struct Biol 138:145–155
17. Meyer B (1998) Self-organizing graphs—a neural network perspective of graph layout. Lect Notes Comput Sci (Graph Drawing) 1547:246–262
18. Podkolodnaya OA, Yarkova EE, Demenkov PS et al (2011) Application of the ANDCell computer system to reconstruction and analysis of associative networks describing potential relationships between myopia and glaucoma. Russ J Genet Appl Res 1:21–28

19. Scheer M, Grote A, Chang A (2011) BRENDA, the enzyme information system in 2011. Nucl Acids Res 39:D670–D676
20. Sommer B (2012) CELLmicrocosmos – integrative cell modeling at the molecular, mesoscopic and functional level. Bielefeld University, Bielefeld
21. Sommer B, Künsemöller J, Sand N et al (2010) CELLmicrocosmos 4.1: an interactive approach to integrating spatially localized metabolic networks into a virtual 3D cell environment. In: Fred A, Filipe J, Gamboa H (eds) BIOINFORMATICS 2010 – proceedings of the 1st international conference on bioinformatics, part of the 3rd international joint conference on biomedical engineering systems and technologies (BIOSTEC 2010), INSTICC, Valencia
22. Sommer B, Tiys E S, Kormeier B et al (2010) Visualization and analysis of a cardio vascular disease- and MUPP1-related biological network combining text mining and data warehouse approaches. J Integr Bioinform 7:148
23. Sommer B, Kormeier B, Demenkov P S et al (2013) Subcellular localization charts: a new visual methodology for the semi-automatic localization of protein-related data sets. J Bioinform Comput Biol 11:1340005
24. Takahashi K, Ishikawa N, Sadamoto Y (2003) E-Cell 2: multi-platform E-Cell simulation system. Bioinformatics 19:1727–1729
25. Webb EC (1992) Enzyme nomenclature – recommendations of the nomenclature committee of the international union of biochemistry and molecular biology on the nomenclature and classification of enzymes, NC-IUBMB. Academic, San Diego
26. Widjaja YY, Pang CNI, Li SS, Wilkins MR et al (2009) The interactorium: visualising proteins, complexes and interaction networks in a virtual 3D cell. Proteomics 9:5309–5315
27. Wurtele ES, Li J, Diao L et al (2003) MetNet: software to build and model the biogenetic lattice of Arabidopsis. Comp Funct Genom 4:239–245
28. Wurtele ES, Bassham DC, Dickerson J et al (2010) Meta!Blast: a serious game to explore the complexities of structural and metabolic cell biology. In: Proceedings of the ASME 2010 world conference on innovative virtual reality, ASME, Ames
29. Yamaguchi R, Lartigue L, Perkins G (2008) Opa1-mediated cristae opening is Bax/Bak and BH3 dependent, required for apoptosis, and independent of Bak oligomerization. Mol Cell 31:557–569

Chapter 11
Bridging Genomics and Phenomics

Dijun Chen, Ming Chen, Thomas Altmann, and Christian Klukas

Abstract Genomics and phenomics are two fundamentally important branches of biological sciences, and they stand at both ends of the multiple "omics" families. A central goal of current biology is to establish complete functional links between the genome and phenome, the so-called genotype–phenotype map. Recent advances in high-throughput and high-dimensional genotyping and phenotyping technologies enable us to uncover the casual networks inside the "black box" that lies between genotypes and phenotypes using the principles of genome-wide association studies (GWAS). Application of GWAS and analogous methodologies and incorporation of multiple omics data begin to unravel the contribution of genetic variation to phenotypic diversity. Integrating "omics" data at broad levels by using the systems-biology approach is paramount to further bridging the gaps between genomics and phenomics and eventually making accurate predictions of phenotypes based on genetic contribution.

D. Chen
Department of Molecular Genetics, Leibniz Institute of Plant Genetics and Crop Plant Research Gatersleben (IPK), Corrensstrasse 3, 06446 Gatersleben, Germany

Department of Bioinformatics, College of Life Sciences, Zhejiang University, Hangzhou, 310058 People's Republic of China
e-mail: chend@ipk-gatersleben.de

M. Chen
College of Life Sciences, Zhejiang University, Hangzhou, People's Republic of China
e-mail: mchen@zju.edu.cn

T. Altmann • C. Klukas (✉)
Department of Molecular Genetics, Leibniz Institute of Plant Genetics and Crop Plant Research Gatersleben (IPK), Corrensstrasse 3, 06446 Gatersleben, Germany
e-mail: altmann@ipk-gatersleben.de; klukas@ipk-gatersleben.de

M. Chen and R. Hofestädt (eds.), *Approaches in Integrative Bioinformatics: Towards the Virtual Cell*, DOI 10.1007/978-3-642-41281-3_11,
© Springer-Verlag Berlin Heidelberg 2014

Keywords High-throughput phenotyping • Next-generation sequencing (NGS) • Genotype–phenotype map (G-P map) • Phenomics • Genome-wide association study (GWAS) • Quantitative trait loci (QTL)

11.1 Introduction

With the rapid advances of high-throughput resequencing and marker genotyping, high-density genetic variation information (such as single-nucleotide polymorphisms, SNPs, and copy-number variants, CNVs) has been collected and need to be linked with functions. Over the past few years, a multitude of genome-wide association studies (GWAS) and related strategies have identified numerous genetic variants associated with complex diseases or other traits in humans and plants, providing valuable insights into their genetic architecture. These findings are definitely enriching our knowledge about the genetic basis of phenotypic variation and provide an opportunity for genetic testing. However, most variants identified so far explain only a small proportion of the causal genetic factors, leaving the remaining "missing" heritability to be explained [1]. Moreover, even with a complete understanding of the genetics of a complex phenotypic trait, it is still challenging to accurately predict phenotypic variation from individual genetic codes. Furthermore, the majority of these disease- or trait-related variants lie within noncoding regions of genomes, complicating their functional evaluation and offering the greatest challenge in the "post-GWAS" era [2].

Globally linking genetic variants to phenotypic diversity is one of the key goals of biology. Our understanding of such a genotype–phenotype map cannot be established without detailed phenotypic data [3]. However, our ability to characterise phenomes – the full set of phenotypes of an individual – largely lags behind our ability to characterise genomes. Hence, phenomics – high-throughput and high-dimensional phenotyping – is emerging as a suit of new technologies to accelerate progress in our understanding of the relationship between genotype and phenotype [3, 4].

In this chapter, we will first review the principle of dissecting genotypes and monitoring phenotypes, usually in high-throughput manners. We also highlight current approaches to obtaining phenomic data and the emerging applications of large-scale phenotyping approaches in the phenomics era. We then outline the current strategies, such as GWAS and analogous methodologies, for globally linking genetic variation to phenotypic diversity. We summarise insights about the complete "genotype–phenotype" map that could be established through integrating "omics" data at broad levels in terms of a systems-biology approach. Related phenome projects and phenomic tools are discussed. Please keep in mind that the results discussed here are mostly based on research in humans and/or plants and that only a subset of published information can be mentioned.

11.2 Defining the Genotype and Phenotype

In this section, we outline the state-of-the-art methods used for the assessment of genotypes and phenotypes and the corresponding mapping approaches for linking genotypes to phenotypes at global levels (Table 11.1). We also present phenomics-related projects that combine rich genomic data with data on quantitative variation in phenotypes and which have recently been launched in both humans and plants (Table 11.2). We highlight many emerging technologies developed for high-throughput phenotyping in plants (Table 11.3).

11.2.1 Genetic Variation: Genotyping

Genotyping technology is referred to as the set of methodologies and protocols used to elucidate the genetic makeup (genotype) of an individual, also known as genotypic assaying. Genotyping is essential in deciphering the genetic causes of complex phenomena, including health, disease, crop yields and evolutionary fitness. Human genetic mapping was initially performed based on restriction fragment length polymorphisms (RFLPs) [5, 6], amplified fragment length polymorphisms (AFLPs) [7] and microsatellite markers (also known as short tandem repeats or simple sequence repeats) [8]. More recently, SNPs, due to their high abundance, low mutation rates and amenability to high-throughput analysis, have become the markers of choice for linkage and linkage disequilibrium (LD) mapping [9, 10]. The usually binary SNP markers are well suited to automated, high-throughput typing. Indeed, it is now feasible to genotype SNPs with high density at the genome-wide scale by utilising array-based [11, 12] or sequencing-based [13, 14] technologies (Table 11.1). Although high-throughput SNP arrays avoid time-consuming cloning and primer design steps, they lack of the discovery process and show bias towards genotyping new populations. Now, with the advent of next-generation sequencing (NGS), new technologies such as reduced-representation libraries (RRLs) [15] or complexity reduction of polymorphic sequences (CRoPS) [16], restriction-site-associated DNA sequencing (RAD-seq) [17] and low-coverage genotyping, including multiplexed shotgun genotyping (MSG) [18] or genotyping by sequencing (GBS) [19], are capable of genome-wide marker discovery for both model organisms and non-model species. Although sequence-level variants have been catalogued more extensively, structural variations – including indels (insertions/deletions), CNVs and inversions – are now investigated for their contribution to complex traits, including many important common diseases [20]. CNVs can be identified with various genome analysis platforms, including array-based comparative genomic hybridisation (CGH), SNP genotyping platforms and NGS.

Table 11.1 Various approaches towards genotype–phenotype map

Approach	Level	Mapping	Information	Related technology	Potential applications
Genomics	DNA	GWAS, linkage mapping	Sequence, SNPs, CNVs, indels	Genome sequencing, array-based CGH, SNP typing (NGS), exon sequencing	High-throughput genotyping and resequencing reference panels Identification of rare alleles High-density genetic markers for QTL detection
Epigenomics	Chromatin	methQTL, EWAS	DNA methylation, histone modifications, siRNAs	Bisulfite sequencing, ChIP-chip, ChIP-seq	High-throughput examination of epigenome High-density epigenetic markers for QTL mapping
Transcriptomics	RNA	eQTLs, TWAS	Transcripts, miRNAs	EST, gene chip, microarray, RNA-seq	High-throughput, unbiased and cost-effective examination of transcriptome
Proteomics	Proteins	pQTLs, PWAS	Proteins	LC-MS/MS, 2D-PAGE	Detecting quantitative and qualitative variation in proteins (abundance and modification)
Metabolomics	Metabolites	mQTLs, MWAS	Metabolites	MS based, NMR, HPLC	Detecting quantitative and qualitative variation in cellular metabolites
Phenomics	Cell, tissue, organism, development, physiology, morphology, behaviour	phQTLs, PheWAS	Biomass, morphology	Automated imaging, noninvasive phenotyping	Phenotyping large samples required for systems genetic analyses Automated noninvasive phenotyping for growth modelling Dissecting quantitative traits for QTL mapping

| Systems biology | Systems | Network mapping | Network | Integrated analysis | Combinational use of genetic and multi-omic data into an integrative analysis (genetical genomics) [160] |

QTLs: quantitative trait loci; methQTLs: QTLs; eQTLs: expression QTLs; pQTLs: protein QTLs; mQTLs: metabolic QTLs; phQTLs: phenotypic QTLs; GWAS: genome-wide association studies; EWAS: epigenome-wide association studies; TWAS: transcriptome-wide association studies; PWAS: proteome-wide association studies; MWAS: metabolome-wide association studies; PheWAS: phenome-wide association studies; SNPs: single-nucleotide polymorphisms; CNVs: copy-number variations; indels: insertion–deletion polymorphisms; siRNAs: small interfering RNAs; miRNAs: microRNAs; NGS: next-generation sequencing; CGH: comparative genome hybridisation; ChIP-chip: chromatin immunoprecipitation (ChIP) followed by microarray technology; ChIP-seq: ChIP followed by high-throughput DNA sequencing; EST: expressed sequence tag; RNA-seq: RNA sequencing; LC-MS/MS: liquid chromatography–tandem mass spectrometry; 2D-PAGE: two-dimensional polyacrylamide gel electrophoresis; MS-based technologies: such as gas chromatography–mass spectrometry (GC-MS) and liquid chromatography–mass spectrometry (LC-MS); NMR: nuclear magnetic resonance; HPLC: high-performance liquid-phase chromatography

Table 11.2 Related projects or resources for phenomics studies

	Project name	Description	URLs
Humans	The International HapMap Project	The International HapMap Project is an organisation that aims to develop a haplotype map (HapMap) of the human genome, which will describe the common patterns of human genetic variation. HapMap is a key resource for researchers to find genetic variants affecting health, disease and responses to drugs and environmental factors. The information produced by the project is made freely available to researchers around the world	http://hapmap.ncbi.nlm.nih.gov/; http://www.hapmap.org/
	The 1000 Genomes Project (TGP)	The 1000 Genomes Project, launched in January 2008, is an international research effort to establish by far the most complete and detailed catalogue of human genetic variations, which in turn can be used for association studies relating genetic variation to disease	http://www.1000genomes.org/
	The Personal Genome Project (PGP)	PGP is a long-term, large cohort study which aims to sequence and publicise the complete genomes and medical records of 100,000 volunteers, in order to enable research into personal genomics and personalised medicine. The project will publish the genotype of the volunteers, along with extensive information about their phenotype: medical records, various measurements, MRI images, etc.	http://www.personalgenomes.org/
	Human Epigenome Project (HEP)	Human Epigenome Project (HEP) is a international science project, with the stated aim to "identify, catalog, and interpret genome-wide DNA methylation patterns of all human genes in all major tissues"	http://www.epigenome.org/
	The Wellcome Trust Case Control Consortium (WTCCC)	The WTCCC aims were to exploit progress in understanding of patterns of human genome sequence variation along with advances in high-throughput genotyping technologies and to explore the utility, design and analyses of genome-wide association (GWA) studies	http://www.wtccc.org.uk/
	Online Mendelian Inheritance in Man (OMIM)	OMIM is a database that catalogues all the known diseases with a genetic component and links them to the relevant genes in the human genome and provides references for further research and tools for genomic analysis of a catalogued gene	http://www.ncbi.nlm.nih.gov/omim; http://www.omim.org/
	The Human Variome Project (HVP)	The Human Variome Project is the global initiative to collect and curate all human genetic variation affecting human health and with the mission is to improve health outcomes	http://www.humanvariomeproject.org/
Plants	IPPN	International Plant Phenomics Network. IPPN is an international consortium that will boost plant phenotyping science by developing novel technologies and concepts used for the application of plant production and the analysis of ecosystem performance	http://www.plantphenomics.com/

EPPN	European Plant Phenotyping Network. This project will establish the network that integrates European plant phenotyping efforts and builds a competitive community to the goal of the understanding of the link between genotype and phenotype as well as their interaction with the environment	http://www.plant-phenotyping-network.eu/
DPPN	German Plant Phenotyping Network. DPPN is a Germany-funded project that partners undertake a joint research programme and share their phenotyping infrastructure within networking activities	http://www.dppn.de/
JPPC	The Jülich Plant Phenotyping Centre. This project is with aims to elucidate the functional role of gene networks under natural conditions with the aid of the development of noninvasive phenotyping tools and methods as well as the existing genetic resources	http://www2.fz-juelich.de/icg/icg-3/jppc/phenotyping/
PHENOME	PHEOME, launched in 2012, is a project funded by French investment for the future. It will provide France with an up-to-date, versatile, high-throughput infrastructure and suite of methods allowing characterisation of panels of genotypes of different species (important crop species) under scenarios associated with climate changes	http://urgi.versailles.inra.fr/Projects/PHENOME/
APPF	The Australian Plant Phenomics Facility. APPF is developed to alleviate the "phenotyping bottleneck" by utilising high-throughput plant phenotyping and "reverse phenomics" approaches with aims to probe and improve plant function and performance	http://www.plantphenomics.org.au/
1001 Genomes Project	A project launched at the beginning of 2008 with a goal to discover the genome-wide genetic variants contributing to adaptation to diverse environments in 1001 strains (accessions) of Arabidopsis. The resulting information has been tested in a GWAS for 107 phenotypes and is ultimately paving the way for a new era of genetics that identifies alleles underpinning phenotypic diversity across the entire genome and the entire species	http://1001genomes.org/
Maize HapMap	The Maize Diversity Project. This project aims to decode the genetic architecture underlying variation in maize (*Zea mays*) quantitative traits and to evaluate our ability to predict phenotype from genotype. Based on this project, a comprehensive "phenomic" association analysis has been tested on several key traits involved in leaf development and disease resistance	http://www.panzea.org/index.html
Rice HapMap	The Rice Haplotype Map (HapMap) Project. Several ecologically or agronomically important traits have been tested for genotype–phenotype associations under this project. The project has a long-term goal to provide a repertoire of the genetic variants in rice that facilitate the genetic mapping of complex traits	http://www.ncgr.ac.cn/RiceHapMap/; http://www.ncgr.ac.cn/RiceHap3/

Table 11.3 Automated or semiautomated plant phenotyping platforms

Name	Description	URL
Analysis software or tools		
WinRHIZO	An image analysis system specifically designed for root measurement in forms of morphology, topology, architecture, colour analyses and so on. WinRHIZO includes both image acquisition components and computer programs to meet different needs and budgets. Other related tools are WinFOLIA and WinSEEDLE for leaf analysis, WinDENDRO for tree ring analysis and WinCELL for wood cell analysis	http://www.regent.qc.ca/index.html
HTPheno	An open source image analysis pipeline based on ImageJ. It can be used for the analysis different phenotypic measurements such as height, width and projected shoot area of the plants. HTPheno is demonstrated to use in an analysis of high-throughput phenotyping data of barley (*Hordeum vulgare*) plants from the LemnaTec system	http://htpheno.ipk-gatersleben.de/
LAMINA	A software for the automated analysis of images of leaves for various plant species. It enables rapid quantification of leaf size (area) and shape (blade dimensions) parameters	http://sourceforge.net/projects/lamina/
HYPOTrace	An image analysis tool that enables automatical extraction hypocotyl growth and shape information from electronic images of Arabidopsis seedlings undergoing photomorphogenesis	http://phytomorph.wisc.edu/software/hypotrace.php
LeafAnalyser	An automated image-processing solution specifically developed for leaf phenotyping (in Arabidopsis), although it can be extended to use in any 2D image-processing application	http://sourceforge.net/apps/trac/leafanalyser/
Integrated Analysis Platform (IAP)	IAP is developed as a comprehensive framework for high-throughput phenotyping in plants, which enables us to extract a high-dimensional list of plant traits from real-time images to quantify the plant growth and performance. IAP is anticipated as an open and extensible platform that will help collaborate between phenomic and genomic research communities	http://iap.ipk-gatersleben.de
Hardware installations (some with their own software)		
LemnaTec	A robotic greenhouse system that uses non-destructive imaging to monitor plant growth under fully controlled conditions in high throughput. The LemnaTec platform aims to visualise and analyse the biology beyond human vision through imaging automatisation	http://www.lemnatec.com/
PHENOPSIS	An automated platform developed by Optimalog (France) for reproducible phenotyping of plant responses to soil water deficit in Arabidopsis (*Arabidopsis thaliana*). The PHENOPSIS platform allows to weight, irrigate precisely and take a picture of more than 500 individual plants in rigorously controlled conditions	http://bioweb.supagro.inra.fr/phenopsis/

11 Bridging Genomics and Phenomics

Phenodyn	A platform to measures growth rate and transpiration rate every minute, together with environmental conditions (current throughput, 480 plants)	http://bioweb.supagro.inra.fr/phenodyn/
GROWSCREEN	An in-house system used in the Jülich Plant Phenotyping Centre (JPPC) to study leaf growth and fluorescence and root architecture in large plant populations	http://www2.fz-juelich.de/icg/icg-3/jppc/growscreen/
TraitMill	A high-throughput gene engineering system developed by CropDesign that enables large-scale plant transformation and automated high-resolution phenotypic evaluation of crop performance in rice (*Oryza sativa*)	http://www.cropdesign.com/tech_traitmill.php
RootReader3D	An imaging and software platform designed for high-throughput 3D analysis of plant roots. RootReader3D was applied to 3D root reconstructions and architecture descriptor quantification during rice seedling development	http://www.plantmineralnutrition.net/rootreader.htm
GROWSCREEN 3D	A pioneered solution developed for 3D analysis of leaves in tobacco (*Nicotiana tabacum*). It enables more accurate measurements of leaf area and extraction of additional volumetric traits	NA
PlantScan 3D meshes	A novel automated screening platform and mesh-based technique developed for high-throughput 3D plant analysis. It was initially used for the analysis of aerial parts in cotton (*Gossypium hirsutum*) and demonstrated highly accurate when comparing with manual measurement data	http://www.plantphenomics.org.au/node/157

NA: not available

Our knowledge regarding human genetic variations is mostly derived from the international effort of the SNP Consortium [21] and the International HapMap Project [22] (Table 11.2). Recent advances in sequencing technology make it possible to comprehensively catalogue genetic variation in population samples. Projects such as the Personal Genome Project (PGP) (e.g. diploid personal genomes [23]), the 1000 Genomes Project (TGP) [24] and exome sequencing projects [25] are under way in an attempt to elucidate the full spectrum of human genetic variations as a foundation to investigate the relationship between genotype and phenotype. For example, the Phase 1 publication of TGP in 2012 included whole-genome sequences of 1,092 individuals from 14 populations. A total of 38 million SNPs, 1.4 million short indels and more than 14,000 larger deletions were identified [26]. Notably, the genome of any apparently healthy individual carries more than 2,500 nonsynonymous variants at conserved regions, 20–40 variants identified as damaging at conserved sites and 150 loss-of-function (LoF) variants in protein-coding genes, some of which are known to cause Mendelian disease [26].

Meanwhile, genome-wide genotyping is extensively performed in plants in recent years (Table 11.2), such as in *Arabidopsis thaliana* [27], rice [28], maize [29, 30], sorghum [31] and barley [32]. These rich resources will ultimately help to explore the genetic basis of plant agriculture-related traits, such as flowering time, growth rate, yield and stress tolerance, and to improve crops and understand plant adaptation.

11.2.2 Phenomics: Multilevel and Multidimensional Assessment of Features

The term phenotype includes the composite of an organism's observable traits or characteristics – such as its morphological, developmental, physiological, pathological or biochemical properties, phenology and behaviour – that can be monitored, quantified and/or visualised by some technical procedure. Phenomics is defined as the study of all the phenotypes of an organism (phenome) that are the result of genetic code (G), environmental factors (E) and their interactions (G × E). In contrast to genotypes, which are essentially single one-dimensional as merely determined by the linear DNA code, phenotypes are usually multi-dimensional and are frequently capricious in different spatial and temporal situations. An important field of research today is trying to improve, both qualitatively and quantitatively, the capacity to measure phenomes. In broad definition, phenome includes epigenomics, transcriptomics, proteomics, metabolomics and many other "omics" data regarding quantitative measurement of biochemical and cellular processes. We have relatively well-developed technologies of measurements, in vivo or in destructive manners, of physiological states and other "internal phenotypes" (endophenotypes), such as gene expression, protein and metabolite levels, whereas our ability to measure "external phenotypes" (exophenotypes) is rapidly evolving.

11 Bridging Genomics and Phenomics

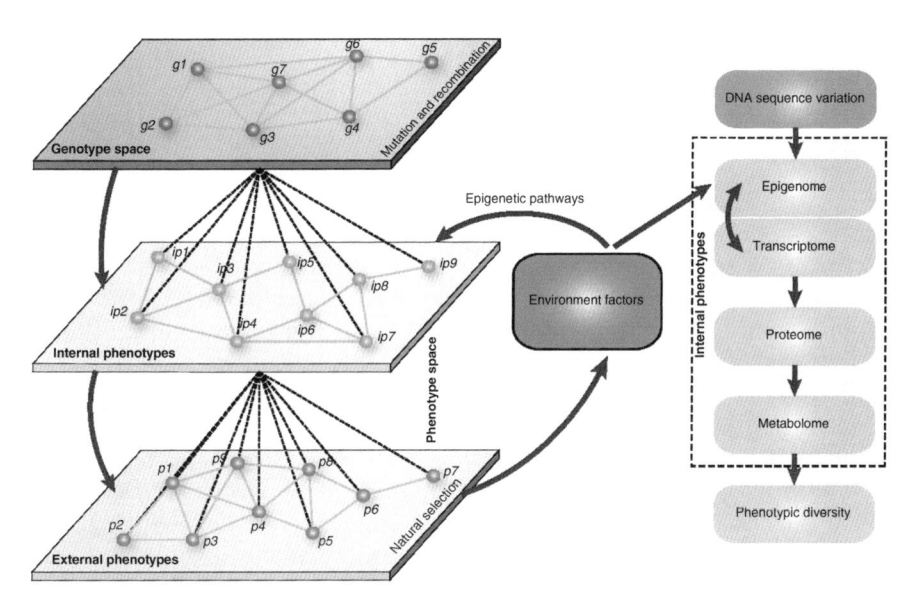

Fig. 11.1 The genotype–phenotype map (G-P map). The *left panel* shows the relationship of the genotype space (G space) and the phenotype space (P space) [3]. The corresponding information that transmits from G space to P space is shown in the *right panel*. Genotypes could gain mutation and recombination over generations. Phenotypes can be broadly classified into internal and external phenotypes. These internal phenotypes include properties from molecular, cellular or tissue levels, which in turn shape external phenotypes such as morphology and behaviour. Upon the environmental stimuli, the epigenetic process creates the phenotypes using genotype information. External phenotypes can in turn shape the environment that an individual occupies, creating complex feedback relationships between genes, environments and phenotypes. Natural selection act in the P space to change the average phenotype of parents away from the average phenotype of the generation. The importance of the environment suggests that we should explicitly broaden the G-P map to the genotype–environment–phenotype (G-E-P) map. *g*: genotype; *p*: phenotype; *ip*: internal phenotype

We will never be able to come even close to a complete characterisation of the phenome due to its highly dynamic and high-dimensional properties. However, increasing the quantitative information obtained by phenotypic measurements is an important goal for phenomics [3]. Phenotypic variation, a fundamental prerequisite and the perpetual force for evolution by natural selection, results from the complex interactions between genotype and environment (G × E). Phenomic-wide data are essential and necessary for enabling us to trace causal links in the genotype–phenotype map (G-P map [33]) as they define the space of all possible phenotypes (P space; Fig. 11.1).

High-throughput automated imaging is the ideal tool for phenomic studies. Owing to the recent increased availability of high-precision robotic handling machinery, many imaging-based technologies that span molecular to organismal spatial scales have been or are being established and enable us to extract multiparametric phenotypic information in great detail. Various detectors using a broad range of the electromagnetic spectrum and magnetic resonance imaging (MRI) with different scales of resolution are widely used imaging techniques for phenotyping [34]. High-dimensional spatiotemporal data on many phenotype classes such as morphology, behaviour, physiological state and locations of proteins and metabolites can be captured by these imaging techniques and analysed via high-performance computing [3]. In recent years, systems for performing high-content microscopy-based assays have become available and are often used to investigate the effects of chemical (such as drugs and small molecules) and genetic (loss-of-function of genes using RNA interference [RNAi]) perturbations on cultured cells [35–42]. Such genome-wide RNAi screens enable us to discover novel gene functions and interrogate their functional relationships based on phenotypic similarity analysis [43, 44]. These screens produced huge amount of high-content image data that can be automatically processed using software tools such as ImageJ [45], EBImage [46], CellProfiler [47] or PhenoRipper [48]. Traditional microscopy is generally used in two-dimensional (2D) imaging. However, high-resolution and dynamic three-dimensional (3D) imaging data can be acquired by confocal laser scanning microscopy (CLSM), X-ray computerised tomography (CT) or MRI.

In plants, the "phenotyping bottleneck" [4] needs to be addressed by high-throughput noninvasive technologies [49]. Thanks to developed new imaging sensors (e.g. high-resolution imaging spectrometers) and the advanced software for image analysis and feature extraction, a range of automated or semiautomated high-throughput plant phenotyping systems (Table 11.3) have been recently developed and applied to assess plant function and performance under controlled conditions [50–58]. One of the pioneer platforms, PHENOPSIS [51], was developed for the dissection of genotype × environment effects on different processes in *Arabidopsis thaliana* with reproducible phenotyping. TraitMill [50, 52], GROWSCREEN [53, 55, 59], LIMINA [54], HYPOTrace [56], HTPheno [57] and LeafAnalyser [58] provide general image-processing solutions for plant morphological measurements (such as plant height, length and width, shape, projected area and biovolume) and colorimetric analysis. Most recently, high-throughput phenotyping has been used for three-dimensional plant analysis [60–64], focusing on a specific organs (e.g. leaves, roots and aerials). However, most of these tools possess the inherent disadvantage that they are designed to address only very specific question [65]. Among the advancing solutions, the state-of-the-art phenotyping platform developed by LemnaTec (http://www.lemnatec.com/) is a robotic greenhouse system that uses non-destructive imaging to monitor plant growth under controlled environmental conditions (such as nutrition, water availability, irradiation and temperature) over a period of time. Several ingenious imaging cameras, such as visible/colour/RGB (red, blue and green) imaging, fluorescence, thermal and near-infrared imaging, have been adopted in this system to assess the physical and physiological status of plants, such us their geometric properties, pigment or fluorophore contents, canopy temperature and tissue water content. LemnaTec systems have now been deployed in growth champers or greenhouses (e.g. at the Leibniz Institute of Plant Genetics and Crop Plant Research [IPK; Germany], the Australian Centre for

Plant Functional Genomics [ACPFG] at the University of Adelaide [Australia], the Aberystwyth University [UK] and the PhenoArch at Institut National de la Recherche Agronomique at Montpellier [France]) for high-throughput phenotyping in *Arabidopsis* [66], wheat [67], barley [57] and maize (unpublished data). The time-lapse phenotypic data from these large-scale phenotyping platforms provide an invaluable opportunity to model and predict plant growth [67, 68]. Also, these data can be used to map quantitative trait loci (QTL) for growth-related traits. Notably, a recent phenotyping application was developed for QTL mapping in pepper plants using phenotypic features such as leaf angle and leaf size from RGB images, resulting in heritabilities of 0.56 and 0.70, respectively [69]. At the same time, however, the huge amounts of imaging data generated from these platforms present a great challenge for data analysis. As one solution, the Integrated Analysis Platform (IAP; http://iap.ipk-gatersleben.de) [70] is being developed as a comprehensive framework for high-throughput phenotyping in plants, which enables us to extract a high-dimensional list of plant features from real-time images to quantify plant growth and performance.

11.2.3 Defining Genotype–Phenotype Relationships

Understanding the interplay between genotype and phenotype (G-P map; Fig. 11.1) is the ultimate goal in both genomics and phenomics research, which will yield insights that are important for predicting disease risk and individual therapeutic treatments in human population, for increasing the speed of selective breeding traits in agriculturally import crops and for predicting adaptive evolution [71]. The interactions between genotypes and phenotypes also inevitably involve the environmental factors [3]. Thus, the interaction between genotype and phenotype has often been conceptualised by the following relationship: genotype (G) + environment (E) + genotype × environment (G × E) → phenotype (P). Since individuals themselves may influence the environment and exert different effects depending on their characteristics, feedback of phenotypes needs to be considered in this concept. Furthermore, the response of a certain genotype to an environmental factor may depend strongly on the phenotypic status of the individual, which is the result of events that occurred in its preceding life history. Towards understanding, the G-P map will provide a framework for the development of personalised medicine and crop breeding [72, 73].

Genomics and other highly parallel technologies – including epigenomics, transcriptomics, proteomics, metabolomics and ionomics – have become the mainstay in biological research. These recently developed technologies commonly termed "omics" permit assessment of the entirety of the components of biological systems at broad levels (Table 11.1). Furthermore, the emerging high-throughput phenotyping technology is moving towards comprehensive, quantitative high-dimensional measurements of individuals (phenome). However, our current knowledge of the genetic basis of complex phenotypic traits probably represents only the tip of the iceberg. Why do even genetically identical twins often substantially differ in phenotypic traits such as disease risk and drug response? Indeed, it is now understood that the differences are to a large extent result of the epigenome and involve chromatin modifications as well as myriads of noncoding RNAs (ncRNAs) [74, 75]. The emerging task is to understand the complex relationships among the genome, the epigenome, the environment and the phenome. The goal of globally linking genotype to phenotype can only be achieved through integrating information from different levels into an integrative model in terms of systems-biology approaches, which makes prediction of phenotypes possible (Fig. 11.2). This model should also consider the complex environmental factors in the real world, which need to be very precisely defined. For example, it is now possible to model rice transcriptome dynamics under fluctuating field conditions [76], rising hopes to predict genome-wide transcriptional responses in the complex real-world settings [77].

11.3 Approaches for Linking the Genome to the Phenome

11.3.1 QTL Detection Through Linkage and Association Mapping: Identifying the Genetic Basis of Complex Traits

Thanks to the advanced high-throughput experimental technologies such as microarray and sequencing, high-density genotyping arrays are available and are widely used recently to establish large-scale genome-wide maps of QTLs for various phenotypes such as human diseases and agricultural traits [20, 79–81]. Genome-wide association studies (GWAS, also called association mapping) are becoming the preferred method to relate genetic variation to phenotypic diversity in populations of unrelated individuals. The most common polymorphic markers used for GWAS are sequence polymorphisms such as SNPs and structural variants such indels and CNVs [20]. GWAS are now preferred over traditional family-based linkage studies (linkage-based QTL mapping; Fig. 11.3) [82], which use interval mapping to estimate the map position and effect of each QTL.

GWAS use dense maps of genetic markers that cover the whole genome to look for allele-frequency differences between cases (e.g. patients with a specific disease or individuals with a certain trait) and controls. Several powerful statistical methods have been established to associate common complex trait with genomic variations, including efficient mixed-model association (EMMA) [83], EMMA expedited (EMMAX) [84], genome-wide EMMA (GEMMA) [85], mixed-model and regression (GRAMMAR) [86], fast linear mixed models (FaST-LMM) [87], general linear model and mixed linear model implemented in TASSEL (Trait Analysis by aSSociation, Evolution and Linkage) [88] and the EIGENSTRAT method [89]. In the past few years, intensive efforts in more than 1,500 GWAS

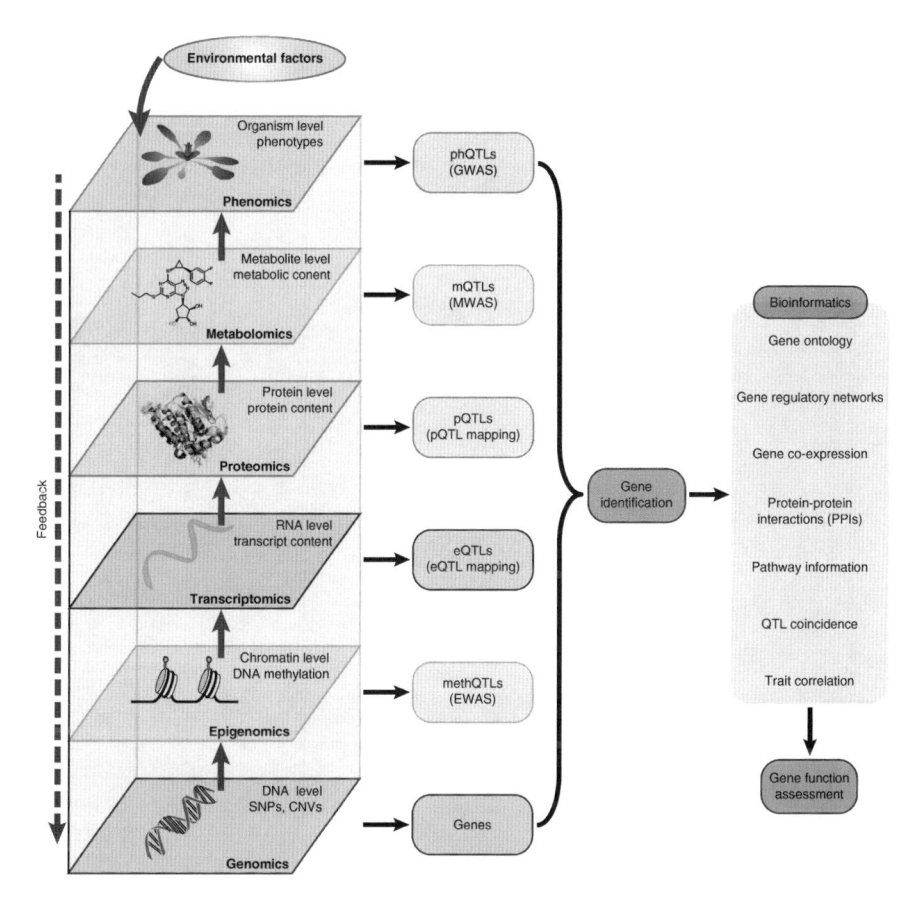

Fig. 11.2 Chart flow of the assessment of gene function using quantitative trait locus (QTL) analyses. Genetic markers (DNA level) such as SNPs and CNVs can be genotyped using next-generation sequencing technology. Quantitative traits, such as DNA methylation level, transcript, protein or metabolite content and biomass can be analysed using different detection methods. The information flow is indicated with *arrows*. Environmental factors are also included. The data generated can be used for mapping to determine the genomic regions (QTLs) responsible for the observed variation. The identification of the causal genes underlying the QTL, and ultimately their functional characterisation, will be facilitated by the combined analysis of the data generated using different profiling techniques and additional information obtained using bioinformatics tools [78]. phQTLs: DNA methylation QTLs; eQTLs: expression QTLs; pQTLs: protein QTLs; mQTLs: metabolic QTLs; phQTLs: phenotypic QTLs; GWAS: genome-wide association studies; EWAS: epigenome-wide association studies; MWAS: metabolome-wide association studies

have uncovered hundreds of genetic variants associated with hundreds of diseases and other traits [90], providing valuable insights into the complexities of genetic architecture of human diseases. Although disease-associated variants in protein-coding regions are expected to be more importantly related to trait/disease diversity, the vast majority (80 %) of variants are found to fall outside coding regions, highlighting the importance of noncoding regions in the search for

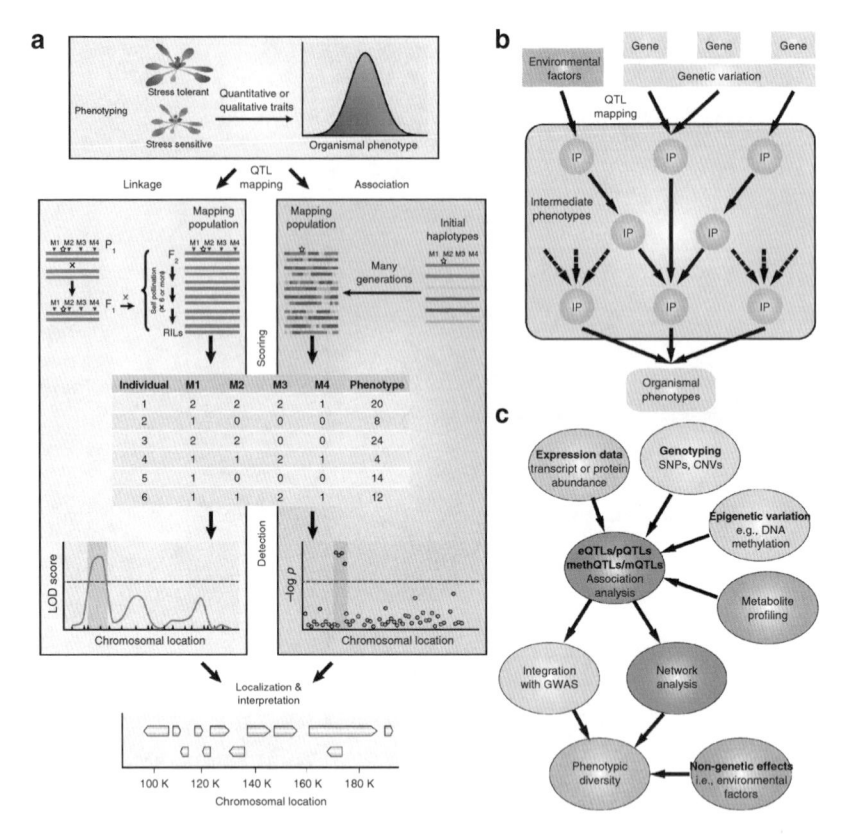

Fig. 11.3 Principle of quantitative trait locus (QTL) mapping. (**a**) Linkage-based mapping versus association mapping. The purpose of QTL mapping is to uncover the genetic basis of quantitative traits of interest. Linkage-based analyses seek to identify segregating genetic markers (M1, M2, M3 and M4) that predict the organismal phenotype, using a population that carries genetic mosaics derived from parental varieties, such as second generation (F_2) plants or recombinant inbred lines (RILs). The relationships of individuals are known ($P_1 \rightarrow F_1 \rightarrow F_2 \rightarrow$ RILs). It should be noted that RILs, rather than the F_2 or F_3 population, are needed to evaluate genotype-by-environment interactions. The region highlighted in yellow indicates the position of a causal locus or QTL. Association mapping (analogous to genome-wide association study [GWAS]) relies on correlations between genetic markers and a phenotype among collections of diverse germplasm. Thus, the recombination used in this strategy is historical. As shown in the figure, the association mapping population is separated by many generations from its progenitors. In linkage-based studies, the haplotype blocks in the mapping population may be large and, as a consequence, the causal locus might only be mapped to a large region. The haplotype blocks in an association mapping population tend to be much smaller, so it might be possible to localise the causal locus to a small genomic region. Within the QTL region, relevant genes may be identified for future studies or candidates may be suggested for targeted sequencing or experimental perturbation. (**b**) Conception of the intermediate phenotype used QTL mapping. The association of genetic variants is strongest with their closest intermediate phenotypes (IPs), such as variation of DNA methylation (methQTLs), transcript (eQTLs) or protein (pQTLs) content and metabolic traits (mQTLs). In some cases, the association of genetic variants with the organismal end point may not even be detectable at a level of genome-wide significance. (**c**) Relationships of GWAS and QTL mapping methodologies in integrative analyses (Part **a** is reproduced, with permission, from Mackay et al. [98], Copyright 2009, Macmillan Publishers Ltd. Part **b** is adapted from Suhre and Gieger [136]. Part **c** is reproduced from Cookson et al. [123])

disease-associated variants [1, 90]. However, the identified loci thus far explain only a small fraction of the phenotypic diversity in humans, raising questions regarding "the missing heritability" [1, 91]. An informative example is the investigation of height in humans, which is 80–90 % heritable, but a list of loci that has been detected in GWAS together accounts for less than 5 % of heritability for height [92]. Several explanations for this missing heritability have been proposed, including rare variants, allelic heterogeneity, epigenetic variation (see the next section), CNVs, gene–gene interactions and, perhaps most importantly, the environmental uncertainty [1, 91]. Intriguingly, GWAS have shown to be even more successful in plants than in humans [93], the key observation being that initial GWAS in plants (e.g. in *Arabidopsis* [94], maize [95, 96] and rice [28]) have explained a much greater proportion of the phenotypic variation. Perhaps the best example is a study in rice [28], in which the authors performed low-coverage resequencing of the genomes of a panel of about 500 rice landraces and identified 80 loci associated with 14 agronomic traits, explaining on average 36 % of the phenotypic variance. Several of these loci matched previously characterised genes. The ongoing development of technologies in both genotyping for detection of CNVs and other structural variants and statistical methods for accurate association testing will help us to examine potential sources of missing heritability and to better illuminate the causality of complex traits/diseases.

Linkage-based QTL mapping approaches have proved to be enormously successful for plant breeding and have identified loci with large effects of genetic variants on complex traits, which include most agriculturally important traits [81, 97]. The primary advantages of QTL mapping in plants are the great feasibility of creating populations of segregating individuals showing measurable phenotypic variation. However, the generation of crosses is time-consuming, and there is the necessity to focus on traits that can be readily and accurately phenotyped. Furthermore, due to the low frequency of recombinations represented in biparental mapping populations, causal loci (QTLs) identified by linkage-based strategies can only be mapped to large chromosomal regions, and tedious fine mapping needs to be carried out to narrow down on candidate genes that can be subjected to targeted sequencing or experimental perturbation [97, 98].

The emergence of a next-generation of mapping populations [97] overcomes many of the limitations of biparental QTL mapping and association mapping. Such experimental designs combine association and linkage analysis as they involve the crossing of multiple parents and advance populations through several generations to increase allelic richness and to improve resolution in genetic mapping. Such designs include the nested association mapping (NAM) [95, 99, 100], the multiparent advanced generation intercross (MAGIC) [101, 102] and the recombinant inbred advanced intercross line (RIAIL) [103, 104] populations.

In a further aspect, it needs to be mentioned that genomic selection (GS) [105], a genomics-based strategy for predicting phenotypes by the use of genome-wide marker data, is receiving considerable attention among (animal and) plant breeders. Similar to linkage and association mapping methods, GS starts with the development of a prediction model on a training population with individuals characterised for genotype and phenotype. Unlike linkage and association mapping approaches, GS models consider all markers as predictors and can thus capture more of the variation due to small-effect QTLs. Most importantly, the training population used in GS is generally closely related to the breeding population under selection. This situation supports the use of GS models for most accurate predictions for breeding [106].

11.3.2 EWAS: Linking Epigenetic Variation and Complex Traits

In addition to genetic variability, epigenetic factors including DNA methylation, histone modifications and ncRNAs (e.g. small interfering RNAs [siRNAs], microRNAs [miRNAs] and large intergenic ncRNAs [lincRNAs]) are considered as the missing part of the underlying molecular control of phenotypic variation (Table 11.1) [71, 75]. DNA methylation is the most studied epigenetic modification, and its variation at a single CpG (cytosine–guanine dinucleotide) site (known as a methylation variable position, MVP), CHG (H = A, T or C) or CHH contexts or a differentially methylated region (DMR) can be considered as the epigenetic equivalent (heritable epigenetic polymorphism) of an SNP in the context of genome [107]. While the DNA-centric model (e.g. GWAS) has allowed scientists to uncover the molecular genetic origins of Mendelian traits and diseases successfully, many complex traits and diseases are non-Mendelian, making them hard to explain. Due to the elasticity and plasticity of epigenetic factors, epigenetics can provide a novel framework for the identification of aetiological factors in complex traits and diseases [108]. The direct evidence that epigenetics could "make the difference" comes from the remarkably different epigenetic profiles, including disease-associated epigenetic differences, in human monozygous (MZ) twins, who share an identical genotype [109–111]. Indeed, with the recent advances in genomic technologies, the large-scale, systematic epigenomic equivalents of GWAS, termed as epigenome-wide association studies (EWAS), are emerging as the promising tool to investigate human disease-associated epigenetic variation [71]. However, it is still challenging in EWAS to distinguish whether epigenetic variation is the cause or functional consequence of the identified effects. In this regard, the sample used in an EWAS should ideally consist of MZ twins, to eliminate the influence of genetic background on the identified epigenetic variation [71] and as recently demonstrated by several studies [112–115]. Analysis of epigenetic variation is likely to be most successful when integrating the analysis of genetic variants (i.e. QTL mapping), leading to the identification of the underlying genetic variants that influence epigenetic state (epigenotype). The loci that harbour genetic variants corresponding to methylation states (e.g. MVPs or DMRs) have thus been termed methylation QTLs (methQTLs) [116]. The most pronounced methQTLs influence epigenetic states in *cis*, and they reside less than 50 bp from the CpG site in question [112]. The notion of methQTLs provides a general idea for integrated GWAS and EWAS (Fig. 11.3) to explore genotypes that exert their function through epigenetic mechanisms, which can be maintained and propagated during cell division, resulting in permanent maintenance of the acquired phenotype [71, 108, 117].

At the same time, there is also evidence from plant research communities that naturally occurring epigenetic changes (i.e. DMRs) in a single gene locus (epiallele) can lead to heritable phenotypic variation [118–122]. The epialleles often show increased cytosine methylation of the promoter and can result in nearby gene expression changes that are sometimes transmitted across generations, thus contributing to heritable phenotypic variation independent of DNA sequence diversity. These outstanding resources will advance our understanding of the relative roles of genetic and epigenetic variation in controlling quantitative trait variation in plants.

11.3.3 Variation in Gene Expression: From eQTLs to Phenotypes

Variation in gene expression is an important mechanism underlying phenotypic variation such as disease susceptibility and drug response. DNA variants may alter transcript abundance and splicing patterns through modification of regulatory elements [123]. Genomic loci responsible for this genetic control are consequently termed expression QTLs (eQTLs). The combination of high-throughput phenotyping and transcriptional profiling has allowed the systematic identification of eQTLs (Fig. 11.3) [98]. In principle, eQTL mapping uses transcript abundance as a phenotypic trait and maps the genomic loci controlling the transcript level, as performed in the same manner of traditional QTL mapping of any other quantitative trait phenotype [124]. According to the genomic context of transcripts, eQTLs can be categorised into *cis* eQTLs if the molecular variants (e.g. SNPs) are mapped to the approximate location (within 100 kb upstream and downstream [112, 125]) of their gene-of-origin transcripts and *trans* eQTLs in other cases. Further statistical analysis revealed a strong enrichment of *cis* eQTLs around transcription start sites (TSSs) and within 250 bp upstream of transcription end sites (TESs) [126]. The *cis*-acting variants are more likely in exonic regions than in intronic regions. Given that genetic variation in the 3′UTR of a gene may create or destroy a miRNA binding site [127], the *cis* effects are likely mediated through miRNA-regulated pathways. Besides this, *cis*-acting variants in promoter or enhancer regions may influence the binding of transcription factors and thus promoter regulation. Nevertheless, it is still not known whether *trans* effects are mediated through transcription factor variants or through other mechanisms [123]. Generally, *cis* eQTLs tend to have stronger influence on target gene regulation than *trans* eQTLs. Moreover, there exist the so-called eQTL hot spots in which the expression levels of many transcripts are associated with the variation.

The resulting comprehensive eQTL maps provide potential insight into a biological basis for complex quantitative trait associations identified through GWAS [123]. Since the expression of transcripts is subject to intensive gene regulation, eQTL data should be interpreted further by the incorporation of additional biological information, such as results from GWAS and EWAS as discussed above, and analysis of regulatory networks, which are discussed below. This kind of integrated analyses has been utilised in several studies [112, 114, 115, 128, 129].

Proteins are mainly responsible for the biological phenotype; they thus should more accurately reflect the cellular physiological state or the changes induced by disease processes, drug treatment or other influences, compared with genetic, epigenetic or transcript variants. Various mechanisms of post-transcriptional regulation can lead to changes in protein abundance in the absence of a corresponding alteration of transcript levels, suggesting that the proteome is expected to provide important biological insights and disease biomarkers that cannot be captured through evaluation of the transcriptome alone [130]. We mention here that association mapping analysis could also be done at the protein level in terms of protein QTL (pQTL or PQL [131]) mapping, in which protein abundance or modification is treated as a phenotypic trait. pQTL mapping, complementary to eQTL mapping, is now becoming feasible with technical advances in mass spectrometry (MS)-based proteomics [130, 132, 133]. The little overlap between pQTLs and eQTLs from the same study [134] indicates that the proteome and the transcriptome give distinct insights into the diversity between different individuals and further highlights the implications for systems-biology approaches that utilise such high-throughput data into integrated analysis.

11.3.4 Genome-Wide Association Studies with Metabolomics: Metabolic QTL Analysis

In addition to genomics, epigenomics, transcriptomics and proteomics, metabolomics is emerging as a complementary approach for globally measuring ideally all endogenous small organic molecules (metabolic traits; normally below 1,500 Da) in a biological sample. However, unlike the transcriptome and to a lesser degree the proteome, the metabolome is much more amenable to variation. The metabolome is much more diverse in terms of chemical structure and function [135]. Metabolite profiles capture important information on the environment (diet, lifestyle, gut microbial activity and bacterial activity) that individuals experience and can give an instantaneous snapshot of the individual's physiological state at that particular time under a particular set of conditions. Some changes in metabolite levels may be a consequence of the phenotypic diversity; therefore, a metabolic trait presents a functional intermediate trait or merely a correlated biomarker [136]. Noninvasive metabolic methodologies include nuclear magnetic resonance (NMR) spectroscopy [137], MS and high-performance liquid-phase chromatography (HPLC). Due to advances in these technologies, quantitative readouts for hundreds of small molecules that are detected in large scale can now be provided. Experimental design concerns the choice of which metabolites to study. While targeted methods provide precise measurements of specific (known) metabolites and are easy to replicate, nontargeted approaches are currently more promising as they provide the opportunity to discover novel associations including hitherto uncharacterised metabolites [136].

In the past few years, GWAS face the challenge that the effect of sizes of genetic association is generally small and information on the underlying biological processes is lacking [136]. These problems can be overcome, at least partially, by association with metabolic traits as functional intermediates [138]. There is the increased interest from the scientific community, and particularly plant biologists, in integrating metabolic approaches into research with the aim to unravel phenotypic diversity and its underlying genetic variation [78]. The combination of high-throughput metabolic phenotyping with general QTL analysis has thus given birth to the emerging field of metabolome-wide association studies (MWAS; Fig. 11.3).

The study of the chemical composition (i.e. the metabolite) of plants has always been of great interest in biological research, in part because metabolic phenotypes (metabotypes) largely reflect the developmental stage of the plant and its interactions with the environment. In plants, the first studies combining metabolic phenotyping with QTL analysis were performed in tomato [139–141] and successfully uncovered loci (metabolite QTLs, mQTLs) regulating plant metabolite composition. In *Arabidopsis* [142–147] and other crops, such as *Brassica napus* [148, 149], potato [150], rice [151] and maize [138, 152], mQTL mapping analyses have also been implemented using targeted and nontargeted metabolic profiling. Metabolite profiling-based approaches furthermore provide important steps towards the goal of hybrid performance prediction [152] and metabolomics-assisted crop breeding [153].

Similar MWAS were later performed in human studies [154–158]. Large panels of metabotypes have been analysed in association with genetic variants, disease-related phenotypes and lifestyle and environmental parameters, allowing dissection of the contribution of these factors to the aetiology of complex diseases [136]. These MWAS have identified genetic factors reliably that influence intermediate traits on phenotypes such as blood pressure [158], cardiometabolic disorder [157] and coronary heart disease [159]. In summary, incorporation of GWAS and metabolomics further refine the G-P map and eventually identify possible prognostic or diagnostic biomarkers of disease risk and biomarkers for predictive plant breeding.

11.3.5 Systems Biology: Genome-Scale Networks That Link Genes to Phenotypes

Associating sequence-level variation (such as SNPs and CNVs) with high-level variation in organismal phenotypes (such as disease susceptibility or crop yield) omits all of the intermediate steps in the chain of causation from genetic perturbation to phenotypic diversity. As mentioned above, intermediate molecular phenotypes (endophenotypes) such as epigenetic variation, transcript/protein abundance and metabolic traits vary genetically in populations and are themselves quantitative traits [98]. These endophenotypes functionally link genetic variation to disease-predisposing (for human) or biomass-predisposing (for plants) factors and then to complex phenotypic end points. Excitingly, the so-called "genetical genomics" approach [160] now enables us to integrate genetic variation, various endophenotypic variation and variation in organismal phenotypes in a linkage or association mapping population in both human [161] and plants [162], allowing to interpret quantitative genetic variation in terms of biologically meaningful causal networks of correlated transcripts.

However, it is becoming clear that each of the intermediate steps in translating biological information from genotype to phenotype does not stand alone [135]. The omics technologies now enable us to understand the biology inside the "black box" that lies between genotype and phenotype in terms of complex interacting networks [135, 163] (Fig. 11.4). Although we are still far away from a holistic understanding of the G-P map, systems biology is an emerging approach that aims to elucidate higher-level behaviour of biological systems and focuses on complex interactions within them, illuminating the path towards this ultimate goal – the complete G-P map. The integrative systems approach tries to link together the single-level omics data (e.g. genome, epigenome, transcriptome, proteome and metabolome) and, over time (if available [164]), to reveal and model the dynamic molecular regulatory networks or pathways from gene-to-function in order to bridge from genomics to phenomics. With the availability of increasingly powerful omics-based technologies, analytical and statistical tools and integrated knowledge bases, it has become possible to establish new links between genes, biological functions and a wide range of human diseases [165–179]. The comprehensive gene-disease associations present important insights that different disease modules (i.e. diseases share common genetic origins) could overlap and perturbations caused by one disease could affect other disease modules [180]. The identification of disease modules leads to the concept of the diseasome [165], which represents disease networks whose nodes are diseases and whose links represent the shared molecular relationships between the disease pairs. The underlying disease-associated cellular components are mostly investigated with protein-coding genes [165, 166, 168, 176, 177], though miRNAs [173, 178, 181], large intergenic noncoding RNAs (lincRNAs) [175] or metabolic pathways [171] are also investigated. Importantly, uncovering such diseasome networks provides hints on how different phenotypes are linked at the molecular level.

Although GWAS and analogous methodologies have presented large numbers of disease-gene candidates, it still has the difficulty to identify the particular gene and the causal mutation [180]. A series of sophisticated strategies have recently been developed to predict potential disease genes (Fig. 11.5). These network-based tools include linkage methods [182], functional module-based or "guilt-by-association" methods [166, 176, 177] and diffusion-based methods [183, 184]. Furthermore, it is

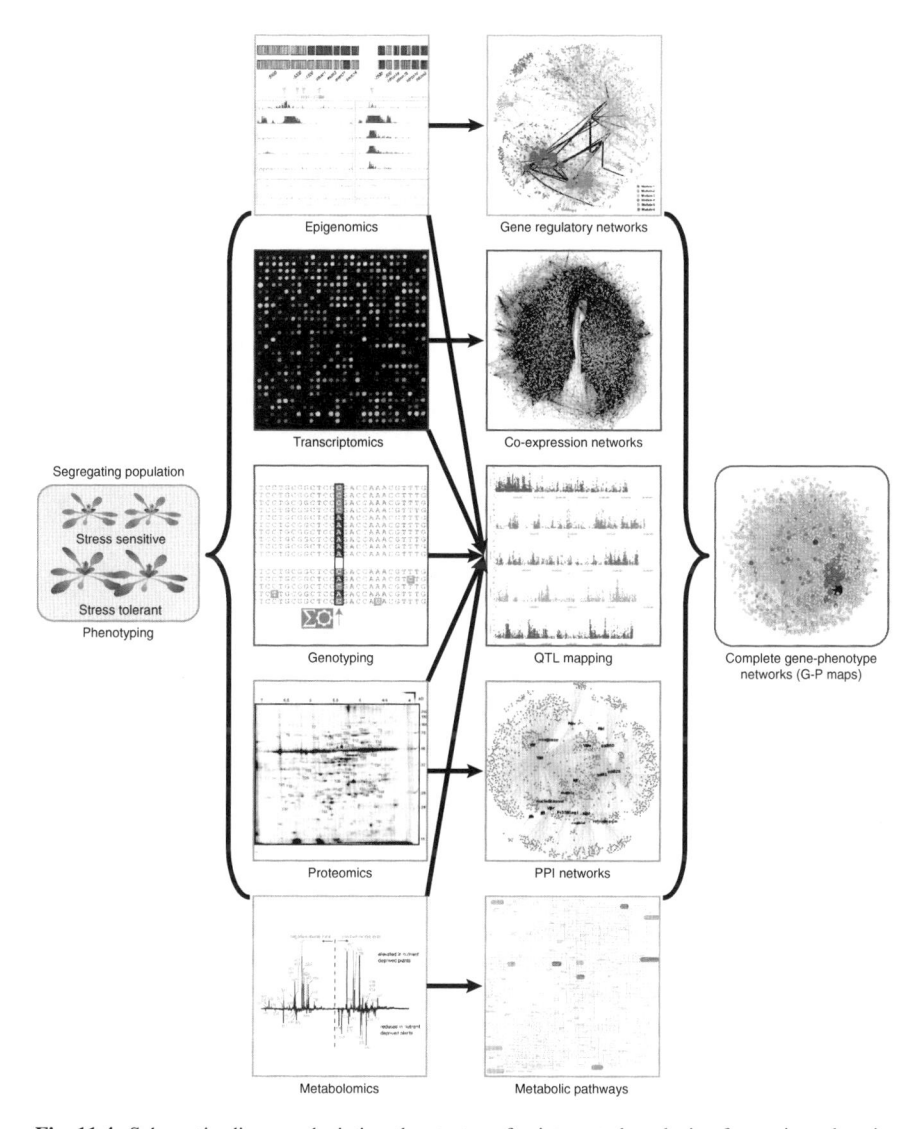

Fig. 11.4 Schematic diagram depicting the strategy for integrated analysis of genetic and omic data. Large-scale genotyping and phenotyping are performed on segregating populations. Quantitative traits can be analysed on different levels to identify responsible loci (QTLs) based on QTL mapping approaches. Retrieved data can also be used in cluster analyses to identify gene-centred networks. The methodology of the combined used of genetic and omic technologies is commonly referred to as "genetical genomics" [160] and enables the elucidation of complex gene–phenotype networks (the G-P maps). This figure extends the work from Keurentjes [135]

believed that genes tend to work in evolutionarily conserved pathways or modules; so the G-P maps can potentially be transferred between different species. Based on this assumption, orthologous phenotypes (phenologs) can be used to systematically

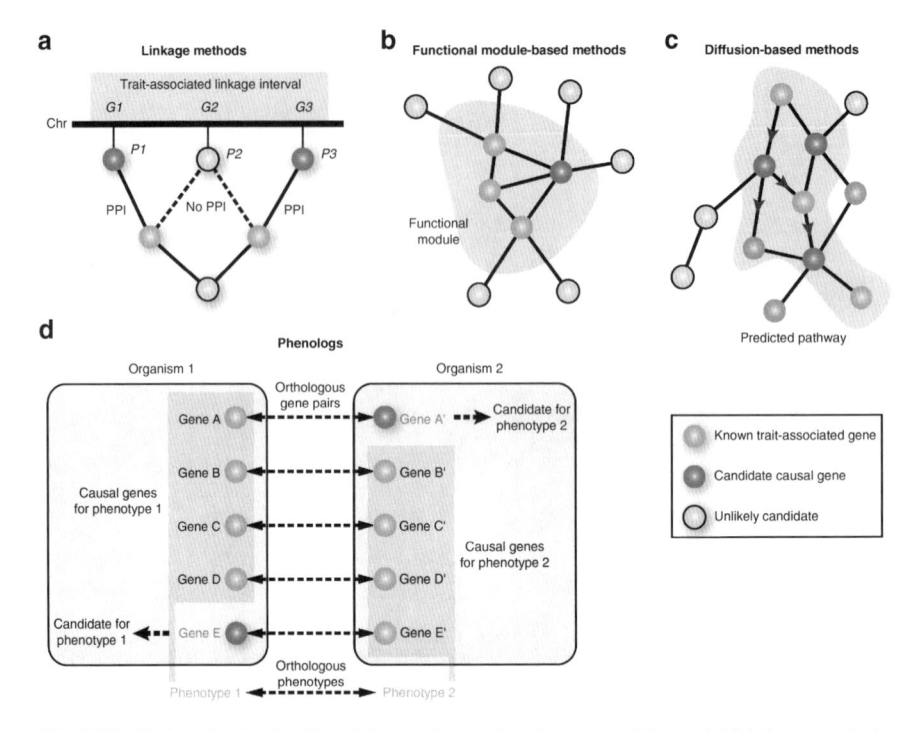

Fig. 11.5 Methodologies for identifying trait-associated gene candidates. (**a**) Linkage methods. These methods combine both the linkage analysis (to determine the linkage interval of a specific trait) and protein–protein interaction (PPI) information. Genes (denoted as G1, G2 and so on) located in the linkage interval whose protein products interact with a known trait-associated protein are considered likely candidate genes. (**b**) Functional module-based or guilt-by-association methods. Function modules are identified from clustering analysis of genome-scale networks. The members of such modules are considered candidate genes linked to specific phenotypes. (**c**) Diffusion-based methods. Starting from proteins that are known to be associated with a phenotype, a random walker visits each node in the interactome with a certain probability. The outcome of this algorithm is a trait-association score that is assigned to each protein, that is, the likelihood that a particular protein is associated with the phenotype. (**d**) Phenologs (orthologous phenotypes). Phenologs is used to map phenotypes between organisms based on significantly overlapping sets of orthologous genes. Perturbation of overlapping modules of orthologous genes may result in one set of phenotypes in one organism but a different set of phenotypes in another organism. The genes in such modules are considered candidates associated with the corresponding phenotypes (Parts **a–c** are modified from Barabasi et al. [180]. Part **d** is modified from McGary et al. [185])

predict genes associated nonobviously with diseases across different organisms using overlapping sets of orthologous genes [185]. In summary, the value of these tools is expected to increase with the wealth of disease gene candidates beyond GWAS. Although most of the initial studies based on these tools were performed in humans, similar strategies can also be applied to the plant biological research [186]. Indeed, networks for *Arabidopsis* [187], rice [188, 189] and maize [189] have been shown to connect thousands of genes accurately to phenotypes.

11.4 Perspectives and Future Challenges

The basic requirements for building an ideal phenomics realm are easy to imagine but still hard to realise. We are facing great opportunities but also great challenges in the areas of both genomics and phenomics. Although technically feasible, extensive and intensive measurement of genetic contents (such as epigenetic modification, gene expression, metabolite content) on large samples of genotypes across the full range of spatial and temporal scales is costly. Furthermore, the high density of genetic markers identified thus far yet awaits to be linked to their consequential phenotypic traits. On the phenomics side, the major challenge resides in the multitudes of phenotypic traits and environmental influences. The cost of a phenome project using current technology is extremely high [3]. High-throughput and high-resolution phenotyping technologies, for detection of both internal and external phenotypes, especially in plants, have started to open new horizons [3, 49]. Extracting as much quantitative information as possible from phenotyping data is a fundamental goal for phenomics. In other words, future phenomic efforts need to focus on comprehensive and quantitative measurements of phenotypes, rather than conventionally low-dimensional and qualitative phenotype categorisations [3]. Developments in phenomics will increase both the number of phenotypic traits that are quantitatively assessed and the sample sizes (number of individuals or genotypes characterised), resulting in major challenges with respect to data analysis. The available state-of-the-art methods, such as partial least squares (PLS) regression, principal component analysis (PCA), random forests (RF) and support vector machines (SVM), can be used to address the high-dimensional phenomic data. Another challenge in new analytics is automated analysis of phenotyping data, since navigating the huge imaging data sets manually is extremely tedious.

Regarding linking genotype to phenotype, many important challenges remain: (a) with respect to the problem of linking genes to traits, according to the observation of vast numbers of associated variants located within noncoding regions of the genome [90]; (b) with respect to epistatic interactions [190]; (c) with respect to gene-environment interactions [191]; (d) with respect to epigenetic influences on phenotypic variation; and (e) with respect to variation in the outcome of mutations among individuals [73]. One promising solution here is to combine data from multiple "omics" technologies in what may be termed "a genome-wide systems-biology approach".

In a nutshell, however, phenomics lags largely behind genomics. In contrast to the situation in humans, in plant organisms it is relatively straightforward to carry out systematic genetic screens and large-scale phenotyping under various controlled environments. This provides unbiased assessment of the genetic complexity of phenotypic traits [73]. The G-P maps are therefore ultimately expected to be more complete and more systematic in plants than they may be in humans. Notably, many ongoing developing or developed phenomics tools will give plant scientists the power to unlock the information coded in genomes (Table 11.3). In the near future, the plant phenotypic landscape will be populated at a faster pace to accelerate research in model organisms and to bridge the gap between genomics and phenomics [3, 49].

Acknowledgement This work was supported by grants from the Federal Office of Agriculture and Food (15/12-13, 530–06.01-BiKo CHN), the Robert Bosch Stiftung (32.5.8003.0116.0) and the Federal Ministry of Education and Research (BMBF – 0315958A).

WWW List in This Chapter

- The NHGRI GWAS Catalogue: http://www.genome.gov/gwastudies/
  A catalogue of published genome-wide association studies (GWAS)
- LemnaTec: http://www.lemnatec.com/
  High-throughput and high-content screening solutions for plant phenomics
- IAP: http://iap.ipk-gatersleben.de/
  An Integrated Analysis Platform (IAP) for plant high-throughput phenotyping data analysis
- Note: Other useful links are listed in Tables 11.2 and 11.3.

References

1. Manolio TA, Collins FS, Cox NJ, Goldstein DB, Hindorff LA, Hunter DJ, McCarthy MI, Ramos EM, Cardon LR, Chakravarti A et al (2009) Finding the missing heritability of complex diseases. Nature 461(7265):747–753
2. Freedman ML, Monteiro AN, Gayther SA, Coetzee GA, Risch A, Plass C, Casey G, De Biasi M, Carlson C, Duggan D et al (2011) Principles for the post-GWAS functional characterization of cancer risk loci. Nat Genet 43(6):513–518
3. Houle D, Govindaraju DR, Omholt S (2010) Phenomics: the next challenge. Nat Rev Genet 11(12):855–866
4. Furbank RT, Tester M (2011) Phenomics–technologies to relieve the phenotyping bottleneck. Trends Plant Sci 16(12):635–644
5. Botstein D, White RL, Skolnick M, Davis RW (1980) Construction of a genetic linkage map in man using restriction fragment length polymorphisms. Am J Hum Genet 32:314–331
6. Kan YW, Dozy AM (1978) Polymorphism of DNA sequence adjacent to human beta-globin structural gene: relationship to sickle mutation. Proc Natl Acad Sci U S A 75(11):5631–5635
7. Vos P, Hogers R, Bleeker M, Reijans M, van de Lee T, Hornes M, Frijters A, Pot J, Peleman J, Kuiper M et al (1995) AFLP: a new technique for DNA fingerprinting. Nucleic Acids Res 23(21):4407–4414
8. Bodmer WF (1986) Human genetics: the molecular challenge. Cold Spring Harb Symp Quant Biol 51 Pt 1:1–13
9. Kruglyak L (1997) The use of a genetic map of biallelic markers in linkage studies. Nat Genet 17(1):21–24
10. Wang DG, Fan JB, Siao CJ, Berno A, Young P, Sapolsky R, Ghandour G, Perkins N, Winchester E, Spencer J et al (1998) Large-scale identification, mapping, and genotyping of single-nucleotide polymorphisms in the human genome. Science 280(5366):1077–1082

11. Gunderson KL, Steemers FJ, Lee G, Mendoza LG, Chee MS (2005) A genome-wide scalable SNP genotyping assay using microarray technology. Nat Genet 37(5):549–554
12. Syvanen AC (2005) Toward genome-wide SNP genotyping. Nat Genet 37(Suppl):S5–S10
13. Davey JW, Hohenlohe PA, Etter PD, Boone JQ, Catchen JM, Blaxter ML (2011) Genome-wide genetic marker discovery and genotyping using next-generation sequencing. Nat Rev Genet 12(7):499–510
14. Nielsen R, Paul JS, Albrechtsen A, Song YS (2011) Genotype and SNP calling from next-generation sequencing data. Nat Rev Genet 12(6):443–451
15. Van Tassell CP, Smith TP, Matukumalli LK, Taylor JF, Schnabel RD, Lawley CT, Haudenschild CD, Moore SS, Warren WC, Sonstegard TS (2008) SNP discovery and allele frequency estimation by deep sequencing of reduced representation libraries. Nat Methods 5(3):247–252
16. van Orsouw NJ, Hogers RC, Janssen A, Yalcin F, Snoeijers S, Verstege E, Schneiders H, van der Poel H, van Oeveren J, Verstegen H et al (2007) Complexity reduction of polymorphic sequences (CRoPS): a novel approach for large-scale polymorphism discovery in complex genomes. PLoS One 2(11):e1172
17. Baird NA, Etter PD, Atwood TS, Currey MC, Shiver AL, Lewis ZA, Selker EU, Cresko WA, Johnson EA (2008) Rapid SNP discovery and genetic mapping using sequenced RAD markers. PLoS One 3(10):e3376
18. Andolfatto P, Davison D, Erezyilmaz D, Hu TT, Mast J, Sunayama-Morita T, Stern DL (2011) Multiplexed shotgun genotyping for rapid and efficient genetic mapping. Genome Res 21(4):610–617
19. Elshire RJ, Glaubitz JC, Sun Q, Poland JA, Kawamoto K, Buckler ES, Mitchell SE (2011) A robust, simple genotyping-by-sequencing (GBS) approach for high diversity species. PLoS One 6(5):e19379
20. Frazer KA, Murray SS, Schork NJ, Topol EJ (2009) Human genetic variation and its contribution to complex traits. Nat Rev Genet 10(4):241–251
21. Thorisson GA, Stein LD (2003) The SNP Consortium website: past, present and future. Nucleic Acids Res 31(1):124–127
22. International HapMap C (2003) The International HapMap project. Nature 426(6968): 789–796
23. Gonzaga-Jauregui C, Lupski JR, Gibbs RA (2012) Human genome sequencing in health and disease. Annu Rev Med 63:35–61
24. Genomes Project C, Abecasis GR, Altshuler D, Auton A, Brooks LD, Durbin RM, Gibbs RA, Hurles ME, McVean GA (2010) A map of human genome variation from population-scale sequencing. Nature 467(7319):1061–1073
25. Yi X, Liang Y, Huerta-Sanchez E, Jin X, Cuo ZX, Pool JE, Xu X, Jiang H, Vinckenbosch N, Korneliussen TS et al (2010) Sequencing of 50 human exomes reveals adaptation to high altitude. Science 329(5987):75–78
26. Genomes Project C, Abecasis GR, Auton A, Brooks LD, DePristo MA, Durbin RM, Handsaker RE, Kang HM, Marth GT, McVean GA (2012) An integrated map of genetic variation from 1,092 human genomes. Nature 491(7422):56–65
27. Atwell S, Huang YS, Vilhjalmsson BJ, Willems G, Horton M, Li Y, Meng D, Platt A, Tarone AM, Hu TT et al (2010) Genome-wide association study of 107 phenotypes in Arabidopsis thaliana inbred lines. Nature 465(7298):627–631
28. Huang X, Wei X, Sang T, Zhao Q, Feng Q, Zhao Y, Li C, Zhu C, Lu T, Zhang Z et al (2010) Genome-wide association studies of 14 agronomic traits in rice landraces. Nat Genet 42(11):961–967
29. Gore MA, Chia JM, Elshire RJ, Sun Q, Ersoz ES, Hurwitz BL, Peiffer JA, McMullen MD, Grills GS, Ross-Ibarra J et al (2009) A first-generation haplotype map of maize. Science 326(5956):1115–1117
30. Lai J, Li R, Xu X, Jin W, Xu M, Zhao H, Xiang Z, Song W, Ying K, Zhang M et al (2010) Genome-wide patterns of genetic variation among elite maize inbred lines. Nat Genet 42(11):1027–1030

31. Zheng LY, Guo XS, He B, Sun LJ, Peng Y, Dong SS, Liu TF, Jiang S, Ramachandran S, Liu CM et al (2011) Genome-wide patterns of genetic variation in sweet and grain sorghum (Sorghum bicolor). Genome Biol 12(11):R114
32. Cockram J, White J, Zuluaga DL, Smith D, Comadran J, Macaulay M, Luo Z, Kearsey MJ, Werner P, Harrap D et al (2010) Genome-wide association mapping to candidate polymorphism resolution in the unsequenced barley genome. Proc Natl Acad Sci U S A 107(50):21611–21616
33. Waddington CH (1968) Towards a theoretical biology. Nature 218(5141):525–527
34. Walter T, Shattuck DW, Baldock R, Bastin ME, Carpenter AE, Duce S, Ellenberg J, Fraser A, Hamilton N, Pieper S et al (2010) Visualization of image data from cells to organisms. Nat Methods 7(3 Suppl):S26–S41
35. Perlman ZE, Slack MD, Feng Y, Mitchison TJ, Wu LF, Altschuler SJ (2004) Multidimensional drug profiling by automated microscopy. Science 306(5699):1194–1198
36. Neumann B, Held M, Liebel U, Erfle H, Rogers P, Pepperkok R, Ellenberg J (2006) High-throughput RNAi screening by time-lapse imaging of live human cells. Nat Methods 3(5):385–390
37. Bakal C, Aach J, Church G, Perrimon N (2007) Quantitative morphological signatures define local signaling networks regulating cell morphology. Science 316(5832):1753–1756
38. Loo LH, Wu LF, Altschuler SJ (2007) Image-based multivariate profiling of drug responses from single cells. Nat Methods 4(5):445–453
39. Jones TR, Carpenter AE, Lamprecht MR, Moffat J, Silver SJ, Grenier JK, Castoreno AB, Eggert US, Root DE, Golland P et al (2009) Scoring diverse cellular morphologies in image-based screens with iterative feedback and machine learning. Proc Natl Acad Sci U S A 106(6):1826–1831
40. Collinet C, Stoter M, Bradshaw CR, Samusik N, Rink JC, Kenski D, Habermann B, Buchholz F, Henschel R, Mueller MS et al (2010) Systems survey of endocytosis by multiparametric image analysis. Nature 464(7286):243–249
41. Fuchs F, Pau G, Kranz D, Sklyar O, Budjan C, Steinbrink S, Horn T, Pedal A, Huber W, Boutros M (2010) Clustering phenotype populations by genome-wide RNAi and multiparametric imaging. Mol Syst Biol 6:370
42. Mukherji M, Bell R, Supekova L, Wang Y, Orth AP, Batalov S, Miraglia L, Huesken D, Lange J, Martin C et al (2006) Genome-wide functional analysis of human cell-cycle regulators. Proc Natl Acad Sci U S A 103(40):14819–14824
43. Carpenter AE, Sabatini DM (2004) Systematic genome-wide screens of gene function. Nat Rev Genet 5(1):11–22
44. Pepperkok R, Ellenberg J (2006) High-throughput fluorescence microscopy for systems biology. Nat Rev Mol Cell Biol 7(9):690–696
45. Schneider CA, Rasband WS, Eliceiri KW (2012) NIH Image to ImageJ: 25 years of image analysis. Nat Methods 9(7):671–675
46. Pau G, Fuchs F, Sklyar O, Boutros M, Huber W (2010) EBImage-an R package for image processing with applications to cellular phenotypes. Bioinformatics 26(7):979–981
47. Carpenter AE, Jones TR, Lamprecht MR, Clarke C, Kang IH, Friman O, Guertin DA, Chang JH, Lindquist RA, Moffat J et al (2006) Cell Profiler: image analysis software for identifying and quantifying cell phenotypes. Genome Biol 7(10):R100
48. Rajaram S, Pavie B, Wu LF, Altschuler SJ (2012) PhenoRipper: software for rapidly profiling microscopy images. Nat Methods 9(7):635–637
49. Fiorani F, Schurr U (2013) Future scenarios for plant phenotyping. Annu Rev Plant Biol 64(1):267–291
50. Reuzeau C, Pen J, Frankard V, de Wolf J, Peerbolte R, Broekaert W (2005) TraitMill: a discovery engine for identifying yield-enhancement genes in cereals. Fenzi Zhiwu Yuzhong (Mol Plant Breed) 3:7534
51. Granier C, Aguirrezabal L, Chenu K, Cookson SJ, Dauzat M, Hamard P, Thioux JJ, Rolland G, Bouchier-Combaud S, Lebaudy A et al (2006) PHENOPSIS, an automated platform for

reproducible phenotyping of plant responses to soil water deficit in Arabidopsis thaliana permitted the identification of an accession with low sensitivity to soil water deficit. New Phytol 169(3):623–635
52. Reuzeau C, Frankard V, Hatzfeld Y, Sanz A, Van Camp W, Lejeune P, De Wilde C, Lievens K, de Wolf J, Vranken E et al (2006) Traitmill™: a functional genomics platform for the phenotypic analysis of cereals. Plant Genet Resour 4(01):20–24
53. Walter A, Scharr H, Gilmer F, Zierer R, Nagel KA, Ernst M, Wiese A, Virnich O, Christ MM, Uhlig B et al (2007) Dynamics of seedling growth acclimation towards altered light conditions can be quantified via GROWSCREEN: a setup and procedure designed for rapid optical phenotyping of different plant species. New Phytol 174(2):447–455
54. Bylesjo M, Segura V, Soolanayakanahally RY, Rae AM, Trygg J, Gustafsson P, Jansson S, Street NR (2008) LAMINA: a tool for rapid quantification of leaf size and shape parameters. BMC Plant Biol 8:82
55. Jansen M, Gilmer F, Biskup B, Nagel KA, Rascher U, Fischbach A, Briem S, Dreissen G, Tittmann S, Braun S et al (2009) Simultaneous phenotyping of leaf growth and chlorophyll fluorescence via GROWSCREEN FLUORO allows detection of stress tolerance in Arabidopsis thaliana and other rosette plants. Funct Plant Biol 36(10–11):902–914
56. Wang L, Uilecan IV, Assadi AH, Kozmik CA, Spalding EP (2009) HYPOTrace: image analysis software for measuring hypocotyl growth and shape demonstrated on Arabidopsis seedlings undergoing photomorphogenesis. Plant Physiol 149(4):1632–1637
57. Hartmann A, Czauderna T, Hoffmann R, Stein N, Schreiber F (2011) HTPheno: an image analysis pipeline for high-throughput plant phenotyping. BMC Bioinform 12:148
58. Weight C, Parnham D, Waites R (2008) LeafAnalyser: a computational method for rapid and large-scale analyses of leaf shape variation. Plant J 53(3):578–586
59. Meyer RC, Kusterer B, Lisec J, Steinfath M, Becher M, Scharr H, Melchinger AE, Selbig J, Schurr U, Willmitzer L et al (2010) QTL analysis of early stage heterosis for biomass in Arabidopsis. Theor Appl Genet Theoretische und angewandte Genetik 120(2):227–237
60. Biskup B, Scharr H, Fischbach A, Wiese-Klinkenberg A, Schurr U, Walter A (2009) Diel growth cycle of isolated leaf discs analyzed with a novel, high-throughput three-dimensional imaging method is identical to that of intact leaves. Plant Physiol 149(3):1452–1461
61. Clark RT, MacCurdy RB, Jung JK, Shaff JE, McCouch SR, Aneshansley DJ, Kochian LV (2011) Three-dimensional root phenotyping with a novel imaging and software platform. Plant Physiol 156(2):455–465
62. Lobet G, Pages L, Draye X (2011) A novel image-analysis toolbox enabling quantitative analysis of root system architecture. Plant Physiol 157(1):29–39
63. Paproki A, Sirault X, Berry S, Furbank R, Fripp J (2012) A novel mesh processing based technique for 3D plant analysis. BMC Plant Biol 12:63
64. Wuyts N, Palauqui JC, Conejero G, Verdeil JL, Granier C, Massonnet C (2010) High-contrast three-dimensional imaging of the Arabidopsis leaf enables the analysis of cell dimensions in the epidermis and mesophyll. Plant Methods 6:17
65. Sozzani R, Benfey PN (2011) High-throughput phenotyping of multicellular organisms: finding the link between genotype and phenotype. Genome Biol 12(3):219
66. Arvidsson S, Perez-Rodriguez P, Mueller-Roeber B (2011) A growth phenotyping pipeline for Arabidopsis thaliana integrating image analysis and rosette area modeling for robust quantification of genotype effects. New Phytol 191(3):895–907
67. Golzarian MR, Frick RA, Rajendran K, Berger B, Roy S, Tester M, Lun DS (2011) Accurate inference of shoot biomass from high-throughput images of cereal plants. Plant Methods 7:2
68. Tardieu F, Tuberosa R (2010) Dissection and modelling of abiotic stress tolerance in plants. Curr Opin Plant Biol 13(2):206–212
69. van der Heijden G, Song Y, Horgan G, Polder G, Dieleman A, Bink M, Palloix A, van Eeuwijk F, Glasbey C (2012) SPICY: towards automated phenotyping of large pepper plants in the greenhouse. Funct Plant Biol 39(10–11):870–877
70. Klukas C, Pape JM, Entzian A (2012) Analysis of high-throughput plant image data with the information system IAP. J Integr Bioinform 9(2):191

71. Rakyan VK, Down TA, Balding DJ, Beck S (2011) Epigenome-wide association studies for common human diseases. Nat Rev Genet 12(8):529–541
72. Tester M, Langridge P (2010) Breeding technologies to increase crop production in a changing world. Science 327(5967):818–822
73. Lehner B (2013) Genotype to phenotype: lessons from model organisms for human genetics. Nat Rev Genet 14(3):168–178
74. Wong AH, Gottesman II, Petronis A (2005) Phenotypic differences in genetically identical organisms: the epigenetic perspective. Hum Mol Genet 14(1):R11–R18
75. Meyer UA, Zanger UM, Schwab M (2013) Omics and drug response. Annu Rev Pharmacol Toxicol 53:475–502
76. Nagano AJ, Sato Y, Mihara M, Antonio BA, Motoyama R, Itoh H, Nagamura Y, Izawa T (2012) Deciphering and prediction of transcriptome dynamics under fluctuating field conditions. Cell 151(6):1358–1369
77. Jaeger PA, Doherty C, Ideker T (2012) Modeling transcriptome dynamics in a complex world. Cell 151(6):1161–1162
78. Carreno-Quintero N, Bouwmeester HJ, Keurentjes JJ (2013) Genetic analysis of metabolome-phenotype interactions: from model to crop species. Trends Genet 29(1):41–50
79. Altshuler D, Daly MJ, Lander ES (2008) Genetic mapping in human disease. Science 322(5903):881–888
80. Donnelly P (2008) Progress and challenges in genome-wide association studies in humans. Nature 456(7223):728–731
81. Takeda S, Matsuoka M (2008) Genetic approaches to crop improvement: responding to environmental and population changes. Nat Rev Genet 9(6):444–457
82. Ott J, Kamatani Y, Lathrop M (2011) Family-based designs for genome-wide association studies. Nat Rev Genet 12(7):465–474
83. Kang HM, Zaitlen NA, Wade CM, Kirby A, Heckerman D, Daly MJ, Eskin E (2008) Efficient control of population structure in model organism association mapping. Genetics 178(3):1709–1723
84. Kang HM, Sul JH, Service SK, Zaitlen NA, Kong SY, Freimer NB, Sabatti C, Eskin E (2010) Variance component model to account for sample structure in genome-wide association studies. Nat Genet 42(4):348–354
85. Zhou X, Stephens M (2012) Genome-wide efficient mixed-model analysis for association studies. Nat Genet 44(7):821–824
86. Aulchenko YS, Ripke S, Isaacs A, van Duijn CM (2007) GenABEL: an R library for genome-wide association analysis. Bioinformatics 23(10):1294–1296
87. Lippert C, Listgarten J, Liu Y, Kadie CM, Davidson RI, Heckerman D (2011) FaST linear mixed models for genome-wide association studies. Nat Methods 8(10):833–835
88. Bradbury PJ, Zhang Z, Kroon DE, Casstevens TM, Ramdoss Y, Buckler ES (2007) TASSEL: software for association mapping of complex traits in diverse samples. Bioinformatics 23(19):2633–2635
89. Price AL, Patterson NJ, Plenge RM, Weinblatt ME, Shadick NA, Reich D (2006) Principal components analysis corrects for stratification in genome-wide association studies. Nat Genet 38(8):904–909
90. Hindorff LA, Sethupathy P, Junkins HA, Ramos EM, Mehta JP, Collins FS, Manolio TA (2009) Potential etiologic and functional implications of genome-wide association loci for human diseases and traits. Proc Natl Acad Sci U S A 106(23):9362–9367
91. Eichler EE, Flint J, Gibson G, Kong A, Leal SM, Moore JH, Nadeau JH (2010) Missing heritability and strategies for finding the underlying causes of complex disease. Nat Rev Genet 11(6):446–450
92. Maher B (2008) Personal genomes: the case of the missing heritability. Nature 456(7218):18–21
93. Brachi B, Morris GP, Borevitz JO (2011) Genome-wide association studies in plants: the missing heritability is in the field. Genome Biol 12(10):232

94. Li Y, Huang Y, Bergelson J, Nordborg M, Borevitz JO (2010) Association mapping of local climate-sensitive quantitative trait loci in Arabidopsis thaliana. Proc Natl Acad Sci U S A 107(49):21199–21204
95. Buckler ES, Holland JB, Bradbury PJ, Acharya CB, Brown PJ, Browne C, Ersoz E, Flint-Garcia S, Garcia A, Glaubitz JC et al (2009) The genetic architecture of maize flowering time. Science 325(5941):714–718
96. Li H, Peng Z, Yang X, Wang W, Fu J, Wang J, Han Y, Chai Y, Guo T, Yang N et al (2013) Genome-wide association study dissects the genetic architecture of oil biosynthesis in maize kernels. Nat Genet 45(1):43–50
97. Morrell PL, Buckler ES, Ross-Ibarra J (2011) Crop genomics: advances and applications. Nat Rev Genet 13(2):85–96
98. Mackay TF, Stone EA, Ayroles JF (2009) The genetics of quantitative traits: challenges and prospects. Nat Rev Genet 10(8):565–577
99. Yu J, Holland JB, McMullen MD, Buckler ES (2008) Genetic design and statistical power of nested association mapping in maize. Genetics 178(1):539–551
100. Kump KL, Bradbury PJ, Wisser RJ, Buckler ES, Belcher AR, Oropeza-Rosas MA, Zwonitzer JC, Kresovich S, McMullen MD, Ware D et al (2011) Genome-wide association study of quantitative resistance to southern leaf blight in the maize nested association mapping population. Nat Genet 43(2):163–168
101. Cavanagh C, Morell M, Mackay I, Powell W (2008) From mutations to MAGIC: resources for gene discovery, validation and delivery in crop plants. Curr Opin Plant Biol 11(2):215–221
102. Kover PX, Valdar W, Trakalo J, Scarcelli N, Ehrenreich IM, Purugganan MD, Durrant C, Mott R (2009) A multiparent advanced generation inter-cross to fine-map quantitative traits in Arabidopsis thaliana. PLoS Genet 5(7):e1000551
103. Rockman MV, Kruglyak L (2008) Breeding designs for recombinant inbred advanced intercross lines. Genetics 179(2):1069–1078
104. Balasubramanian S, Schwartz C, Singh A, Warthmann N, Kim MC, Maloof JN, Loudet O, Trainer GT, Dabi T, Borevitz JO et al (2009) QTL mapping in new Arabidopsis thaliana advanced intercross-recombinant inbred lines. PLoS One 4(2):e4318
105. Jannink JL, Lorenz AJ, Iwata H (2010) Genomic selection in plant breeding: from theory to practice. Brief Funct Genomics 9(2):166–177
106. Hamblin MT, Buckler ES, Jannink JL (2011) Population genetics of genomics-based crop improvement methods. Trends Genet 27(3):98–106
107. Rakyan VK, Hildmann T, Novik KL, Lewin J, Tost J, Cox AV, Andrews TD, Howe KL, Otto T, Olek A et al (2004) DNA methylation profiling of the human major histocompatibility complex: a pilot study for the human epigenome project. PLoS Biol 2(12):e405
108. Petronis A (2010) Epigenetics as a unifying principle in the aetiology of complex traits and diseases. Nature 465(7299):721–727
109. Fraga MF, Ballestar E, Paz MF, Ropero S, Setien F, Ballestar ML, Heine-Suner D, Cigudosa JC, Urioste M, Benitez J et al (2005) Epigenetic differences arise during the lifetime of monozygotic twins. Proc Natl Acad Sci U S A 102(30):10604–10609
110. Javierre BM, Fernandez AF, Richter J, Al-Shahrour F, Martin-Subero JI, Rodriguez-Ubreva J, Berdasco M, Fraga MF, O'Hanlon TP, Rider LG et al (2010) Changes in the pattern of DNA methylation associate with twin discordance in systemic lupus erythematosus. Genome Res 20(2):170–179
111. Nguyen A, Rauch TA, Pfeifer GP, Hu VW (2010) Global methylation profiling of lymphoblastoid cell lines reveals epigenetic contributions to autism spectrum disorders and a novel autism candidate gene, RORA, whose protein product is reduced in autistic brain. Fed Am Soc Exp Biol 24(8):3036–3051
112. Gibbs JR, van der Brug MP, Hernandez DG, Traynor BJ, Nalls MA, Lai SL, Arepalli S, Dillman A, Rafferty IP, Troncoso J et al (2010) Abundant quantitative trait loci exist for DNA methylation and gene expression in human brain. PLoS Genet 6(5):e1000952
113. Shoemaker R, Deng J, Wang W, Zhang K (2010) Allele-specific methylation is prevalent and is contributed by CpG-SNPs in the human genome. Genome Res 20(7):883–889

114. Bell JT, Pai AA, Pickrell JK, Gaffney DJ, Pique-Regi R, Degner JF, Gilad Y, Pritchard JK (2011) DNA methylation patterns associate with genetic and gene expression variation in HapMap cell lines. Genome Biol 12(1):R10
115. Bell JT, Tsai PC, Yang TP, Pidsley R, Nisbet J, Glass D, Mangino M, Zhai G, Zhang F, Valdes A et al (2012) Epigenome-wide scans identify differentially methylated regions for age and age-related phenotypes in a healthy ageing population. PLoS Genet 8(4):e1002629
116. Zhang D, Cheng L, Badner JA, Chen C, Chen Q, Luo W, Craig DW, Redman M, Gershon ES, Liu C (2010) Genetic control of individual differences in gene-specific methylation in human brain. Am J Hum Genet 86(3):411–419
117. Kilpinen H, Dermitzakis ET (2012) Genetic and epigenetic contribution to complex traits. Hum Mol Genet 21(R1):R24–R28
118. Vaughn MW, Tanurdzic M, Lippman Z, Jiang H, Carrasquillo R, Rabinowicz PD, Dedhia N, McCombie WR, Agier N, Bulski A et al (2007) Epigenetic natural variation in Arabidopsis thaliana. PLoS Biol 5(7):e174
119. Becker C, Hagmann J, Muller J, Koenig D, Stegle O, Borgwardt K, Weigel D (2011) Spontaneous epigenetic variation in the Arabidopsis thaliana methylome. Nature 480(7376):245–249
120. Eichten SR, Swanson-Wagner RA, Schnable JC, Waters AJ, Hermanson PJ, Liu S, Yeh CT, Jia Y, Gendler K, Freeling M et al (2011) Heritable epigenetic variation among maize inbreds. PLoS Genet 7(11):e1002372
121. Chodavarapu RK, Feng S, Ding B, Simon SA, Lopez D, Jia Y, Wang GL, Meyers BC, Jacobsen SE, Pellegrini M (2012) Transcriptome and methylome interactions in rice hybrids. Proc Natl Acad Sci U S A 109(30):12040–12045
122. Weigel D (2012) Natural variation in Arabidopsis: from molecular genetics to ecological genomics. Plant Physiol 158(1):2–22
123. Cookson W, Liang L, Abecasis G, Moffatt M, Lathrop M (2009) Mapping complex disease traits with global gene expression. Nat Rev Genet 10(3):184–194
124. Carlborg O, De Koning DJ, Manly KF, Chesler E, Williams RW, Haley CS (2005) Methodological aspects of the genetic dissection of gene expression. Bioinformatics 21(10):2383–2393
125. Dixon AL, Liang L, Moffatt MF, Chen W, Heath S, Wong KC, Taylor J, Burnett E, Gut I, Farrall M et al (2007) A genome-wide association study of global gene expression. Nat Genet 39(10):1202–1207
126. Veyrieras JB, Kudaravalli S, Kim SY, Dermitzakis ET, Gilad Y, Stephens M, Pritchard JK (2008) High-resolution mapping of expression-QTLs yields insight into human gene regulation. PLoS Genet 4(10):e1000214
127. Ryan BM, Robles AI, Harris CC (2010) Genetic variation in microRNA networks: the implications for cancer research. Nat Rev Cancer 10(6):389–402
128. Quon G, Lippert C, Heckerman D, Listgarten J (2013) Patterns of methylation heritability in a genome-wide analysis of four brain regions. Nucleic Acids Res 41(4):2095–2104
129. Emilsson V, Thorleifsson G, Zhang B, Leonardson AS, Zink F, Zhu J, Carlson S, Helgason A, Walters GB, Gunnarsdottir S et al (2008) Genetics of gene expression and its effect on disease. Nature 452(7186):423–428
130. Foss EJ, Radulovic D, Shaffer SA, Ruderfer DM, Bedalov A, Goodlett DR, Kruglyak L (2007) Genetic basis of proteome variation in yeast. Nat Genet 39(11):1369–1375
131. Damerval C, Maurice A, Josse JM, de Vienne D (1994) Quantitative trait loci underlying gene product variation: a novel perspective for analyzing regulation of genome expression. Genetics 137(1):289–301
132. Picotti P, Clement-Ziza M, Lam H, Campbell DS, Schmidt A, Deutsch EW, Rost H, Sun Z, Rinner O, Reiter L et al (2013) A complete mass-spectrometric map of the yeast proteome applied to quantitative trait analysis. Nature 494(7436):266–270
133. Gstaiger M, Aebersold R (2009) Applying mass spectrometry-based proteomics to genetics, genomics and network biology. Nat Rev Genet 10(9):617–627

134. Ghazalpour A, Bennett B, Petyuk VA, Orozco L, Hagopian R, Mungrue IN, Farber CR, Sinsheimer J, Kang HM, Furlotte N et al (2011) Comparative analysis of proteome and transcriptome variation in mouse. PLoS Genet 7(6):e1001393
135. Keurentjes JJ (2009) Genetical metabolomics: closing in on phenotypes. Curr Opin Plant Biol 12(2):223–230
136. Suhre K, Gieger C (2012) Genetic variation in metabolic phenotypes: study designs and applications. Nat Rev Genet 13(11):759–769
137. Borisjuk L, Rolletschek H, Neuberger T (2012) Surveying the plant's world by magnetic resonance imaging. Plant J Cell Mol Biol 70(1):129–146
138. Riedelsheimer C, Lisec J, Czedik-Eysenberg A, Sulpice R, Flis A, Grieder C, Altmann T, Stitt M, Willmitzer L, Melchinger AE (2012) Genome-wide association mapping of leaf metabolic profiles for dissecting complex traits in maize. Proc Natl Acad Sci U S A 109(23):8872–8877
139. Schauer N, Semel Y, Roessner U, Gur A, Balbo I, Carrari F, Pleban T, Perez-Melis A, Bruedigam C, Kopka J et al (2006) Comprehensive metabolic profiling and phenotyping of interspecific introgression lines for tomato improvement. Nat Biotechnol 24(4):447–454
140. Tieman DM, Zeigler M, Schmelz EA, Taylor MG, Bliss P, Kirst M, Klee HJ (2006) Identification of loci affecting flavour volatile emissions in tomato fruits. J Exp Bot 57(4):887–896
141. Schauer N, Semel Y, Balbo I, Steinfath M, Repsilber D, Selbig J, Pleban T, Zamir D, Fernie AR (2008) Mode of inheritance of primary metabolic traits in tomato. Plant Cell 20(3):509–523
142. Steinfath M, Strehmel N, Peters R, Schauer N, Groth D, Hummel J, Steup M, Selbig J, Kopka J, Geigenberger P et al (2010) Discovering plant metabolic biomarkers for phenotype prediction using an untargeted approach. Plant Biotechnol J 8(8):900–911
143. Meyer RC, Steinfath M, Lisec J, Becher M, Witucka-Wall H, Torjek O, Fiehn O, Eckardt A, Willmitzer L, Selbig J et al (2007) The metabolic signature related to high plant growth rate in Arabidopsis thaliana. Proc Natl Acad Sci U S A 104(11):4759–4764
144. Keurentjes JJ, Fu J, de Vos CH, Lommen A, Hall RD, Bino RJ, van der Plas LH, Jansen RC, Vreugdenhil D, Koornneef M (2006) The genetics of plant metabolism. Nat Genet 38(7):842–849
145. Lisec J, Meyer RC, Steinfath M, Redestig H, Becher M, Witucka-Wall H, Fiehn O, Torjek O, Selbig J, Altmann T et al (2008) Identification of metabolic and biomass QTL in Arabidopsis thaliana in a parallel analysis of RIL and IL populations. Plant J Cell Mol Biol 53(6):960–972
146. Lisec J, Steinfath M, Meyer RC, Selbig J, Melchinger AE, Willmitzer L, Altmann T (2009) Identification of heterotic metabolite QTL in Arabidopsis thaliana RIL and IL populations. Plant J Cell Mol Biol 59(5):777–788
147. Rowe HC, Hansen BG, Halkier BA, Kliebenstein DJ (2008) Biochemical networks and epistasis shape the Arabidopsis thaliana metabolome. Plant Cell 20(5):1199–1216
148. Lou P, Zhao J, He H, Hanhart C, Del Carpio DP, Verkerk R, Custers J, Koornneef M, Bonnema G (2008) Quantitative trait loci for glucosinolate accumulation in Brassica rapa leaves. New Phytol 179(4):1017–1032
149. Feng J, Long Y, Shi L, Shi J, Barker G, Meng J (2012) Characterization of metabolite quantitative trait loci and metabolic networks that control glucosinolate concentration in the seeds and leaves of Brassica napus. New Phytol 193(1):96–108
150. Carreno-Quintero N, Acharjee A, Maliepaard C, Bachem CW, Mumm R, Bouwmeester H, Visser RG, Keurentjes JJ (2012) Untargeted metabolic quantitative trait loci analyses reveal a relationship between primary metabolism and potato tuber quality. Plant Physiol 158(3):1306–1318
151. Matsuda F, Okazaki Y, Oikawa A, Kusano M, Nakabayashi R, Kikuchi J, Yonemaru J, Ebana K, Yano M, Saito K (2012) Dissection of genotype-phenotype associations in rice grains using metabolome quantitative trait loci analysis. Plant J Cell Mol Biol 70(4):624–636
152. Riedelsheimer C, Czedik-Eysenberg A, Grieder C, Lisec J, Technow F, Sulpice R, Altmann T, Stitt M, Willmitzer L, Melchinger AE (2012) Genomic and metabolic prediction of complex heterotic traits in hybrid maize. Nat Genet 44(2):217–220

153. Fernie AR, Schauer N (2009) Metabolomics-assisted breeding: a viable option for crop improvement? Trends Genet 25(1):39–48
154. Gieger C, Geistlinger L, Altmaier E, Hrabe de Angelis M, Kronenberg F, Meitinger T, Mewes HW, Wichmann HE, Weinberger KM, Adamski J et al (2008) Genetics meets metabolomics: a genome-wide association study of metabolite profiles in human serum. PLoS Genet 4(11):e1000282
155. Illig T, Gieger C, Zhai G, Romisch-Margl W, Wang-Sattler R, Prehn C, Altmaier E, Kastenmuller G, Kato BS, Mewes HW et al (2010) A genome-wide perspective of genetic variation in human metabolism. Nat Genet 42(2):137–141
156. Suhre K, Shin SY, Petersen AK, Mohney RP, Meredith D, Wagele B, Altmaier E, Cardio-Gram, Deloukas P, Erdmann J et al (2011) Human metabolic individuality in biomedical and pharmaceutical research. Nature 477(7362):54–60
157. Kettunen J, Tukiainen T, Sarin AP, Ortega-Alonso A, Tikkanen E, Lyytikainen LP, Kangas AJ, Soininen P, Wurtz P, Silander K et al (2012) Genome-wide association study identifies multiple loci influencing human serum metabolite levels. Nat Genet 44(3):269–276
158. Holmes E, Loo RL, Stamler J, Bictash M, Yap IK, Chan Q, Ebbels T, De Iorio M, Brown IJ, Veselkov KA et al (2008) Human metabolic phenotype diversity and its association with diet and blood pressure. Nature 453(7193):396–400
159. Shah SH, Bain JR, Muehlbauer MJ, Stevens RD, Crosslin DR, Haynes C, Dungan J, Newby LK, Hauser ER, Ginsburg GS et al (2010) Association of a peripheral blood metabolic profile with coronary artery disease and risk of subsequent cardiovascular events. Circ Cardiovasc Genet 3(2):207–214
160. Jansen RC, Nap JP (2001) Genetical genomics: the added value from segregation. Trends Genet 17(7):388–391
161. de Koning DJ, Haley CS (2005) Genetical genomics in humans and model organisms. Trends Genet 21(7):377–381
162. Fu J, Keurentjes JJ, Bouwmeester H, America T, Verstappen FW, Ward JL, Beale MH, de Vos RC, Dijkstra M, Scheltema RA et al (2009) System-wide molecular evidence for phenotypic buffering in Arabidopsis. Nat Genet 41(2):166–167
163. Keurentjes JJ, Koornneef M, Vreugdenhil D (2008) Quantitative genetics in the age of omics. Curr Opin Plant Biol 11(2):123–128
164. Chen R, Mias GI, Li-Pook-Than J, Jiang L, Lam HY, Chen R, Miriami E, Karczewski KJ, Hariharan M, Dewey FE et al (2012) Personal omics profiling reveals dynamic molecular and medical phenotypes. Cell 148(6):1293–1307
165. Goh KI, Cusick ME, Valle D, Childs B, Vidal M, Barabasi AL (2007) The human disease network. Proc Natl Acad Sci U S A 104(21):8685–8690
166. Wu X, Jiang R, Zhang MQ, Li S (2008) Network-based global inference of human disease genes. Mol Syst Biol 4:189
167. Lamb J, Crawford ED, Peck D, Modell JW, Blat IC, Wrobel MJ, Lerner J, Brunet JP, Subramanian A, Ross KN et al (2006) The connectivity map: using gene-expression signatures to connect small molecules, genes, and disease. Science 313(5795):1929–1935
168. Schuster-Bockler B, Bateman A (2008) Protein interactions in human genetic diseases. Genome Biol 9(1):R9
169. Ramsey SA, Gold ES, Aderem A (2010) A systems biology approach to understanding atherosclerosis. EMBO Mol Med 2(3):79–89
170. Inouye M, Kettunen J, Soininen P, Silander K, Ripatti S, Kumpula LS, Hamalainen E, Jousilahti P, Kangas AJ, Mannisto S et al (2010) Metabonomic, transcriptomic, and genomic variation of a population cohort. Mol Syst Biol 6:441
171. Lee DS, Park J, Kay KA, Christakis NA, Oltvai ZN, Barabasi AL (2008) The implications of human metabolic network topology for disease comorbidity. Proc Natl Acad Sci U S A 105(29):9880–9885
172. Park J, Lee DS, Christakis NA, Barabasi AL (2009) The impact of cellular networks on disease comorbidity. Mol Syst Biol 5:262

173. Jiang Q, Hao Y, Wang G, Juan L, Zhang T, Teng M, Liu Y, Wang Y (2010) Prioritization of disease microRNAs through a human phenome-microRNAome network. BMC Syst Biol 4(Suppl 1):S2
174. Butte AJ, Kohane IS (2006) Creation and implications of a phenome-genome network. Nat Biotechnol 24(1):55–62
175. Ning S, Wang P, Ye J, Li X, Li R, Zhao Z, Huo X, Wang L, Li F, Li X (2013) A global map for dissecting phenotypic variants in human lincRNAs. Eur J Hum Genet 21(10):1128–1133
176. Lee I, Blom UM, Wang PI, Shim JE, Marcotte EM (2011) Prioritizing candidate disease genes by network-based boosting of genome-wide association data. Genome Res 21(7):1109–1121
177. Lage K, Karlberg EO, Storling ZM, Olason PI, Pedersen AG, Rigina O, Hinsby AM, Tumer Z, Pociot F, Tommerup N et al (2007) A human phenome-interactome network of protein complexes implicated in genetic disorders. Nat Biotechnol 25(3):309–316
178. Iliopoulos D, Malizos KN, Oikonomou P, Tsezou A (2008) Integrative microRNA and proteomic approaches identify novel osteoarthritis genes and their collaborative metabolic and inflammatory networks. PLoS One 3(11):e3740
179. Pujana MA, Han JD, Starita LM, Stevens KN, Tewari M, Ahn JS, Rennert G, Moreno V, Kirchhoff T, Gold B et al (2007) Network modeling links breast cancer susceptibility and centrosome dysfunction. Nat Genet 39(11):1338–1349
180. Barabasi AL, Gulbahce N, Loscalzo J (2011) Network medicine: a network-based approach to human disease. Nat Rev Genet 12(1):56–68
181. Lu M, Zhang Q, Deng M, Miao J, Guo Y, Gao W, Cui Q (2008) An analysis of human microRNA and disease associations. PLoS One 3(10):e3420
182. Oti M, Snel B, Huynen MA, Brunner HG (2006) Predicting disease genes using protein-protein interactions. J Med Genet 43(8):691–698
183. Kohler S, Bauer S, Horn D, Robinson PN (2008) Walking the interactome for prioritization of candidate disease genes. Am J Hum Genet 82(4):949–958
184. Vanunu O, Magger O, Ruppin E, Shlomi T, Sharan R (2010) Associating genes and protein complexes with disease via network propagation. PLoS Comput Biol 6(1):e1000641
185. McGary KL, Park TJ, Woods JO, Cha HJ, Wallingford JB, Marcotte EM (2010) Systematic discovery of nonobvious human disease models through orthologous phenotypes. Proc Natl Acad Sci U S A 107(14):6544–6549
186. Bassel GW, Gaudinier A, Brady SM, Hennig L, Rhee SY, De Smet I (2012) Systems analysis of plant functional, transcriptional, physical interaction, and metabolic networks. Plant Cell 24(10):3859–3875
187. Lee I, Ambaru B, Thakkar P, Marcotte EM, Rhee SY (2010) Rational association of genes with traits using a genome-scale gene network for Arabidopsis thaliana. Nat Biotechnol 28(2):149–156
188. Lee I, Seo YS, Coltrane D, Hwang S, Oh T, Marcotte EM, Ronald PC (2011) Genetic dissection of the biotic stress response using a genome-scale gene network for rice. Proc Natl Acad Sci U S A 108(45):18548–18553
189. Ficklin SP, Feltus FA (2011) Gene coexpression network alignment and conservation of gene modules between two grass species: maize and rice. Plant Physiol 156(3):1244–1256
190. Lehner B (2011) Molecular mechanisms of epistasis within and between genes. Trends Genet 27(8):323–331
191. Thomas D (2010) Gene–environment-wide association studies: emerging approaches. Nat Rev Genet 11(4):259–272

Part V
Biocompution

Chapter 12
Parallel Computing for Gene Networks Reverse Engineering

Jaroslaw Zola

Abstract Gene networks provide a mathematical representation of gene interactions that govern biological processes in every living organism. Given a gene expression data, the goal of network inference is to reconstruct the underlying regulatory network. The problem is challenging owing to the convoluted nature of biological interactions and imperfection of experimental data. In many cases, the resulting computational models are too complex to execute on a sequential computer and require scalable parallel approaches. In this chapter, we describe network inference methods based on information theory and show a parallel algorithm that enables whole-genome networks reconstruction.

12.1 Introduction

Biological processes in every living organism are governed by complex interactions between thousands of genes, gene products, and other molecules. Genes that are encoded in the DNA are transcribed and translated to form multiple copies of gene products including proteins and various types of RNAs. These gene products coordinate to execute cellular processes or to regulate the expression of other genes depending on the signals carried by, e.g., small molecules. Gene regulatory networks are an attempt to develop a system-level model of these complex interactions, using observations of gene expression.

Gene regulatory networks are typically expressed as graphs with vertices representing genes and edges representing regulatory interactions between genes (see Fig. 12.1). The functioning of a gene regulatory network in an organism determines the expression levels of various genes to help carry out a biological process.

J. Zola (✉)
Rutgers Discovery Informatics Institute, Rutgers University, Piscataway, NJ 08854, USA
e-mail: jaroslaw.zola@rutgers.edu

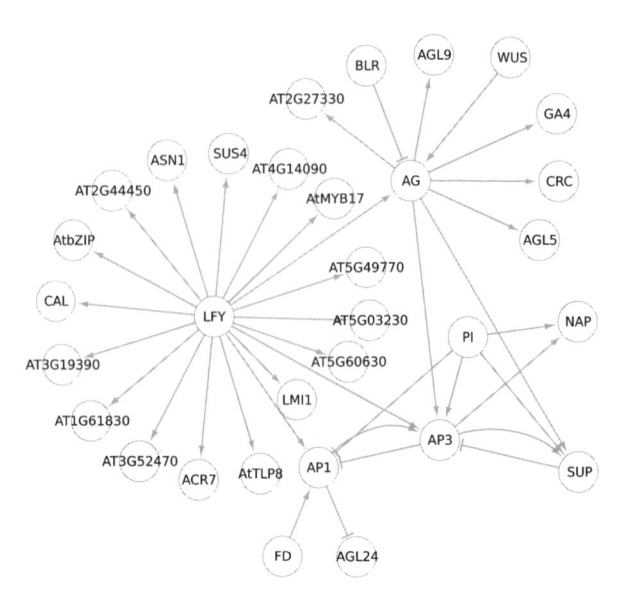

Fig. 12.1 Example gene regulatory network. Here nodes represent genes, T-edges denote regulation in which a source gene inhibits expression of the target gene, and *arrow edges* denote regulation in which a source gene induces expression of the target gene

Network inference, or reverse engineering, is the problem of predicting the underlying network from multiple observations of gene expressions (outputs of the network). To infer a network, one relies on experimental data from high-throughput technologies such as microarrays, or short-read sequencing, which measure a snapshot of all gene expression levels under a particular condition or in a time series. The problem of gene network inference is challenging for several reasons:

- Functioning of any complex organism involves thousands of genes, and usually it is impossible to limit analysis to only a subset of them. In fact, in many cases, the opposite situation takes place – gene networks are used to limit the number of genes that should be target of a biological analysis.
- Despite the rapid progress in high-throughput biotechnology, the number of available expression measurements often falls significantly short of what is required by the underlying computational methods. At the same time, expression data is inherently noisy and significantly influenced by experiment-specific attributes. Consequently in many cases the number of genes in the network significantly outnumbers the number of available expression measurements.
- Finally, our understanding of regulatory mechanisms (e.g., posttranscriptional effects) is still limited, leading to many simplifications in the existing models of regulation.

Due to its importance, gene network inference is an intensely studied problem for which many techniques have been developed. Relevance networks [3, 7], Gaussian graphical models [6, 19], information theory-based methods [2, 9, 25], Bayesian networks [10, 24] and dynamic models [12] are just some examples of the existing approaches. At the same time, two key problems remain with the current methods. The first one is the quality of reconstructed networks. In a recent comprehensive study of 29 network inference methods, Marbach et al. concluded that many do poorly on an absolute basis and 11 do no better than random guessing [15]. The second challenge is computational complexity of the methods and their ability to encompass data from organisms with thousands of genes and large number of expression observations. The computational cost of network inference grows at least as square of the number of the genes and at least linearly with the number of experiments analyzed. Furthermore, statistical methods used to assess significance of the inference, such as bootstrapping, add an extra layer of computational complexity.

To overcome the above limitations and to improve accuracy of network inference while scaling to large expression data, parallel methods for gene networks reverse engineering have been recently proposed [18, 21, 25]. Thanks to the emergence of inexpensive multi and many core processors, and almost ubiquitous adoption of parallel computing, these methods become a solution of choice when large or complex expression data has to be analyzed. More importantly, parallel methods can be used to reconstruct genome-level interactions without sacrificing accuracy, which is where sequential methods fall short.

In this chapter, we focus on application of parallel computing for reverse engineering of gene regulatory networks. First, we explain how the sequential inference process works and then we show how it can be scaled to large distributed memory systems. We base our presentation on information-theoretic methods, a popular class of inference algorithms. Finally, we discuss how to validate inference algorithms *in silico*, and we demonstrate applicability of parallel methods in reconstructing genome-level regulatory networks.

12.2 Network Inference Using Information Theory

In this section, we introduce a more formal statement of the network reconstruction problem, and we present an inference procedure based on information theory. We explain concepts of mutual information and data processing inequality and show how mutual information can be estimated and its significance assessed. We start however with a brief description of a general network inference process.

12.2.1 From Experiment to Network

The goal of network inference is to provide a qualitative, and if possible quantitative, explanation of the observed expression data. The quality of the reconstructed

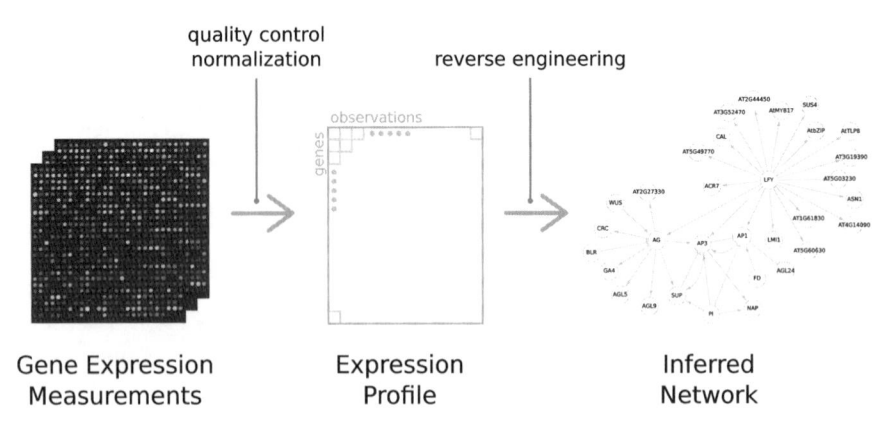

Fig. 12.2 Typical process of reconstructing gene regulatory network

network and its information content are affected not only by the inference method but also by the input data. Gene expression can be measured using several methods, for instance, quantitative PCR, microarrays, or more recently RNA-seq. However, irrespective of which method we select to measure genes expression, three important questions must be answered: first, what should be the set of experimental conditions, how the expression data should be preprocessed to obtain an expression profile suitable for network inference, and finally, which inference method should be used taking into account the two above.

Figure 12.2 illustrates an example inference process. We start with the data acquisition. This step is determined by the underlying scientific hypothesis, which in turn involves careful design of the biological experiment. Note that the data gathered in the experiment can be, and usually is, extended with data deposited in the public repositories, such as Gene Expression Omnibus (GEO) [17]. In general, in this stage, we want to ensure that the collected expression data is sufficient to obtain accurate predictions in the inference process. The following step is to convert the aggregated data into an expression profile. The choice of method depends purely on the experimental platform. For example, processing microarrays will typically require sophisticated signal-calling procedures, followed by filtering and normalization, while RNA-seq will most likely depend on reads mapping and reads counting to obtain a digital expression. This stage is crucial to minimize noise impact, to eliminate low-quality data, and to render different experiments comparable. Only when the expression profile is ready a network can be reconstructed. As we already mentioned, multiple inference methods exist and which method should be used depends on many factors, including size of the expression data and type of queries that the inferred network is meant to answer. In this chapter, however, we consider an information-theoretic approach and its parallel realization.

12.2.2 Problem Formulation

Let us consider the following situation. We performed a set of m experiments, e.g., microarray tests, and obtained expression measurements for n genes. We will represent these genes as a set $G = \{g_1, g_2, \ldots, g_n\}$, where g_i is a gene. Furthermore, we will assume that the expression measurements have been post-processed and normalized accordingly and taking into account properties of the underlying technology. Collectively, we can represent such expression data as a profile matrix $Y_{n \times m}$, where $Y[i, j]$ is an expression of gene g_i in experiment j. The core assumption of virtually all network inference methods is to represent expression of a gene g_i as a random variable $X_i \in \mathcal{X}$, $\mathcal{X} = \{X_1, \ldots, X_n\}$, with marginal probability p_{X_i} derived from some unknown joint probability characterizing the entire system. This random variable is described by a vector of observations $\langle x_{i,1}, \ldots, x_{i,m} \rangle$, where $x_{i,j} = Y[i, j]$. In this form, the network inference problem becomes that of finding a model that best explains the data in Y. The problem can be approached using a variety of methods, including Bayesian networks and Gaussian graphical models. However, one class of methods that have been widely adopted due to their effectiveness uses the concept of mutual information [2, 9, 25]. These methods operate under the assumption that correlation of expression implies co-regulation. Although not always true, the assumption is broadly accepted, especially when analyzing microarray data.

Inference methods based on information theory usually proceed in two phases. First, correlations between pairs of genes are detected. If expression of two genes shows strong correlation, we can assume that they are interacting in the regulatory processes and hence should be connected in the network. Unfortunately, looking solely into pairwise correlations is insufficient to capture more complex regulatory patterns. Consider a scenario where gene g_x regulates gene g_y, which in turn regulates g_z. If we analyze expression of all three genes, it is very likely that g_x and g_z will be significantly correlated, even though they should not be directly connected in the network. To account for such situations, in the second phase, information-theoretic strategies perform additional check to detect and remove indirect interactions.

12.2.2.1 Mutual Information

Let us now focus on how correlation between expression profile of two genes is established. Recall that we represent expression of each gene as a random variable. Although we are given observations of that variable, we do not know its actual distribution. Moreover, the expression observations are delivered from inherently noisy experiments and thus are not perfect. Consequently, to establish whether two expression profiles are correlated, we have to account for potentially complex, e.g., nonlinear, patterns of correlations. This can be achieved using the concept of mutual information [4].

Mutual information is arguably the best measure of correlation between two random variables. It is defined based on the entropy in the following way:

$$\mathcal{I}(X_i; X_j) = \mathcal{H}(X_i) + \mathcal{H}(X_j) - \mathcal{H}(X_i, X_j), \tag{12.1}$$

where entropy $\mathcal{H}$ is given as

$$\mathcal{H}(X) = -\sum p_X(x) \log(p_X(x)), \tag{12.2}$$

p_X defines the probability distribution of X, and $\sum$ is replaced by integral if X is continuous. Intuitively, mutual information $\mathcal{I}(X_i; X_j)$ quantifies information that both variables provide about each other. If two variables are correlated, then their joint entropy is smaller than the sum of their individual entropies, and hence greater is their mutual information. Note that from Eqs. (12.1) and (12.2), we can write mutual information as

$$\mathcal{I}(X_i; Y_i) = \sum_{x_i}\sum_{x_j} p_{X_i X_j}(x_i, x_j) \log\left(\frac{p_{X_i X_j}(x_i, x_j)}{p_{X_i}(x_i) p_{X_j}(x_j)}\right), \tag{12.3}$$

which is equivalent of the Kullback-Leibler divergence between distribution of X_i and X_j when both variables are dependent (i.e., $p_{X_i X_j}$) and when they are independent (i.e., $p_{X_i X_j} = p_{X_i} p_{X_j}$).

Mutual information is a symmetric, nonnegative function and is equal to zero if and only if two random variables are independent. Consequently, to connect two genes in the reconstructed network, we have to check if mutual information between their expression profiles is greater than zero.

12.2.2.2 Data Processing Inequality

Having defined a correlation measure, we are left with the task of identifying indirect interactions between genes. One popular approach to address this problem, which first has been introduced in the ARACNe method [2], is to rely on the data processing inequality principle, or DPI for short. Briefly, DPI states that if three random variables X_i, X_j, X_k form a Markov chain in that order (i.e., conditional probability of X_k depends only on X_j and is independent of X_i), then $\mathcal{I}(X_i; X_k) \leq \mathcal{I}(X_i; X_j)$, which implies also that $\mathcal{I}(X_i; X_k) \leq \mathcal{I}(X_j; X_k)$. In other words, X_k cannot provide more information about X_i than X_j provides about X_i. The DPI reasoning can be used to detect indirect interactions between genes: each time the pair (X_i, X_k) satisfies both inequalities, the corresponding connection between genes g_i and g_k can be removed from the network. Note that the above procedure is based on the assumption that DPI implies independence of X_i and X_k given X_j. This is not always true: in some situations the inequalities may hold even though

X_i and X_k are dependent (consider for example, binary variables, where X_i and X_j are uniform and X_k is a XOR function of X_i and X_j). Nevertheless, DPI has been shown to perform very well in practice.

12.2.2.3 Inference Algorithm

So far we defined two information-theoretic concepts that can be used to infer gene regulatory networks – mutual information and DPI. Algorithm 12.1 shows how the two are combined into a working solution.

We represent the network using adjacency matrix D. Although gene regulatory networks are usually very sparse, initially we have to compute mutual information between all $\binom{n}{2}$ pairs of genes (line 1). Then, we remove edges between genes that are not significantly correlated (lines 2–4) and proceed with the DPI phase (lines 5–9).

Algorithm 12.1 Network inference using information theory

Input: Expression profile $Y_{n \times m}$, mutual information threshold $\mathcal{I}^0$
Output: Adjacency matrix $D_{n \times n}$
1: $D[i, j] = $ Estimate $\mathcal{I}(X_i; X_j)$ from $(Y[i, \cdot], Y[j, \cdot])$
2: **if** $D[i, j] < \mathcal{I}^0$ **then**
3: $D[i, j] = 0$
4: **end if**
5: **for all** (i, k) **do**
6: **if** $\exists j$ s.t. $D[i, k] \leq D[i, j]$ and $D[i, k] \leq D[j, k]$ **then**
7:   $D[i, k] = 0$
8: **end if**
9: **end for**

The main component of the algorithm is estimation of mutual information from the expression data. Observe that although in Eq. (12.3) we express mutual information through probability distributions, we do not know the distribution that governs gene expression. Consequently, we have to estimate mutual information from observations provided by the expression profile matrix Y. Fortunately, because mutual information is a widely used concept, there are several estimators available. Here, we will describe the B-spline-based estimator that has been proposed by Daub et al. for analyzing expression data [5].

The estimator works by discretizing observations into b categories, but with the assumption that given observation can be assigned simultaneously to k categories with different weights. The weights are obtained using B-spline functions of order k, defined over b uniformly spaced knot points. Note that knot points define bins (categories) to which each continuous observation can be assigned. Let $\mathcal{B}_k^b$ be a B-spline function of order k defined over b knot points. For a continuous observation, this function returns a vector of size b with k nonnegative weights that indicate to which bins the observation should be assigned. Given two random variables X_i and X_j with m observations, we can discretize them into variables A and B with probabilities:

$$p_A = \frac{1}{m} \sum_{l=1}^{m} \left(\mathcal{B}_k^b(x_{i,l}) \right), \qquad (12.4)$$

and

$$p_{AB} = \frac{1}{m} \sum_{l=1}^{m} \left(\mathcal{B}_k^b(x_{i,l}) \times \mathcal{B}_k^b(x_{j,l}) \right), \qquad (12.5)$$

where in our case $x_{i,j} = Y[i, j]$. By plugging the resulting probabilities directly into entropy calculations, we can compute marginal and joint entropy for A and B and then approximate $\mathcal{I}(X_i; X_j) \approx \mathcal{I}(A; B)$. A nice property of the B-spline estimator is that it can be very efficiently implemented, and its complexity is of order $O(m)$.

The last element of the inference algorithm is the choice of the threshold value $\mathcal{I}^0$ to decide when correlation is significant. Recall that two random variables are independent only if their mutual information is equal zero. However, because we are estimating mutual information, it would be unrealistic to expect precise results. Therefore, it is a common practice to assume that mutual information lower than the carefully chosen cutoff $\mathcal{I}^0$ implies independence. There are different ways $\mathcal{I}^0$ can be selected, and we describe one particular solution next.

12.2.2.4 Testing Significance of Mutual Information

As we already mentioned, using mutual information requires deciding when its estimate implies independence. This can be regarded as assessing statistical significance of the quantity $\mathcal{I}(X_i; X_j)$ itself. This assessment can be done through permutation testing.

Let $\pi(X_i) = \pi(\langle x_{i,1}, x_{i,2}, \ldots, x_{i,m} \rangle)$ denote a permutation of the vector of m observations of X_i. If there exists dependency between X_i and X_j, it is expected that $\mathcal{I}(X_i; X_j)$ is significantly higher than $\mathcal{I}(\pi(X_i); X_j)$. The permutation testing method involves computing $\mathcal{I}(\pi(X_i); X_j)$ for all $m!$ permutations of $\langle x_{i,1}, x_{i,2}, \ldots, x_{i,m} \rangle$ and accepting the dependency between X_i and X_j to be statistically significant only if $\mathcal{I}(X_i; X_j) > \mathcal{I}(\pi(X_i); X_j)$ for at least a fraction $(1 - \epsilon)$ of the $m!$ permutations tested, for some small constant $\epsilon > 0$. As testing all $m!$ permutations is computationally prohibitive for large m, a large sampling of the permutation space is considered adequate in practice.

Ideally, permutation testing should be conducted for assessing the significance of each pair $\mathcal{I}(X_i; X_j)$ using a large number of random permutations. Clearly, this is computationally prohibitive. However, we proposed a simple solution to overcome this limitation, by using only a few random permutations per pair, while collectively obtaining statistically meaningful results for all pairs [25].

Mutual information has the property of being invariant under homeomorphic transformations:

$$\mathcal{I}(X_i; X_j) = \mathcal{I}(f(X_i); h(X_j)), \qquad (12.6)$$

for any homeomorphisms f and h. Consider replacing the vector of observations for X_i, i.e., $\langle x_{i,1}, x_{i,2}, \ldots, x_{i,m} \rangle$ with the vector $\langle \text{rank}(x_{i,1}), \text{rank}(x_{i,2}), \ldots, \text{rank}(x_{i,m}) \rangle$, where $\text{rank}(x_{i,l})$ denotes the rank of $x_{i,l}$ in the set $\{x_{i,1}, x_{i,2}, \ldots, x_{i,m}\}$; i.e., we replace each observation with its rank in the set of all observations. The transformation, which is termed *rank transformation*, while not continuous, is considered a good approximation to homeomorphism [13]. In our case, instead of computing mutual information of pairs of gene expression vectors directly, we equivalently compute the mutual information of their rank transformed counterparts. With this change, each observation vector is now a permutation of $\{1, 2, \ldots, m\}$. Therefore, a permutation $\pi(X_i)$ corresponds to some permutation of the observation vector of any other random variable X_j. More formally, consider applying permutation testing to a specific pair (X_i, X_j) by computing $\mathcal{I}(\pi(X_i); X_j)$ for some randomly chosen permutation π. For any other pair (X_k, X_l), $\exists \pi', \pi''$ such that $\pi(X_l) = \pi'(X_j)$ and $\pi(X_i) = \pi''(\pi'(X_k))$. Since π is a random permutation, so is π'' and $\mathcal{I}(\pi(X_i); X_j)$ is a valid permutation test for assessing the statistical significance of $\mathcal{I}(X_k; X_l)$ as well. Thus, each permutation test is a valid test for all $\binom{n}{2}$ pairs of observations.

Using the above procedure, we can easily find the threshold value $\mathcal{I}^0$. When estimating mutual information for each rank transformed pair (X_i, X_j), we perform additional q permutation tests. Then, $\mathcal{I}^0$ is the rth largest value among all values generated by the permutation test, where $r = (1 - \epsilon) \cdot q \cdot \binom{n}{2}$ and ϵ is a small constant describing the significance level.

12.3 Parallel Method for Networks Inference

In the previous section, we explained how concepts from information theory and statistics can be used for reverse engineering gene regulatory networks. Although the resulting method is considered to be very accurate, it becomes limited for reconstructing genome-level networks with thousands of genes. This is because the whole-genome gene network reconstruction is both compute and memory intensive. Memory consumption arises from the $\Theta(nm)$ size of input data and from the $\Theta(n^2)$ dense initial network generated in the first phase of the reconstruction algorithm. Computational cost is dominated by the $O(n^2)$ computations of mutual information, where the complexity of a mutual information estimator is at least $O(m)$, but it can be $O(m^2)$ for, e.g., Gaussian kernel estimator. As a result, large-scale network construction is out of the scope of sequential methods, and scalable parallel approach becomes necessary.

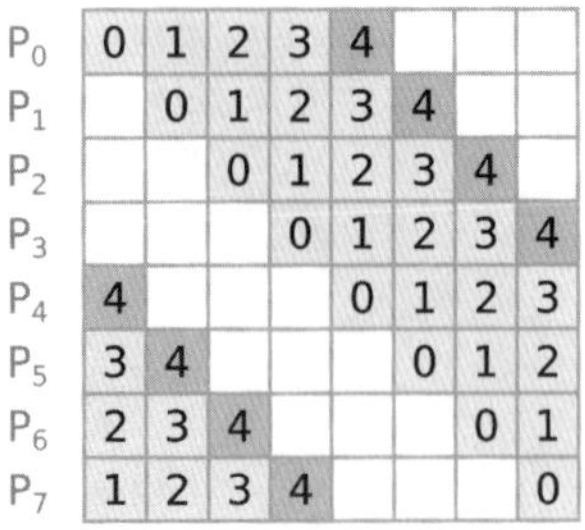

Fig. 12.3 An example of partitioning of matrix D for eight processors. For each block, the iteration number in which it will be processed is marked

P_0	0	1	2	3	4			
P_1		0	1	2	3	4		
P_2			0	1	2	3	4	
P_3				0	1	2	3	4
P_4	4				0	1	2	3
P_5	3	4				0	1	2
P_6	2	3	4				0	1
P_7	1	2	3	4				0

Let p denote the number of processors in a parallel computer. Recall that we represent the network using the standard adjacency matrix $D_{n \times n}$. To store the matrix in the distributed memory, we use row-wise data distribution where each processor stores up to $\lceil \frac{n}{p} \rceil$ consecutive rows of the expression profile Y and the same number of rows of the corresponding gene network adjacency matrix D. To begin, each processor reads and parses its block of input data and then applies rank transformation. The algorithm proceeds in three phases: in the first phase, mutual information is computed for each of the $\binom{n}{2}$ pairs of genes and q randomly chosen permutations per pair. Note that the total number of permutations used in the test is $Q = q \cdot \binom{n}{2}$, allowing a small constant value of q for large n. In the second phase, the threshold value $\mathcal{I}^0$ is computed. In the final phase, indirect interactions are detected using DPI and removed.

12.3.1 Computing Mutual Information

Without loss of generality, we can assume that n is a multiple of p. To compute mutual information between all pairs of genes, we first partition D into $p \times p$ blocks of submatrices $D_{i,j}$ ($0 \leq i, j < p$), each of $\frac{n}{p} \times \frac{n}{p}$ size. Then we proceed in $\lceil \frac{p+1}{2} \rceil$ iterations. In each iteration, a processor is assigned a submatrix. Its task is to compute mutual information for each position in the submatrix, along with mutual information of q random permutations for each position. Observe that to do so, it requires the expression profile vectors of all genes representing rows or columns in the submatrix. For blocks on the main diagonal, the same genes represent both rows and columns. For other blocks, the row genes and column genes are distinct. Because the matrix D is symmetric, we need to compute only half of it, i.e., as $D_{i,j} = D_{j,i}^T$ only one of them needs to be computed. We call a set of $\frac{p \cdot (p+1)}{2}$ submatrices containing only one of $D_{i,j}$ or $D_{j,i}$ for each pair (i, j), to be the complete set of unique submatrices. The assignment of submatrices to processors is as follows: in iteration i, processor with rank j is assigned the submatrix $D_{j,(j+i) \bmod p}$ (see Fig. 12.3 for an illustration). It is easy to argue that this scheme computes all unique submatrices.

The assignment of submatrices to processors creates the same workload with the following exceptions: in iteration 0, the submatrices assigned are diagonal, for which we only need the lower (or upper) triangular part. As all processors are dealing with diagonal submatrices in the same iteration, it simply means that this iteration will take roughly half the compute time as others. The other exception may occur during the last iteration. To see this, consider that the processors collectively compute p submatrices in each iteration. The total number of unique submatrices is $\frac{p \cdot (p+1)}{2}$. The following two cases are possible:

1. p is odd. In this case, the number of iterations is $\lceil \frac{p+1}{2} \rceil = \frac{p+1}{2}$. The total number of submatrices computed is $\frac{p \cdot (p+1)}{2}$, which is the same as the total number of unique submatrices. Since the algorithm guarantees that all unique submatrices are computed, each unique submatrix is computed only once.
2. p is even. In this case, the number of iterations is $\lceil \frac{p+1}{2} \rceil = \frac{p}{2} + 1$, causing the total number of submatrices computed to be $p \cdot \left(\frac{p}{2} + 1 \right)$, which is $\frac{p}{2}$ more than the number of unique submatrices. It is easy to show that this occurs because in the last iteration, half the processors are assigned submatrices that are transpose counterparts of the submatrices assigned to the other half (marked with darker color in Fig. 12.3).

When p is even, we can optimize the computational cost by recognizing this exception during the last iteration and having each processor compute only half of the submatrix assigned to it, so that the processor which has the transpose counterpart computes the other half. Note that this will save half an iteration, significant only if p is small. For large p, we can ignore this cost and run the last iteration similar to others.

Let us now compute the parallel runtime of the above method assuming a simple point-to-point communication model with latency τ and bandwidth $\frac{1}{\mu}$ (adequate for most distributed memory parallel systems). Under this model, the first phase takes $O\left(\frac{qn^2m}{p}\right)$ compute time and $O\left(p\tau + \mu n m\right)$ time for communication. Thus, we can scale p linearly with n while maintaining parallel compute time as the dominant factor in runtime.

12.3.2 Computing $\mathcal{I}^0$

Having computed the adjacency matrix D, we have to now remove edges that are not statistically significant, i.e., their mutual information is lower than $\mathcal{I}^0$. Recall that the threshold is computed by finding the element with rank $r = (1-\epsilon) \cdot q \cdot \binom{n}{2}$ among the $q \cdot \binom{n}{2}$ mutual information values computed as part of permutation testing. As each processor stores at most $\frac{qn^2}{p}$ results of the permutation test, we can find the threshold using a parallel selection algorithm. However, ϵ is a very small positive constant close to zero, and hence, the threshold value is close to the largest value in the sorted order of the computed mutual information values. Hence, we can proceed as follows: each processor sorts its $\Theta\left(\frac{qn^2}{p}\right)$ values and selects the r largest values. Then, a parallel reduction operation is applied using the r largest values in each processor as input. The reduction operator performs linear merging of two samples of size r and retains the r largest elements. Once the rth globally largest value is found and broadcast, each processor eliminates edges from its local adjacency matrix that are below the threshold. This phase takes $O\left(\frac{qn^2 \log n}{p} + r \log p\right)$ parallel compute time and $O((\tau + \mu r) \log p)$ parallel communication time. Assuming $\epsilon < \frac{1}{p}$, we can expect linear scaling.

12.3.3 Removing Indirect Interactions

In the final phase of the algorithm, we have to apply DPI to prune indirect interactions. To decide if a given edge $D[i, j]$ should be removed, we have to compare it with all values $D[i, k]$ and $D[j, k]$. Consequently, complete information about rows i and j is needed. Because matrix D is stored row-wise, we have to stream row j to the processor responsible for row i. Moreover, because matrix D is symmetric, it is sufficient to analyze its upper (or lower) triangular part. We can achieve this in $p - 1$ communication rounds, where in round i only processors with ranks $0, 1, \ldots, p - i$ participate in communication and processing. The parallel runtime of this phase is $O\left(\frac{n^3}{p} + \tau p + \mu n^2\right)$. While this worst case analysis indicates this to be the most compute-intensive phase of the algorithm, it is not so. This is because DPI requires no significant computation and just single memory write, and it needs to be applied only to current existing edges in the network, and the network is expected to be significantly sparse.

12.3.4 Testing Parallel Scalability

The parallel approach described in the previous sections has been implemented in the software package TINGe [1,25]. TINGe is a C++ software based on the Message Passing Interface (MPI) that can be executed on large distributed memory machines. It implements several low-level optimizations to exploit SIMD instructions of modern processors during mutual information computations and uses MPI I/O routines to handle large input and output data. TINGe employs the B-spline-based mutual information estimator.

To demonstrate scalability of the parallel inference method, we executed TINGe on the IBM Blue Gene/L system with $p = 1{,}024$ processors and analyzed four different expression profiles with varying number of genes ($n = 2{,}048$ and $n = 4{,}096$) and varying number of observations ($m = 911$ and $m = 2{,}996$). We tested

Table 12.1 TINGe runtime in seconds for different number of genes n and different number of expression observations m

	$m = 911$		$m = 2,996$	
p	$n = 2,048$	$n = 4,096$	$n = 2,048$	$n = 4,096$
32	382	1,525	1,489	5,932
64	193	766	752	2,986
128	98	385	378	1,495
256	50	196	193	762
512	27	101	101	386
1,024	17	55	56	203

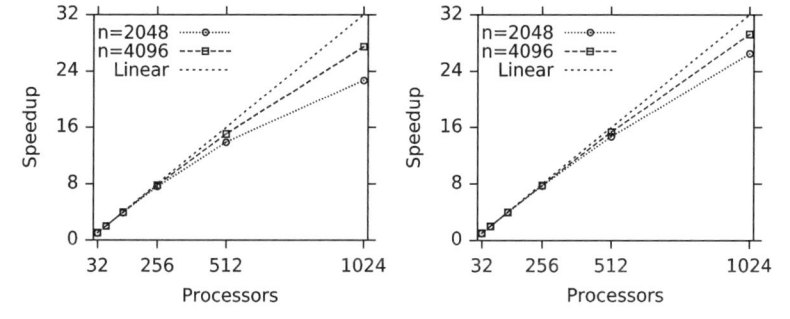

Fig. 12.4 Relative speedup of TINGe as a function of number of processors for the data sets with 911 observations (*left*) and 2,996 observations (*right*)

how runtime changes with the number of processors and what is the relative speedup of the software. Results are summarized in Table 12.1 and Fig. 12.4.

Note that scalability is a crucial characteristic of any parallel software. If a parallel application is scalable, we can decrease its runtime proportionally by executing it on the larger number of processors, and we can solve larger problems. On the other hand, software that scales poorly is of little use as it does not benefit from the parallel hardware.

As we can see, TINGe maintains almost linear scalability up to 1,024 processors, that is, with the increasing number of processors, its runtime decreases linearly. The runtime grows as square function of the number of genes n and linearly with the number of observations m. This is what we expected taking into account the $O(m)$ complexity of the B-spline-based mutual information estimator and the fact that computations are dominated by the first phase, i.e., computing mutual information between all pairs of genes.

12.4 Example Applications

So far we described a parallel information theory-based method for gene network inference. We also demonstrated that the method scales very well on distributed memory parallel systems and hence can be used to process large expression data.

One remaining question that has to be addressed is the accuracy and applicability of the method. In this section, we show how to assess quality of network inference methods. Then, we analyze performance of TINGe, and we explain how it has been used to reconstruct a genome-level regulatory network of the model plant *Arabidopsis thaliana*.

12.4.1 Assessing Quality of Network Inference Methods

Computational methods for regulatory networks reverse engineering are necessarily error-prone, owing to simplifications in the underlying models. This is not surprising taking into account our limited understanding of regulatory processes. When designing a new inference method, we would like to meet three main quality criteria: sensitivity, specificity, and precision. Let TP denote the number of true positives, i.e., the number of correctly predicted gene interactions (network edges); FP be the number of false positives, i.e., incorrectly predicted edges; TN be the number of correctly avoided edges; and, finally, FN be the number of incorrectly avoided interactions. Then, sensitivity relates to the ability of the method to identify positive results Sensitivity $= \frac{TP}{TP+FN}$, and it is a fraction of correct interactions predicted. Likewise, specificity relates to the ability of the method to identify negative results Specificity $= \frac{TN}{TN+FP}$, and it is a fraction of missing edges correctly classified. Finally, precision describes predictive power of the method Precision $= \frac{TP}{TP+FP}$.

Having defined quality criteria, the question is how can we assess performance of a particular method. Naturally, performing a biological experiment to confirm predictions in the inferred network is the most desired approach. However, in most cases, it is infeasible because of the cost and technical limitations (not all interactions can be easily validated). To address this challenge, several researchers proposed methods to perform quality assessment using a synthetic data [12, 14, 20, 23]. The basic idea of the approach is illustrated in Fig. 12.5.

The process starts with two input components – some known network structure and a model of expression dynamics, which usually involves a set of differential equations describing how regulators affect expression of the target gene [23]. The input network is sampled to obtain a benchmark synthetic network. By combining the synthetic network and the expression model, we can generate a synthetic gene expression profile that next can be used as an input to the tested inference method. Observe that the above process guarantees that we know both expression data and the network from which this data has been derived. Consequently, we can easily assess the quality of the inference method by comparing our prediction with the synthetic network. This approach is very flexible – different models of regulation and different sampling strategies can be used to generate synthetic data, and hence to capture various properties of the real-life biological systems. There are several tools that implement such strategy (see, for example, GNW [20], SynTReN [23],

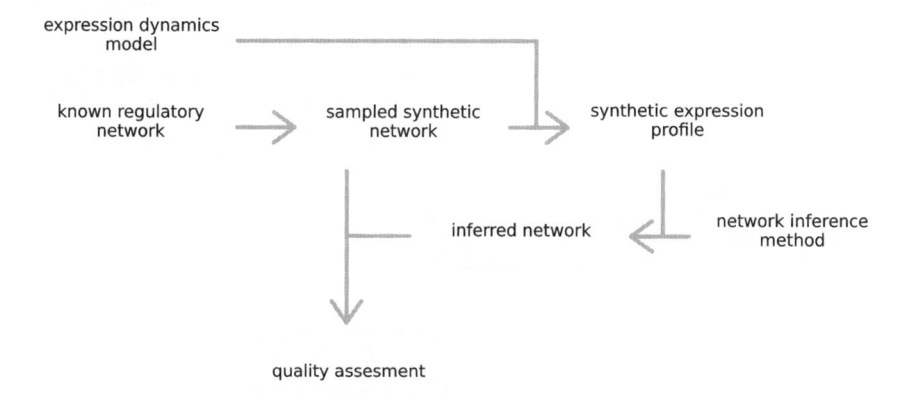

12 Parallel Computing for Gene Networks Reverse Engineering

Fig. 12.5 Quality assessment of a network inference method using synthetic data

Table 12.2 Quality of ARACNe and TINGe using a 250-gene synthetic network with two different sets of expression profiles

	$m = 500$		$m = 900$	
	ARACNe	TINGe	ARACNe	TINGe
Time (s)	473	24	1,866	40
Specificity	0.99	0.99	0.99	0.99
Sensitivity	0.42	0.42	0.44	0.44
Precision	0.52	0.52	0.56	0.57
TP	181	180	190	187
TN	30,535	30,538	30,553	30,562
FP	166	163	148	139
FN	243	244	234	237

or COPASI [12]). These tools provide a functionality to create benchmark data sets with desired number of genes and observations and can be readily used to assess quality of inference methods.

We used SynTReN to assess quality of two mutual information-based inference methods: ARACNe and TINGe. Both tools use the same underlying inference algorithm; however, ARACNe uses different methods to estimate mutual information and to establish the threshold $\mathcal{I}^0$. We generated two synthetic regulatory networks, each consisting of $n = 250$ genes, but differing in the number of expression observations ($m = 500$ and $m = 900$, respectively). Table 12.2 shows that both methods preserve very good precision and sensitivity, while TINGe outperforms ARACNe in terms of the runtime. Increasing the number of observations improves performance of both methods, which is expected as we gain more information with more observations.

While synthetic data provides a convenient way to assess quality of inference methods, we should keep in mind that it is not an ultimate quality indicator. This is because the way synthetic data is generated is model dependent, and hence, it is subject to similar limitations as inference methods. Nevertheless, if a given inference method performs well when tested with synthetic data, it is very likely that it will perform well in practice. On the contrary, methods that perform poorly will fail when analyzing real-life data.

12.4.2 Reconstructing Whole-Genome Network of Arabidopsis

Arabidopsis thaliana is the model plant to study plants' biology and hence is of great practical importance. Its genome contains estimated 27,000 genes, and hence, constructing genome-level regulatory network becomes challenging both computationally and in terms of assembling a sufficiently reach expression profile. Consequently, reconstructing *Arabidopsis* network demonstrates applicability and necessity of parallel inference methods.

We used TINGe to reconstruct the gene regulatory network of *Arabidopsis* [1]. We started reconstruction by obtaining a total of 3,546 nonredundant Affymetrix ATH1 microarray observations, grouped into 197 experiments. Here, each experiment contained several gene expression measurements related to the same biological process or condition. The data was aggregated from the main *Arabidopsis* repositories at NASC [16], GEO [17], ArrayExpress [8], and AtGenExpress [22]. It covers different plant development stages and various treatment experiments, and collectively it provides a broad overview of expression profiles in *Arabidopsis*.

To accommodate for the variability in this highly diverse collection, we developed the following pipeline to obtain the final expression profile. We first removed microarrays which did not pass a rigorous quality control (e.g., exhibited problems, with RNA hybridization). For this we depended on several existing quality indicators offered by the Affymetrix platform. The screening process returned 3,137 microarrays that were subject to normalization: we transformed expression measures into $\log_2$ space and changed to $Y[i, j] = S[i, j] - \overline{S}_i$, where $S[i, j]$ represents $\log_2$-transformed expression of gene i in observation j and $\overline{S}_i$ is the average expression of gene i across all the microarray chips in the experiment containing chip j. Finally, the resulting expression was quantile normalized, and to guarantee that the expression profile of every gene covers a wide range of expression levels, expression profiles with interquartile range of expression lower than 0.65 were removed. As a result, we obtained the final expression matrix with $m = 3,137$ observations and $n = 15,495$ genes.

Using this data, TINGe constructed a whole-genome network in 30 min on the IBM Blue Gene/L with $p = 2,048$ processors. I/O operations took 1 min, finding threshold value $\mathcal{I}^0$ required 1 s, and application of DPI ran in 16 s. Analysis of this network enabled several important insights into biological processes in plants, for instance, the carotenoid biosynthesis. More importantly, this experiment demonstrates that thanks to application of parallel computing, mutual information methods can be used to reconstruct genome-level regulatory networks.

12.5 Final Remarks

The problem of gene regulatory networks inference is one of many in the broad area of computational systems biology. In this chapter, we covered information-theoretic approach to the network inference, together with its scalable parallel implementation. We also demonstrated how application of parallel computing can be used to reconstruct some of the largest gene regulatory networks. Recently, several other parallel reverse engineering methods have been proposed [11, 18, 21]. These methods use different criteria to model gene interactions, e.g. based on Bayesian networks, or different approaches to parallelization, e.g., with GPU accelerators. In spite of that parallel processing only recently attracted attention of systems biology researchers. Together with the rapid progress in high-throughput biological technologies, we can expect accumulation of massive and diverse expression data, which will enable more complex and realistic models of regulation. Clearly, these models will require large computational power offered by parallel systems.

References

1. Aluru M, Zola J, Nettleton D et al (2013) Reverse engineering and analysis of large genome-scale gene networks. Nucl Acids Res 41(1):e24
2. Basso K, Margolin A, Stolovitzky G et al (2005) Reverse engineering of regulatory networks in human B cells. Nat Genet 37(4):382–390
3. Butte AJ, Kohane IS (1999) Unsupervised knowledge discovery in medical databases using relevance networks. In: Proceedings of the American medical informatics association symposium, Washington, DC, pp. 711–715
4. Cover TM, Thomas JA (2006) Elements of information theory, 2nd edn. Wiley, Hoboken
5. Daub CO, Steuer R, Selbig J et al (2004) Estimating mutual information using B-spline functions – an improved similarity measure for analysing gene expression data. BMC Bioinform 5:118
6. de la Fuente A, Bing N, Hoeschele I et al (2004) Discovery of meaningful associations in genomic data using partial correlation coefficients. Bioinformatics 20(18):3565–3574
7. D'haeseleer P, Wen X, Fuhrman S et al (1998) Mining the gene expression matrix: inferring gene relationships from large scale gene expression data. In: Information processing in cells and tissues. Plenum Press, New York
8. EMBL-EBI ArrayExpress. http://www.ebi.ac.uk/microarray-as/aer/
9. Faith JJ, Hayete B, Thaden JT et al (2007) Large-scale mapping and validation of Escherichia coli transcriptional regulation from a compendium of expression profiles. PLoS Biol 5(1):e8
10. Friedman N, Linial M, Nachman I et al (2000) Using Bayesian networks to analyze expression data. J Comput Biol 7:601–620
11. Gregoretti F, Belcastro V, di Bernardo D et al (2010) A parallel implementation of the network identification by multiple regression (NIR) algorithm to reverse-engineer regulatory gene networks. PLoS One 5(4):e10179
12. Hoops S, Sahle S, Gauges R, et al (2006) COPASI – a complex pathway simulator. Bioinformatics 22(24):3067–3074
13. Kraskov A, Stogbauer H, Grassberger P (2004) Estimating mutual information. Phys Rev E 69(6 Pt 2):066138

14. Long J, Roth M (2008) Synthetic microarray data generation with RANGE and NEMO. Bioinformatics 24(1):132–134
15. Marbach D, Prill RJ, Schaffter T et al (2010) Revealing strengths and weaknesses of methods for gene network inference. PNAS 107(14):6286–6291
16. NASC European Arabidopsis Stock Centre. http://www.arabidopsis.info/
17. NCBI Gene Expression Omnibus. http://www.ncbi.nlm.nih.gov/geo/
18. Nikolova O, Zola J, Aluru S (2013) Parallel globally optimal structure learning of Bayesian networks. J Parallel Distrib Comput 73(8):1039–1048. ISSN 0743-7315, http://dx.doi.org/10.1016/j.jpdc.2013.04.001
19. Schafer J, Strimmer K (2005) An empirical Bayes approach to inferring large-scale gene association networks. Bioinformatics 21(6):754–764
20. Schaffter T, Marbach D, Floreano D (2011) GeneNetWeaver: in silico benchmark generation and performance profiling of network inference methods. Bioinformatics 27(16):2263–2270
21. Shi H, Schmidt B, Liu W et al (2011) Parallel mutual information estimation for inferring gene regulatory networks on GPUs. BMC Res Notes 4:189
22. TAIR. http://www.arabidopsis.org/
23. van den Bulcke T, Van Leemput K, Naudts B et al (2006) SynTReN: a generator of synthetic gene expression data for design and analysis of structure learning algorithms. BMC Bioinform 7:43
24. Yu H, Smith A, Wang P et al (2002) Using Bayesian network inference algorithms to recover molecular genetic regulatory networks. In: Proceedings of the international conference on systems biology, Edmonton
25. Zola J, Aluru M, Sarje A et al (2010) Parallel information-theory-based construction of genome-wide gene regulatory networks. IEEE Trans Parall Distrib Syst 21(12):1721–1733

Chapter 13
Computational Biomarker Discovery

Fan Zhang, Xiaogang Wu, and Jake Y. Chen

Abstract The advent of omics technologies as genomics and proteomics has brought the hope of discovering novel biomarkers that can be used to diagnose, predict, and monitor progress of disease. The importance of computational biomarker discovery for diagnostic classification and prognostic assessment in the context of microarray and proteomic data has been increasingly recognized. We present an overview of computational methods and their applications to biomarker

---

F. Zhang
Department of Academic and Institutional Resources and Technology, University of North Texas Health Science Center, Fort Worth, TX 76107, USA

Institute of Biopharmaceutical Informatics and Technology, Wenzhou Medical College, Wenzhou, Zhejiang, China
e-mail: Fan.Zhang@unthsc.edu

X. Wu
School of Informatics, Indiana University, Indianapolis, IN 46202, USA

Indiana Center for Systems Biology and Personalized Medicine, Indianapolis, IN 46202, USA

Institute of Biopharmaceutical Informatics and Technology, Wenzhou Medical College, Wenzhou, Zhejiang, China
e-mail: wu33@IUPUI.edu

J.Y. Chen (✉)
School of Informatics, Indiana University, Indianapolis, IN 46202, USA

Department of Computer and Information Science, School of Science, Purdue University, Indianapolis, IN 46202, USA

Indiana Center for Systems Biology and Personalized Medicine, Indianapolis, IN 46202, USA

Institute of Biopharmaceutical Informatics and Technology, Wenzhou Medical College, Wenzhou, Zhejiang, China
e-mail: Jakechen@IUPUI.edu

discovery with particular focus on genomics and proteomics data. One case study is exemplarily presented, and relevant computational biomarker discovery terminology and techniques are explained.

Keywords Biomarker discovery • Data mining • Breast cancer

13.1 Molecular Biomarkers: What and Why

13.1.1 Definition

A biomarker as defined by the National Cancer Institute is "a biological molecule found in blood, other body fluids, or tissues that is a sign of a normal or abnormal process, or of a condition or disease." A biomarker may be used to see how well the body responds to a treatment for a disease or condition. A biomarker is also called molecular marker and signature molecule.

13.1.2 Application Types

Biomarkers can be used clinically to screen for, diagnose, or monitor the activity of diseases and to guide molecularly targeted therapy or assess therapeutic response. In the biopharmaceutical industry, biomarkers define molecular taxonomies of patients and diseases and serve as surrogate endpoints in early-phase drug trials. Molecular biomarkers can be much more sensitive than traditional lab tests. From a clinical perspective, biomarkers may have a variety of functions, which correspond to different stages (Table 13.1) [1] in disease development, such as in the progression in cancer or cardiovascular disease. Biomarkers can be used to detect and treat early-state

Table 13.1 Rationale and objectives for use of clinical application of cancer biomarkers

Type of biomarker	Objective for use
Screening	To detect and treat early-state (pre)cancers
Diagnostic	To definitively establish the presence of cancer
Prognostic	To portend disease outcome at the time of diagnosis without reference to any specific therapy
Predictive	To predict outcome of a particular therapy
Monitoring	To measure response to treatment and early detect disease progression or relapse
Risk profiling	To determine the risk profile
Companion	To lead to companion diagnostic development
Toxicity	To provide important, compound-specific information regarding toxic drug side effects

(pre)cancers in the asymptomatic patients (screening biomarkers), to definitively establish the presence of cancer for those who are suspected to have the disease (diagnostic biomarkers), or to portend disease outcome at the time of diagnosis without reference to any specific therapy for those with overt disease (prognostic biomarkers) for whom therapy may or may not have been initiated. Biomarkers can be also used to predict outcome of a particular therapy (predictive biomarkers) or to measure response to treatment and early detect disease progression or relapse (monitoring biomarkers) [2]. In addition, biomarkers can be used to evaluate the disease risk profile [3] (risk profiling), to lead to companion diagnostic development (companion biomarker), and to provide important, compound-specific information regarding toxic drug side effects (toxicity biomarker).

13.1.3 Clinical Application from a Historical Perspective

Biomarkers can be classified based on different parameters. They can be classified based on their characteristics such as imaging biomarkers or molecular biomarkers. Imaging biomarkers are measurable characteristics obtained by imaging that indicates a specific biological process is occurring in the body. Imaging-based biomarkers employ a variety of technologies to capture images of anatomical and physiological changes in the body, for example, X-ray, computed tomography (CT), positron emission tomography (PET), single-photon emission computed tomography (SPECT), and magnetic resonance imaging (MRI). Molecular biomarkers have been defined as biomarkers that can be discovered using basic and acceptable platforms such as genomics and proteomics. Apart from genomics and proteomics platforms, biomarker assay techniques, metabolomics, lipidomics, and glycomics are also the most commonly used as techniques in identification of biomarker.

A genomic biomarker is defined by FDA as a measurable DNA and/or RNA characteristic that is an indicator of normal biological processes, pathogenic processes, and/or response to therapeutic or other interventions. Similarly, the definitions can be extended to other approaches such as proteomics, metabolomics, lipidomics, and glycomics. The common genomic approach includes northern blot, gene expression, SAGE, DNA microarray, and next-generation sequencing. Technologies involved in proteomic biomarker research include 2D-GE, LS/MS, SELDI-TOF, Ab microarray, and tissue microarray. Metabolomics approach is to characterize metabolite differences between altered, stressed, or otherwise abnormal physiological states by extracting, identifying, and quantifying all of the small molecule compounds (e.g., metabolites). Lipidomics approach refers to the analysis of lipids. Three key platforms used for lipid profiling include mass spectrometry, chromatography, and nuclear magnetic resonance. Glycans have unique characteristics that are significantly different from nucleic acids and proteins in terms of biosynthesis, structures, and functions. Disease development and progression are usually associated with alternations in glycosylation on tissue proteins and/or blood proteins. Glycans released from tissue/blood proteins hence provide a valuable source of biomarkers. Three common platforms used for glycosylation analysis include lectin microarray, MALDI-TOF MS/tandem MS, and HPLC/capillary electrophoresis.

13.2 Genome Technologies for Biomarker Development

13.2.1 High-Throughput Multiplexing Assays

High-throughput genomics technologies (e.g., gene expression microarrays) have been tremendously changing biomedical research nowadays, which allow researchers to simultaneously monitor the expression of tens of thousands of genes [4]. Microarray data analysis has also become a common practice in many experimental laboratories. Numerous literatures describe the innovative insights within microarray data analysis [5, 6]. It has been widely applied into many medical areas, including distinguishing disease subtypes [7], identifying candidate biomarkers [8], and revealing the underlying molecular mechanisms of disease [9] or drug response [10].

Gene expression microarrays can take a snapshot of all the transcriptional activity in a biological sample, while it also generates a huge amount of data with intrinsic noise (sample or instrument noise), which is still a quite challenging task to interpret it even by exploiting modern computational and statistical tools [6, 11, 12]. This challenge no longer lies in the acquisition of gene expression profiles, but rather in the interpretation for the results to gain insights into biological mechanisms [13]. In many cases, crucial genes show relatively slight changes, and many genes selected are also poorly annotated [5]. From a biological perspective, functionally related genes often display a coordinated expression to accomplish their roles in the cell [14]. Hence, to translate such lists of differentially expressed genes into a functional profile able to understand the underlying biological phenomena, one approach to aid interpretation is to look for changes in a group of genes with a common function (gene cluster) [5].

Accordingly, gene set analysis (GSA) methods aim to test the activity of such gene clusters instead of testing the activity of individual genes—individual gene analysis (IGA) [15]. In recent years, GSA approach has received a great deal of attention, since it is free from the problems of the "cutoff-based" methods. In this direction, GSA methods enable the understanding of cellular processes as an intricate network of functionally related components [14]. Among these GSA methods, gene set enrichment analysis (GSEA) is one of the most widely used methods [13]. GSEA analyzes predefined gene sets based on prior biological knowledge to determine whether this gene set as a whole exhibits differential expression. GSEA has many advantages as it does not employ an arbitrary cutoff to select significant genes. Instead, it uses all the information about every gene involved in the experiment [11]. However, GSEA does rely on predefined gene sets (without gene interaction information), making IGA more beneficial when not much is known about the biological function being considered [6]. Furthermore, GSEA still assumes that more differentially expressed genes are more crucial to the biology, which is not always true [11]. In many cases, extensive upstream data processing, comprehensive gene selection statistics, and downstream pathway/network analysis cannot be replaced by GSEA [11]. Therefore, gene expression signature analysis and pathway analysis (using tools such as DAVID [16]) remain two separate processes.

Network-based gene expression analysis is proposed for candidate biomarker discovery by integrating disease susceptibility genes, their gene expressions, and their gene/protein interaction network [17, 18]. In 2007, Marc Vidal's group at Harvard constructed a protein interaction network for breast cancer susceptibility using various "omics" datasets and identified HMMR as a new susceptibility locus for the disease [17]. Later, Trey Ideker's group at UCSD integrated protein network and gene expression data to improve the prediction of metastasis formation in patients with breast cancer [18]. The two studies marked the exciting beginning of a new paradigm which suggests protein interaction networks and pathway, although drafty, error-prone, and incomplete can serve as a molecular-level conceptual roadmap to guide future microarray analysis.

13.2.2 Next-Generation Sequencing

Over the past 4 years, the application of automated Sanger sequencing for genome analysis has been shift away. The automated Sanger method is considered as a "first-generation" technology, and newer methods are referred to as next-generation sequencing (NGS). These newer technologies constitute various strategies that rely on a combination of template preparation, sequencing and imaging, and genome alignment and assembly methods.

A great deal of NGS effort today centers on cancer, but other basic research areas stand to benefit as well (e.g., immunogenetic studies, neurological and psychiatric diseases, infectious diseases, metagenomics, evolution). NGS has empowered the growth of epigenomics; several approaches exist, but bisulfite-enabled methyl-seq currently dominates the scene.

NGS is also becoming increasingly popular for applications once dominated by microarrays. ChIP-seq's improved data quality compared to microarrays permits greater accuracy in identifying protein-binding DNA targets. RNA-seq provides an alternative to microarrays in assessing cell transcriptomes and is well on its way to becoming the dominant mode in transcriptomics. Other applications of NGS include microRNA-seq, targeted sequencing (sequence capture), de novo sequencing of small genomes, whole-genome sequencing, resequencing of any genome, whole exome sequencing, cancer genome sequencing, methylation, mutation and structural DNA analysis, single molecule sequencing, SNP analysis, microbial/viral genome sequencing, and bisulfite sequencing.

Next-generation sequencing has been applied to biomarker discovery, validation, and characteristics. For example, two studies published recently showed convincingly that whole-genome sequencing of individual patients or affected families can reveal the one gene out of some 25,000 in the human genome bearing a deleterious mutation.

Baylor College of Medicine's Jim Lupski, Richard Gibbs, and colleagues showed that by sequencing the whole genome of an affected individual—in this case Lupski himself—it is possible to identify the rogue gene for a recessive disease by filtering the variations in the coding genes to focus on just those that are novel and predicted to cause a significant phenotypic change [19]. Meanwhile, researchers at the Institute of Systems Biology (ISB) identified the gene for a rare Mendelian disorder called Miller syndrome by sequencing a family of four (parents and two children) [20].

13.2.3 Clinical Proteomics

Clinical proteomics is the application of proteomic techniques to the field of medicine with the aim of solving a specific clinical problem within the context of a clinical study. In the past year significant commitments from research institute and development of clinical proteomics have been witnessed. The application of clinical proteomic research is growing rapidly in the field of biomarker discovery, especially in the area of cancer diagnostics. Clinical proteomics holds the potential of taking a snapshot of the total protein complement of a cell, or body fluid, and identifying proteins as potential biomarkers for the differentiation of disease and health [21]. The study of clinical proteomic may provide us with opportunities in more effective strategies for early disease detection and monitoring, more effective therapies, and developing a better understanding of disease pathogenesis [22]. Such studies may aim at earlier or more accurate diagnosis, improvement of therapeutic strategies, and better evaluation of prognosis and/or prevention of the disease. Although clinical proteomics currently mainly focuses on diagnostics and biomarker discovery, it includes the identification of new therapeutic targets, drugs, and vaccines for better therapeutic outcomes and successful disease prevention. In addition, success for a clinical proteomics requires the communication among clinicians, statisticians, bioinformatists, and biologists [23].

Recent advances in clinical proteomics technology, particularly liquid chromatography coupled with tandem mass spectrometry (LC-MS/MS), have enabled biomedical researchers to characterize thousands of proteins in parallel in biological samples. Using LC-MS/MS, it has become possible to detect complex mixtures of proteins, peptides, carbohydrates, DNA, drugs, and many other biologically relevant molecules unique to disease processes [24]. A modern mass spectrometry (MS) instrument consists of three essential modules: an anion source module that can transform molecules to be detected in a sample into ionized fragments: a mass analyzer module that can sort ions by their masses, charges, or shapes by applying electric and magnetic fields: and a detector module that can measure the intensity or abundance of each ion fragment separated earlier. Tandem mass spectrometry (MS/MS) has the additional analytical modules for bombarding peptide ions into fragment peptide ions by pipeline two MS modules together, and therefore can provide peptide sequencing potentials for selected peptide ions in real time. Recent developments of new generations of mass spectrometers and improvements in the field of chromatography have revolutionized protein analytics. Particularly the combination of liquid chromatography as a separation tool for proteins and peptides with tandem mass spectrometry as an identification tool referred to as LC-MS/MS has generated a powerful and broadly used technique in the field of proteomics [25]. LC-MS/MS proteomics has been used to identify candidate molecular biomarkers in a diverse range of samples, including cells, tissues, serum/plasma, and other types of body fluids. For example, Flaubert et al. discovered highly secreted protein biomarkers which changed significantly in abundance, corresponding with aggressiveness by using LC-MS/MS to analyze the secreted proteomes from a series of isogenic breast cancer cell lines varying in aggressiveness: non-tumorigenic MCF10A, premalignant/tumorigenic MCF10AT, tumorigenic/locally invasive MCF10 DCIS.com, and tumorigenic/metastatic MCF 10CA cl. D. They obtained proteomes from conditioned serum-free media, analyzed the tryptic peptide digests of the secreted proteins using a Waters' capillary liquid chromatography coupled to the nanoflow electrospray source of a Waters' Q-TOF Ultima API-US mass spectrometer, and separated peptide on a C18 reversed-phase column [26].

Figure 13.1 describes the typical workflow for identifying a biomarker from LC-MS/MS: (1) protein separation, (2) enzyme digestion, (3) peptide separation, (4) mass spectrometry, (5) database search, and (6) statistics analysis and pathway analysis for biomarker.

13.3 Computational Method for Biomarker Identification

13.3.1 Computational Experimental Design

Experimental design is the process of planning a study to meet specified objectives. The parts of an experiment design are as follows:

Hypothesis: A statement that predicts the outcome of testing the relationship between the two groups as specified in the problem.

Materials and Procedure: A recipe for conducting the experiment. It consists of a list of materials/equipment followed by step-by-step instructions.

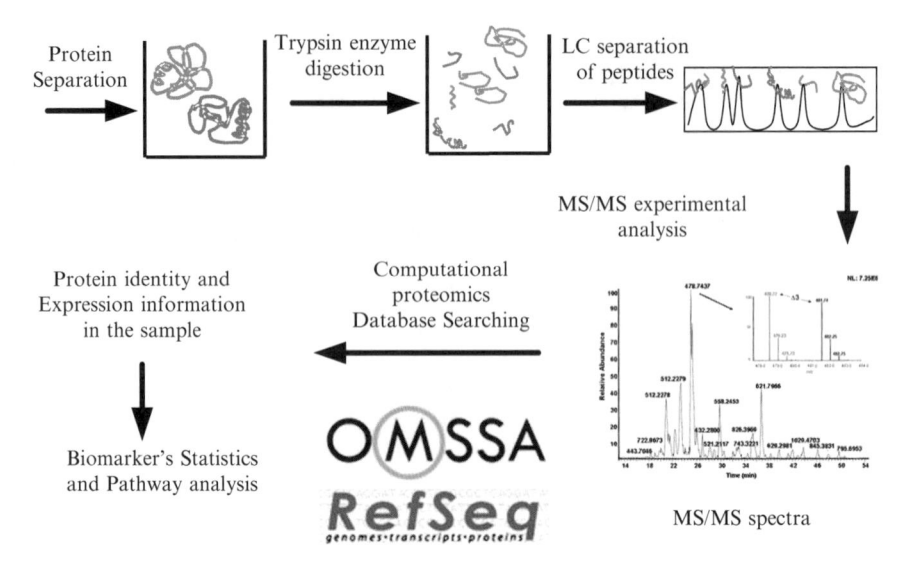

Fig. 13.1 Proteomics analysis using liquid chromatography coupled with tandem mass spectrometry (LC/MS-MS) techniques

Statistical Analysis of Data

Measure of Variation: For qualitative data, a frequency table or histogram can be used. For quantitative data, the range and standard deviation should be used.

Regression Analysis: Using an equation or graph to show the relationship of variables; finding the line of best fit is often used.

Sample size: Sample size estimation is an important aspect of experimental design, because without these calculations, sample size may be too high or too low. If sample size is too low, the experiment will lack the precision to provide reliable answers to the questions it is investigating. If sample size is too large, time and resources will be wasted, often for minimal gain.

Variance Analysis: Each design can be analyzed by using a specific analysis of variance (ANOVA) that is designed for that experimental design. An effect is a change in the response due to a change in a factor level. There are different types of effects. One objective of an experiment is to determine if there are significant differences in the responses across levels of a treatment (a fixed effect) or any interaction between the treatment levels. If this is always the case, the analysis is usually easily manageable, given that the anomalies in the data are minimal (outliers, missing data, homogeneous variances, unbalanced sample sizes, and so on). A random effect exists when the levels that are chosen represent a random selection from a much larger population of equally usable levels. This is often thought of as a sample of interchangeable individuals or conditions. The chosen levels represent arbitrary realizations from a much larger set of other equally acceptable levels.

Prospective Study: A study in which the subjects are identified and then followed forward in time.
Retrospective Study: In medicine, a study that looks backward in time, usually using medical records and interviews with patients who are already known to have a disease.

13.3.2 Statistical Data Analysis

13.3.2.1 t-test

The *t*-test is a statistical hypothesis test in which the test statistic follows a Student's *t* distribution if the null hypothesis is supported and assesses whether the means of two groups are statistically different from each other. Two-sample *t*-statistics and Welch's *t*-test statistics are used to calculate the *p*-value of null hypothesis that the means of two normally distributed populations are equal, for equal variance and unequal variance, respectively.

For equal variance, a 2-sample *t*-test statistics is calculated as

$$t = \frac{\overline{X_1} - \overline{X_2}}{S_{X_1 X_2} \cdot \sqrt{\frac{1}{n_1} + \frac{1}{n_2}}} \tag{13.1}$$

where

$S_{X_1 X_2} = \sqrt{\frac{(n_1-1)S_{X_1}^2 + (n_2-1)S_{X_2}^2}{n_1+n_2-2}}$. $S_{X_1 X_2}$ is an estimator of the common standard deviation of the two samples. The degree of freedom for this test is $n_1 + n_2 - 2$.

For unequal variance, Welch's *t*-test statistics is calculated as

$$t = \frac{\overline{X_1} - \overline{X_2}}{S_{\overline{X_1}-\overline{X_2}}} \tag{13.2}$$

where

$S_{\overline{X_1}-\overline{X_2}} = \sqrt{\frac{S_1^2}{n_1} + \frac{S_2^2}{n_2}}$. For use in significance testing, the distribution of the test statistic is approximated as being a Student's *t* distribution with the degrees of freedom calculated using

$$df = \frac{\left(S_1^2/n_1 + S_2^2/n_2\right)^2}{\left(S_1^2/n_1\right)^2/(n_1-1) + \left(S_2^2/n_2\right)^2/(n_2-1)}. \tag{13.3}$$

13.3.2.2 GSEA

Though there are many variations on the gene set enrichment analysis (GSEA) method, we describe here the version of the algorithm developed by Subramanian and colleagues [13], which is the most widely used form of the GSEA method.

Suppose microarray data are given from samples belonging to two phenotypes, phenotype 1 and phenotype 2 (e.g., control vs. experimental). In the microarray data, each gene and each sample are given a gene expression value. Suppose a gene set S is also given, usually derived from some common biological category. The question here is whether the gene set S shows differential expression between the two phenotypes.

First, an association score is calculated for each gene that measures the difference of that gene's expression in the two phenotypes using any suitable metric. For example, we may compute for each gene an independent two-sample t-statistic between phenotype 1 and phenotype 2 or the difference between signal-to-noise ratios (mean divided by variance) in each phenotype. Second, all the N genes are put into a list $L = \{g_1, g_2, \ldots, g_N\}$ and the list is sorted by each gene's association score r_i from most positive to most negative. Genes that appear toward the top of the list are more expressed in phenotype 1 and genes that appear toward the bottom of the list are more expressed in phenotype 2. Third, walk down the gene list and compute a running sum. Each time a gene is hit in the gene set S, the sum is increased, and each time a gene is not hit in the gene set S, the sum is decreased. The degree to which the sum is increased or decreased is weighted and normalized so that the total sum after going through all the genes is 0. Finally, let the enrichment score (ES) be the maximum deviation of the running sum from 0. More specifically, for some weighting parameter p, usually $p = 1$, let

$$P_{\text{hit}}(S, i) = \sum_{g_j \in S, j \leq i} \frac{|r_j|^p}{N_R}, \quad \text{where } N_R = \sum_{g_j} |r_j|^p$$

$$P_{\text{miss}}(S, i) = \sum_{g_j \notin S, j \leq i} \frac{1}{N - N_S}, \quad \text{where } N_S = \text{num of genes in } S.$$

Then ES is the maximum deviation of $P_{\text{hit}} - P_{\text{miss}}$ from 0.

In order to determine the significance of the ES, a number of permutations are created and the ES for each permutation is recalculated. Permutations of the phenotypes in the original microarray data are preferred over permutations of the genes in the gene list, since this preserves the structure between genes. The ESs of the permutations generate a null distribution, and a nominal p-value is given by the number of permutations with a larger ES than the original data. This nominal p-value is then used to help identify whether this gene set is associated with the difference between the gene expression levels in the samples of the two phenotypes.

13.3.2.3 Bayesian Classification

A simple Bayesian classifier is the naive Bayes classifier which is a probabilistic classifier based on applying Bayes' theorem with strong (naive) independence assumptions.

Assuming in general that Y is any discrete-valued variable and the attributes $X_1, \ldots, X_n$ are any discrete- or real-valued attributes, the naive Bayes classification rule is

$$Y^* = \arg\max_{y \in \{y_1,\ldots,y_m\}} P(Y|X_1,\ldots,X_n). \tag{13.4}$$

Using Bayes' theorem and assuming the attributes $X_1, \ldots, X_n$ are all conditionally independent of one another given Y, the Eq. (13.4) can be rewritten as

$$\begin{aligned} Y^* &= \arg\max_{y \in \{y_1,\ldots,y_m\}} P(Y|X_1,\ldots,X_n) \\ &= \arg\max_{y \in \{y_1,\ldots,y_m\}} \frac{P(X_1,\ldots,X_n|Y)P(Y)}{P(X_1,\ldots,X_n)} \\ &\vdots \\ &= \arg\max_{y \in \{y_1,\ldots,y_m\}} \frac{1}{P(X_1,\ldots,X_n)} P(Y) \prod_{i=1}^{n} P(X_i|Y). \end{aligned} \tag{13.5}$$

where $P(Y)$ is the prior probability, $\prod_{i=1}^{n} P(X_i|Y)$ is likelihood, and $P(X_1, \ldots, X_n)$ is evidence.

13.3.2.4 Bayesian Network

Bayesian networks are a probabilistic graphical model that represents a set of random variables and their conditional independencies via a directed acyclic graph (DAG) whose nodes represent random variables (observable quantities, latent variables, unknown parameters, or hypotheses) and edges represent conditional dependencies.

Let $G = (V, E)$ be a directed acyclic graph (or DAG), and let $X = \{X_1, X_2, \ldots, X_n\}$ be a set of random variables. Suppose that each variable is conditionally independent of all its non-descendants in the graph given the value of all its parents. Then X is a Bayesian network with respect to G. Its joint probability density function (with respect to a product measure) can be written as a product of the individual density functions, conditional on their parent variables as follows [27]:

$$P(X_1,\ldots,X_n) = \prod_{i=1}^{n} P(X_i|\text{parents}(X_i)), \tag{13.6}$$

where parents(X_i) is the set of parents of X_i.

For any set of random variables, the probability of any member of a joint distribution can be calculated from conditional probabilities using the chain rule as follows [27]:

$$P(X_1 = x_1, \ldots, X_n = x_n) = \prod_{i=1}^{n} P(X_i = x_i | X_{i+1} = x_{i+1}, \ldots, X_n = x_n)$$

(13.7)

13.3.3 Computational Data Analysis

13.3.3.1 Artificial Neural Network

Neural networks have several unique advantages and characteristics as research tools for the cancer prediction problems [28–32]. A very important feature of these networks is their adaptive nature, where "learning by examples" replaces conventional "programming by different cases" in solving problems.

A generalized feed-forward neural network has three layers: input layer, hidden layer, and output layer, and is trained using a back propagation supervised training algorithm. The input is used as activation for the input layer and is propagated to the output layer. The received output is then compared to the desired output and an error value is calculated for each node in the output layer. The weights on edges going into the output layer are adjusted by a small amount relative to the error value. This error is propagated backwards through the network to correct edge weights at all levels.

13.3.3.2 Support Vector Machine

Support vector machine (SVM), a supervised learning method that analyzes data and recognizes patterns for classification and regression analysis, performs classification by constructing an N-dimensional hyperplane that optimally separates the data into two categories. For example, the classification problem in biomarkers can be restricted to consideration of the two-class problem without loss of generality (functional biomarker and unfunctional biomarker). In this problem the goal is to separate the two classes by a function which is induced from available samples. In order to build a classifier to infer functional biomarker, we first perform an extensive literature curation to determine the constituents of functional biomarker and record the experiment condition of each functional biomarker, and build the training pattern set. Each pattern contains several features such as species, tissue, platform, time, and conditions. Then we will use support vector machine (SVM)-based methods [33] to develop the classifier for functional biomarker. And then, we will use randomly assigned patterns for validation. Last, we will apply the classifier to predict functional biomarker from the mapping results of clinical proteomics.

Consider the problem of separating the set of training patterns belonging to two separate classes (1, functional biomarker; −1, unfunctional biomarker):

$$D = \{(x_1, y_1), \ldots, (x_l, y_l)\}, x \in \mathbb{R}^n, y \in \{-1, 1\} \quad (13.8)$$

with a hyperplane

$$<w, x> + b = 0. \quad (13.9)$$

The set of patterns is said to be optimally separated by the hyperplane if it is separated without error and the distance between the closest pattern and the hyperplane is maximal. Without loss of generality it is appropriate to consider a canonical hyperplane [34], where the parameters w, b are constrained by

$$\min_i |<w, x_i> + b| = 1. \quad (13.10)$$

That is, the norm of the weight vector should be equal to the inverse of the distance of the nearest point in the dataset to the hyperplane. A separating hyperplane in canonical form must satisfy the following constraints:

$$y_i [<w, x_i> + b] \geq 1 - e_i, \quad i = 1, \ldots, l. \quad (13.11)$$

Therefore, according to the structural risk minimization inductive principle, the training of an SVM is to minimize the guaranteed risk bound

$$\min_{w,b,e} \varphi(w, b, e) = \frac{1}{2} w^T w + \frac{1}{2} C \sum_{i=1}^{l} e_i^2, \quad (13.12)$$

subject to the constraints

$$y_i [<w, x_i> + b] \geq 1 - e_i, \quad i = 1, \ldots, l. \quad (13.13)$$

The above optimization problem can be used in a linear recognition problem, but in this case, the classification problem is nonlinear. To solve the nonlinear classification problem, we can map first the training data to another dot product space (called the feature space) F via a nonlinear map $\phi : \mathbb{R}^n \to F$ and then perform the above computations in F. For example, we can use Gaussian radius basis function (RBF) kernels function for SVM.

13.3.3.3 Leave-One-Out Cross-Validation Method

Some features may not be relevant to the prediction of functional biomarker. Accordingly, we should use the leave-one-out cross-validation method to reduce the dimension of the features to overcome the risk of "overfitting" and determine kernel feathers. The leave-one-out cross-validation method is described as follows:

Input: $D = \{(x_1, y_1), \ldots, (x_1, y_1)\}, x \in \mathbb{R}^n, y \in \{-1, 1\}$
For $t = 1$ to n {
$X_t = \{x_{l,m} | m \neq t\}$//exclude t-th feature;
Train t-th SVM using [35];
Compute LOSS(t) using LOSS $(y_p - y) = \begin{cases} 0 & \text{if } y_p = y \\ 1 & \text{if } y_p \neq y \end{cases}$ //the loss of t-th SVM;}
Output: $p = \arg\max_t \text{LOSS}(t)$//find the worst feature.

where l is the number of the specimen used for training, n is the number of the used features during the procedure, and y_i is the label of i-th specimen. The above algorithms performed until the necessary number of features are reached.

13.3.3.4 Genetic Algorithms

Genetic algorithms are implemented in a computer simulation in which a population of abstract representations (called chromosomes or the genotype of the genome) of candidate solutions (called individuals, creatures, or phenotypes) to an optimization problem evolves toward better solutions. As you can guess, genetic algorithms are inspired by Darwin's theory about evolution. Simply said, solution to a problem solved by genetic algorithms is evolved.

Chromosomes

All living organisms consist of cells. In each cell there is the same set of *chromosomes*. Chromosomes are strings of DNA and serve as a model for the whole organism. A chromosome consists of *genes*, blocks of DNA. Each gene encodes a particular protein. Basically, it can be said that each gene encodes a *trait*, for example, color of the eyes. Possible settings for a trait (e.g., blue, brown) are called *alleles*. Each gene has its own position in the chromosome. This position is called *locus*.

Complete set of genetic material (all chromosomes) is called *genome*. Particular set of genes in genome is called *genotype*. The genotype is with later development after birth base for the organism's *phenotype*, its physical and mental characteristics, such as eye color, intelligence etc.

Reproduction

During reproduction, first occurs *recombination* (or *crossover*). Genes from parents form in some way the whole new chromosome. The newly created offspring can then be mutated. *Mutation* means that the elements of DNA are a bit changed. These changes are mainly caused by errors in copying genes from parents.

The *fitness* of an organism is measured by success of the organism in its life.

Genetic algorithms are inspired by Darwin's theory about evolution. Solution to a problem solved by genetic algorithms is evolved.

Algorithm is started with a *set of solutions* (represented by *chromosomes*) called *population*. Solutions from one population are taken and used to form a new population. This is motivated by a hope, that the new population will be better than the old one. Solutions which are selected to form new solutions (*offspring*) are selected according to their fitness—the more suitable they are, the more chances they have to reproduce.

This is repeated until some condition (e.g., number of populations or improvement of the best solution) is satisfied.

The basic genetic algorithm is outlined as follows:

1. **[Start]** Generate random population of n chromosomes (suitable solutions for the problem).
2. **[Fitness]** Evaluate the fitness $f(x)$ of each chromosome x in the population.
3. **[New population]** Create a new population by repeating the following steps until the new population is complete:

    (a) **[Selection]** Select two parent chromosomes from a population according to their fitness (the better fitness, the bigger chance to be selected).
    (b) **[Crossover]** With a crossover probability, cross over the parents to form a new offspring (children). If no crossover was performed, the offspring is an exact copy of parents.
    (c) **[Mutation]** With a mutation probability, mutate new offspring at each locus (position in chromosome).
    (d) **[Accepting]** Place new offspring in a new population.

4. **[Replace]** Use newly generated population for a further run of algorithm.
5. **[Test]** If the end condition is satisfied, **stop**, and return the best solution in current population.
6. **[Loop]** Go to step **2**.

13.3.3.5 Decision Trees

A *decision tree* (or *tree diagram*) is a decision support tool that uses a treelike graph or model of decisions and their possible consequences, including chance event outcomes, resource costs, and utility. These decisions generate rules for the classification of a dataset. Specific decision tree methods include classification and regression trees (CART) and chi-squared automatic interaction detection (CHAID). CART and CHAID are decision tree techniques used for classification of a dataset. They provide a set of rules that you can apply to a new (unclassified) dataset to predict which records will have a given outcome. CART segments a dataset by creating 2-way splits while CHAID segments using chi square tests to create multi-way splits. CART typically requires less data preparation than CHAID.

13.3.3.6 Graph Theory

A graph refers to a collection of vertices or nodes and a collection of edges that connect pairs of vertices and can be abstracted mathematically as a *graph* $G(V, E)$. The *vertex set* of G is usually denoted by $V(G)$, and the *edge set* of G is usually denoted by $E(G)$. The *degree*, or *valency*, $d_G(v)$ of a vertex v in a graph G is the number of edges incident to v, with loops being counted twice. The *vertex connectivity* or *connectivity* $\kappa(G)$ of a graph G is the minimum number of vertices that need to be removed to disconnect G. The graph theory has been widely applied to the analysis of molecular interaction networks such as protein-protein interaction networks, gene-gene co-expression networks, genetic interaction networks, molecular co-annotation networks, literature co-occurrence networks, and molecular entity association networks, where a vertex can represent gene, protein, or pathway and the edge can represent interaction, similarity, or distance.

13.3.3.7 Nearest Neighbor Method

Nearest neighbor method is a technique that classifies each record in a dataset based on a combination of the classes of the k record(s) most similar to it in a historical dataset (where $k \geq 1$). This is sometimes called the k-nearest neighbor technique.

13.3.4 Biological Data Analysis

13.3.4.1 Gene Prioritization Based on Networks

Identification of disease genes is important to better understand gene functions. Recently, a number of computational approaches have been developed to predict or prioritize candidate disease genes. Network-based approaches have also been employed to infer new candidate disease genes based upon network linkages with known disease genes. Typically, these methods first construct a gene-gene or protein-protein association network based on one or more types of genomic and proteomic data and then rank candidate genes based on network proximity to know disease-associated genes. The influx of human molecular interaction data, e.g., high-throughput protein-protein interaction (PPI) data, has led to many recent studies that aim to connect disease-modifying genes with other molecular interacting entities [36–38]. In the chapter, we primarily focus on molecular interaction network topological information that model physical interactions and functional relationship between proteins and use them to predict and prioritize disease genes or disease biomarkers. A common assumption in these studies is that candidate disease genes may be found in close topological proximity of known disease genes in the molecular interaction network [36–40]. A seeding strategy is often used, e.g., to incorporate prior knowledge of disease genes from public databases such as OMIM (Online Mendelian Inheritance in Man) [41] or literature curation and generate disease-specific context for subsequent network topological function analysis. These methods usually correlate known disease genes with candidate disease genes by relating their local topological features, such as *node degrees* (the number of PPI connections to a node protein), *closeness* (average distance to disease genes), and *betweenness* (average neighborhood overlapping with disease genes) [39].

Recent progress has been made to prioritize genes in molecular interaction networks using PageRank [42] and HITS [43], which were inspired by algorithms to rank Web pages through Web links. These methods calculate scores to measure global similarity between functionally known genes and unknown genes, outperforming local topological association methods [37]. For example, Sebastian et al. used a random walk (RW) algorithm to rank candidate genes to known members of a disease-gene family [37]. Wu et al. introduced the ant colony optimization (ACO) algorithm into ranking yeast PPI network annotated with lethality information, which revealed intriguing patterns [44]. Jing Chen et al. compared PageRank, HITS with Priors, and K-step Markov method to estimate the relative importance of candidate disease genes in PPI network to known disease genes [36], which found that the three methods and their modifications yielded similar results. Huang et al. evaluated how the quality of seed genes and PPI data affect the ranking results and suggested that disease specific prior knowledge should be included in prioritizing candidate disease genes whenever possible [45].

13.3.4.2 Pathway Analysis

A biological system is very dynamical and complex. Systems biology results show that genes and proteins do not function in isolation [46]; instead, they work in interconnected pathways and molecular networks [47]. A mere study of individual molecules such as genes or proteins (which has been the traditional approach for many years) cannot help in decoding the mystery of a system. Since we do not completely understand the entire system globally, there is no cure for several diseases or disorders such as cancer, AIDS etc. Also, there are certain processes which we do not understand yet such as regeneration. Hence, obtaining a genome or a proteome is just half the story. Studying the interactions between these molecules is a step ahead to understand a biological system. As a part of this process, it is also essential to visualize how a system behaves in response to various compounds such as drugs. Only if we know all the interactions of a drug, it can be used as a better cure.

With the advent of high-throughput technology over the last decade, vast amounts of data have been generated which has led to the development of systems biology approaches that interrelate the elements of biological processes, such as mRNAs, micro RNAs, and proteins, revealing higher-level pathways and networks of organization. Systems biology first describes the elements of the system then the biological networks that interrelate these elements and finally characterizes the flow of information that links these elements and their networks to an emergent biological process. A pathway network can be defined as a set of molecular interactions, and in this set, a subset of genes that coordinate to achieve a specific task forms a pathway. Interactions among several genes and gene products lead to biological processes. Most of the effort these days is concerted on recognizing some patterns among these networks and pathways. However, it is not enough to analyze patterns; we need to understand their dynamical nature, that is, how they evolve and change. It is important to understand the transition from DNA sequence to disease symptoms to their therapeutic targets.

A biological pathway tries to understand a specific biological process. A pathway is a series of interactions leading to a cellular process/molecular function. Mathematically, pathways are described as graphs consisting of a node and edges. A node represents a biological entity which can be a gene, protein, or any compound, while edges reflect the relationship between two entities which it connects. The relationship can be activation, inhibition, chemical modification, or undefined. Also these edges can be directional or nondirectional. Pathways can be classified into several categories. Some of the important biological categories are metabolic, regulatory, protein function, and disease.

Biological pathway construction either follows a data-driven objective (DDO) or knowledge-driven objective (KDO). DDO is derived from experimental data such as genomic or proteomic data while KDO is constructed by considering a particular domain of interest, such as disease, system etc. Literature is the main source of information in KDO.

There are some pathway visualization tools such as CellDesigner, Cytoscape, and Ingenuity etc. There are various tools available for manually curating the pathways: Pathway Editor, Knowledge Editor, Map Editor, etc. Several tools such as Pathway Studio, Pathway Finder, and PubGene use natural language processing (NLP) to identify associations from literature which can be built into pathways. There are some important pathway databases described below.

KEGG (Kyoto Encyclopedia of Genes and Genomes http://www.genome.jp/kegg): It is a free, online, open source pathway database. Developed by Kanehisa Laboratories, it is an integrated database resource consisting of 16 main databases which contain systems and genomic and chemical information. KEGG consists of pathways stored as pathway maps. They cover various domains such as metabolism (the most popular domain in KEGG), genetic information processing (transcription, translation, etc.), environmental information processing (cell growth, cell motility, etc.), cellular processes (immune system, nervous system, etc.), and human diseases (neurodegenerative, circulatory, etc.). It contains 343 pathway maps, 114 human diseases, 9,149 drugs, 5,135,391 genes in high-quality genomes, and 16,055 metabolites and other small molecules [48].

Reactome (http://www.reactome.org): It is also a free, online, curated, open source pathway database. It cross-references several databases such as UniProt (http://www.uniprot.org) and NCBI (http://www.ncbi.nlm.nih.gov). It contains information about 23 species. For humans, it contains information on 3,916 proteins, 2,955 complexes, 3,541 reactions, and 1,045 pathways. It is an effort of collaboration between Cold Spring Harbor Laboratory, Ontario Institute for Cancer Research, European Bioinformatics Institute, Gene Ontology Consortium, and New York University. It also offers tools for pathway analysis [49].

Pathguide (http://www.pathguide.org): It is a meta database which contains information on various biological pathway resources. It contains information about 310 resources which include a listing of protein-protein interactions, metabolic pathway, signaling pathways, pathway diagrams, transcription factors/gene regulatory networks, protein compound interactions, genetic interactions, and proteins sequences. It contains this information about 24 different species [50].

BioCyc (http://biocyc.org): It is a collection of 505 pathway or genome databases, and each database describes genome and metabolic pathways of a single organism. It contains many tools such as for a comparative analysis, visual analysis (editing of pathways, etc.), genome browser, and display of individual metabolic maps. BioCyc is organized into three tiers: tier 1 contains an intensively curated database, tier 2 contains computationally derived databases subject to moderate curation, and tier 3 contains computationally derived databases with no curation [51].

NCI pathway interaction database (http://pid.nci.nih.gov): It is very well structured and curated collection of biomolecular interactions and key cellular processes pulled together into signaling pathways. It is a collaborative effort between National Cancer Institute (NCI) and Nature Publishing Group (NPG). It contains 100 human pathways with 6,298 interactions curated by NCI nature. And 392 human pathways with 7,418 interactions imported from Reactome/BioCarta [52].

13.3.5 Validation

The technical validation of biomarkers depends on all aspects of the analytic method, including specificity, sensitivity, bias, and robustness. *Specificity*, a measure of true negatives among all negatives, refers to the probability that a validation technique will indicate a positive test result when biomarker is positive. *Sensitivity*, a measure of true positives among all positives, refers to the probability that a validation technique will indicate a negative test result when biomarker is negative. *Bias* refers to lack of representation of all cases covered due to lack in sample size. *Robustness* refers to consistency of test results performed in different conditions.

13.4 Case: Breast Cancer Plasma Protein Biomarker Discovery by Coupling LC-MS/MS Proteomics and Systems Biology

Use our breast cancer example (only published results). It should cover genome technology, experiment design, statistical/computational analysis, biological analysis, validation, etc.

Breast cancer is worldwide the second most common type of cancer after lung cancer. According to the American Cancer Society, this year in the USA, approximately 192,370 women will be diagnosed with breast cancer, and about 40,170 women will die from the disease.

Biomarkers are important clinical tools for breast cancer screening and diagnosis and also can be used by doctors to tailor patients' treatments. There have been a lot of researches about breast cancer biomarker identification. The researchers, led by Prakash Rao, studied 102 women; 52 had breast cancers and 50 were women who either had leukemia or fibrocystic breast disease or who had no cancer. They found increased levels of riboflavin carrier protein (RCP) in the blood of the women with breast cancer. Blood levels of RCP were more than nine times higher in women with breast cancer than in women without the disease [53]. Dua used an intraductal approach to identify breast cancer biomarker [54]. Ou used integrative proteomic and gene expression mapping to identify breast cancer biomarkers [55]. Also proteomic analysis of breast nipple aspirate fluid (NAF) was used to identify candidate markers of breast cancer [56].

The majority of current breast cancer biomarker identification is conducted using established breast cancer cell lines [26, 57–60]. Cell lines are widely used in many aspects of laboratory research and particularly as *in vitro* models in cancer research. They have a number of advantages. But many researchers supported that the plasma proteome profiling might have a higher chance to indentify biomarkers than proteins present in other medium [61].

On the other hand, a protein biomarker or set of biomarkers that identify patients with cancer from a single type of samples has proven elusive for most forms of the disease. Therefore, comparing protein change in plasma proteome sample with other types of sample may help to identify with a higher confidence a candidate set of protein biomarker.

Moreover, as it is becoming increasingly apparent that genes do not function alone but through complex biological pathways with more information revealed through large-scale "omics" techniques, extensive pathway, network, and function analysis of those identified protein biomarkers allowed us to discover their proteomic signatures in plasma of patients at high risk for cancer disease.

In the case study, we showed how to apply "systems biology" approach to the study of panel biomarker discovery problem in breast cancer proteomics data study. First we used a *t*-statistics and permutation procedure to identify initial protein biomarker candidates. Then an extensive literature-mining curation enabled us to determine the final protein biomarkers. Last, focusing on these final protein biomarkers, we used gene ontology analysis and ingenuity pathway analysis to validate the list and unravel the intricate pathways, networks, and functional contexts in which genes or proteins function. Our results showed that the systems biology approach is essential to the understanding molecular mechanisms of panel protein biomarkers.

13.4.1 Experiment Design

Plasma protein profiles were collected in two batches, which we refer to as Study A and B. Study A and B were processed in the same laboratory but at different times. Each sample was analyzed in a single batch by mass spectrometry. In either study, 80 plasma samples were collected (40 samples collected from women with breast cancer and 40 from healthy volunteer women who served as controls). The demography and clinical distribution of breast cancer stages/subtypes for Study A and B are comparable.

We compared our results with 4 previously published proteomic studies of breast cancer cell lines. Their methods and results presented in peer-reviewed journals [26, 57, 59, 60] have established a higher reliability. A total of 3,085 protein biomarkers were identified from five breast cell lines, MCF-10A, BT474, MDA-MB-468, MD-MB-468, and T47D/MCF7, in their papers.

13.4.2 Protein Identification and Quantification

For protein identification, tryptic peptides were analyzed using Thermo-Finnigan linear ion trap mass spectrometer (LTQ) coupled with an HPLC system. Peptides were eluted with a gradient from 5 to 45 % acetonitrile developed over 120 min and data were collected in the *triple-play* mode (MS scan, zoom scan, and MS/MS scan). The acquired raw peak list data were generated by Xcalibur (version 2.0) using default parameters and further analyzed by the label-free identification and quantitative algorithm using default parameters described by Higgs et al. [62]. MS database searches were performed against the combined protein dataset from International Protein Index (IPI; version 3.60) and the nonredundant NCBI-nr human protein database (updated 2009), which totaled 22,180 protein records. Various data processing filters for protein identification were applied to control false-discovery rate at below 5 % levels.

For protein quantification, first, all extracted ion chromatograms (XICs) were aligned by retention time. Each aligned peak were matched by precursor ion, charge state, fragment ions from MS/MS data, and retention time within a 1-min window. Then, after alignment, the area under the curve (AUC) for each individually aligned peak from each sample was measured, normalized, and compared for relative abundance—all as described in [62]. Here, a linear mixed model generalized from individual ANOVA (analysis of variance) was used to quantify protein intensities. In principle, the linear mixed model considers three types of effects when deriving protein intensities based on weighted average of quantile-normalized peptide intensities: (1) *group effect*, which refers to the fixed nonrandom effects caused by the experimental conditions or treatments that are being compared; (2) *sample effect*, which refers to the random effects (including those arising from sample preparations) from individual biological samples within a group; and (3) *replicate effect*, which refers to the random effects from replicate injections from the same sample preparation.

13.4.3 "Systems Biology" Analysis

We applied "systems biology" approach to the study of panel biomarker discovery problem in breast cancer proteomics data study in this study. Our strategies for analyzing potentially noisy proteomics dataset are threefold. First, we used a *t*-statistics and permutation procedures to calculate *p*-value for proteins changed in all samples, instead of fold change or *t*-test for a given sample that are commonly used in previous studies. This allowed us to enhance the statistical power to filter the proteomics results. Second, we used extensive literature-mining curation to focus on breast-cancer-relevant differentially expressed proteins only. This literature curation step enabled us to concentrate on breast-cancer-relevant signals, with generally noisy proteomics datasets. Third, we used gene ontology analysis and ingenuity pathway analysis to identify and validate correlated changes due to cancer cell signaling that may, individually, elude the detection.

13.4.3.1 *t*-statistics and Permutation Process

Our test statistic is a mean of 40 values (protein intensities in health samples) minus the mean of another 40 values (protein intensities in cancer samples). A permutation procedure was used to determine the *p*-value for each protein, representing the chance of observing a test statistic at least as large as the value actually obtained. The 80 samples for each protein were permuted 100,000 times and the complete set of *t*-tests was performed for each permutation. The permutation *p*-value for a particular protein is the proportion of the permutations in which the permuted test statistic exceeds the observed test statistic in absolute values. We chose a significance level $\alpha = 0.001$ to select proteins where we estimated significant differences in the health and cancer sampled. The corresponding "per-family Type 1 error rate, PFER," that is, the expected number of false positives for such a multiple test procedure, is PFER = number of genes $\times$ 0.001. Alternatively, the nominal "false-discovery rate, FDR," or expected proportion of false positive among the genes declared differentially expressed, is FDR = PFER/number of genes declared expressed.

13 Computational Biomarker Discovery   377

Table 13.2 Top networks involved

Primary network functions	Computed score	Molecules in network
Cancer, cell-to-cell signaling and interaction, hepatic system disease	48	25
Genetic disorder, hematological disease, ophthalmic disease	43	24
Endocrine system disorders, skeletal and muscular system development and function, tissue morphology	22	15
Cancer, cell cycle, reproductive system disease	22	14
Drug metabolism, small molecule biochemistry, cancer	18	12

Reprinted from Ref. [67], with kind permission from BMC Genomics

13.4.3.2 Pathway, Network, and Function Annotation Analysis

Ingenuity pathway analysis was used for building pathway and network. DAVID database was used to study levels 2 and 5 of biological process in gene ontology.

13.4.4 Results

13.4.4.1 Pathway Analysis and Gene Ontology Categorization of Significant Proteins

A total of 4,832 peptides in Study A are mapped to 1,422 proteins by searching against IPI database. Using a t-statistics and permutation process described in the method section and setting a p-value cutoff (0.001) after initial ANOVA analysis of mass spectra data, we identified 254 statistically significant differentially expressed proteins (PFER = 1.422, FDR = 0.0056), among which 208 are overexpressed and 46 are underexpressed in breast cancer plasma. Compared to the result of traditional statistical test (PFER = 2.5596, FDR = 0.01), our result shows that the coupled statistical process outperforms the sensitivity of a parametric traditional statistical test that requires strong and sometimes untenable data assumptions since it is nonparametric and requires no assumption about the distribution under the null hypothesis.

A comparison of the set of 254 proteins with published findings from proteomic analysis of human breast cancer cell lines yielded 26 proteins with differentially expressed in human and cancer samples that were identified in breast cancer cell lines. Top networks and canonical pathways were identified with ingenuity pathway analysis (Tables 13.2 and 13.3, and Fig. 13.2). And level 2 of the biological process in gene ontology is mainly studied (Table 13.4). An interesting finding from pathway analysis is that those top networks and pathways shown in Tables 13.2 and 13.3, and Fig. 13.2, especially the top 1 network (cancer, cell-to-cell signaling and interaction, hepatic system disease) and top 1 pathway (acute phase response

Table 13.3 Top pathways involved

Pathway	−Log(p-value)
Acute phase response signaling	1.20E+01
Complement system	1.05E+01
Coagulation system	4.55E+00
PPAR signaling	1.90E+00
Glutathione metabolism	1.49E+00

Reprinted from Ref. [67], with kind permission from BMC Genomics

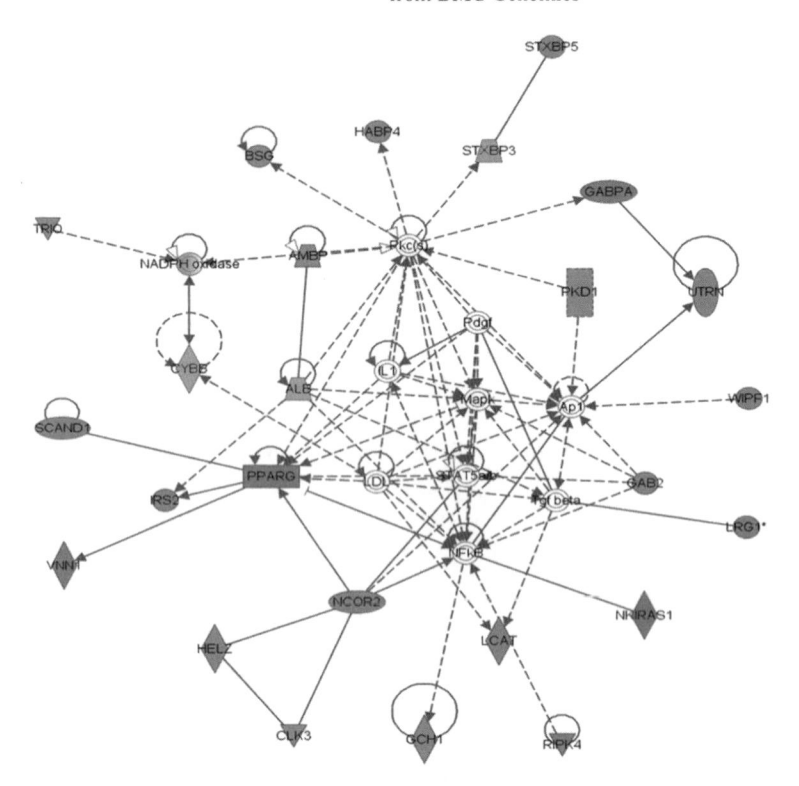

Fig. 13.2 The 26 proteins are involved in a single cancer signaling network (Reprinted from Ref. [67], with kind permission from BMC Genomics)

signaling), are validated by Study B dataset and the 26 candidate protein biomarkers and are similar to previously reported works [63–65]. Another interesting finding from gene ontology is the role of cellular metabolic process, and response to external stimulus (especially proteolysis and acute inflammatory response in level 5) in Table 13.4 in breast cancer was also reported by other authors. For example, cancer, like other diseases, is accompanied by strong metabolic disorders [58]. And It also was reported that stress and external stimulus such as microbial infections, ultraviolet radiation, and chemical stress from heavy metals and pesticides affect the progression of breast cancer [66].

Table 13.4 Gene ontology biological processes enrichment analysis for 26 protein biomarkers

GO term	Percentage
Response to external stimulus	50
Response to stress	46
Primary metabolic process	38
Defense response	38
Regulation of biological process	35
Establishment of localization	35
Regulation of biological quality	31
Anatomical structure development	27
Cellular metabolic process	27
Multicellular organismal development	27
Cell communication	27
Transport	27
Cellular component organization and biogenesis	23
Macromolecule metabolic process	23
Regulation of cellular process	19

Reprinted from Ref. [67], with kind permission from BMC Genomics

13.4.4.2 Cross-Validation of Candidate Biomarkers

In order to validate the computational results, the same methods and procedures as we used in Study A were applied to Study B. Forty-eight candidate protein biomarkers were identified, of which 13 were found in common with the 26 protein biomarkers we identified in Study A. Fisher's exact test shows that our methods are feasible and reliable (Fisher's exact test, p-value $= 1.074\text{e}{-}09$). Using ingenuity pathway analysis and DAVID GO analysis, we also found that the 48 candidate protein biomarkers identified from Study B have the similar pathway, network, and function as the 26 candidate protein biomarkers identified from Study A.

13.5 Discussion: Biomarkers Toward Systems Biology

Research interest in identifying novel biomarkers has grown significantly in recent years. Evolvement of biomarker concept toward systems biology [68] is shown in the Fig. 13.3.

13.5.1 Single Biomarker

Known breast cancer susceptibility genes such as *P53*, *BRCA1*, *BRCA2*, *ERBB2*, and *PTEN* account only for 15–20 % of the familial risk for breast cancer [69]. Identification of these genes and locus [70, 71], while extremely valuable, is only

Fig. 13.3 Concept evolution on biomarkers towards systems biology

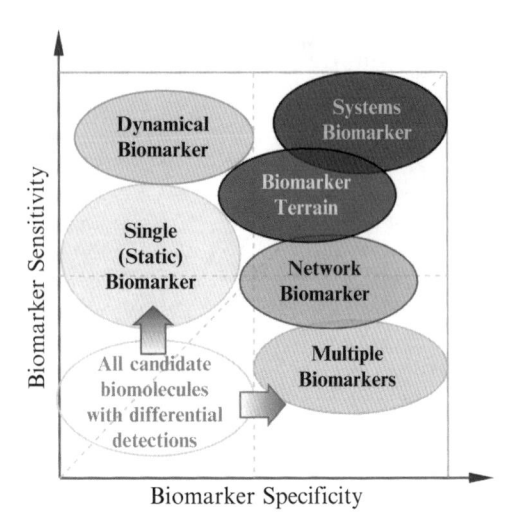

the first step to the development of molecular diagnosis and prognosis solutions. Viewed from an emerging network biology perspective [72], these genes do not function in isolation [46].

13.5.2 Dynamical Biomarker

Dynamical biomarker [73–75] was first introduced on a speech by A.L. Goldberger in 2006 [76], which can be seen as an initiation of using nonlinear dynamical properties as biomarkers, although this concept has not extended to the area of molecular networks. It is found that both biological shape [77, 78] and physiological signals [73, 75] have chaotic and/or fractal characteristics [79], which indicate that many biological systems and networks could be analyzed effectively by applying nonlinear dynamical approaches involving chaos, fractal, bifurcation, pattern formation, and complex systems [74].

13.5.3 Network Biomarker

Network biomarker [18, 80, 81] is a new concept for candidate biomarker discovery by integrating cancer susceptibility genes, their gene expressions, and their protein interaction network. In 2007, Marc Vidal's group at Harvard constructed a protein interaction network for breast cancer susceptibility using various "omics" datasets and identified HMMR as a new susceptibility locus for the disease [17]. Later, Trey Ideker's group at UCSD integrated protein network and gene expression data to improve the prediction of metastasis formation in patients with breast cancer [18]. The two studies marked the exciting beginning of a new paradigm which suggests protein interaction networks and pathway, although drafty, error-prone, and incomplete can serve as a molecular-level conceptual road map to guide future breast cancer biomarkers studies [82].

13.5.4 Systems Biomarker

Systems biomarker, as an innovative concept shown in the Fig. 13.3, derives from the marriage of network biomarkers and dynamical biomarkers. Although many methods have been presented in network biology, including network-based gene ranking for molecular biomarker discovery [83], and graph clustering for functional module discovery [84], it is still hard to find systems biomarkers hidden in disease-specific molecular interaction networks. Although network biomarker can be successfully applied in a small-scale breast cancer centered protein network, it still lacks capability to analyze large-scale protein networks, which are always visualized as "ugly" hairballs on 2D network layout. While the large-scale disease-specific protein network modeling with gene expression profiles is a key step toward systems biomarkers.

13.5.5 Network Biomarker Terrain

Using the concept of terrain to visualize gene expression profiles started from a work by Stuart K. Kim et al. in 2001 [85]. They assembled data from *C. elegans* DNA microarray experiments and visualized grouped co-regulated genes in a 3D expression map that displays correlations of gene expression profiles as distances in two dimensions and gene density in the third dimension. In a followed study at 2008, Qian You et al. visualized an Alzheimer's disease (AD)-specific protein interaction network as a 3D terrain (GeneTerrain) and successfully characterized the differentials between the three distinct stages of AD [86]. Network biomarker terrain bridges the gap between network biomarker and dynamical biomarker, which will finally implement the concept of systems biomarker.

13.6 Conclusions

This chapter has discussed some of the technologies that are available for computational biomarker discovery for diagnostic classification and prognostic assessment, has explored how these technologies can be applied to the discovery of different types of biomarkers, and, finally, has worked through many of the basic, but critical, issues associated with the generation and analysis of the microarray and proteomic expression data. The bioinformatics discussed herein are required to ensure that the data being produced are of high quality, that the experiments are appropriately designed, that the methods are correctly performed, and that the results are analyzed appropriately to identify biomarkers in a systematical biology way. We hope that this will be useful in assisting the computational biomarker discovery efforts of current and future genomics and proteomics researchers.

Acknowledgements This work was supported in part by a grant from the National Cancer Institute (U24CA126480-01), part of NCI's Clinical Proteomic Technologies Initiative (http://proteomics.cancer.gov), awarded to Dr. Fred Regnier (PI) and Dr. Jake Chen (co-PI).

References

1. Soreide K (2009) Receiver-operating characteristic curve analysis in diagnostic, prognostic and predictive biomarker research. J Clin Pathol 62(1):1–5
2. Jaffe CC (2009) Pathology and imaging in biomarker development. Arch Pathol Lab Med 133(4):547–549
3. Rhodes DR, Sanda MG, Otte AP, Chinnaiyan AM, Rubin MA (2003) Multiplex biomarker approach for determining risk of prostate-specific antigen-defined recurrence of prostate cancer. J Natl Cancer Inst 95(9):661–668
4. Allison DB, Cui X, Page GP, Sabripour M (2006) Microarray data analysis: from disarray to consolidation and consensus. Nat Rev Genet 7(1):55–65
5. Reimers M (2010) Making informed choices about microarray data analysis. PLoS Comput Biol 6(5):e1000786
6. Slonim DK, Yanai I (2009) Getting started in gene expression microarray analysis. PLoS Comput Biol 5(10):e1000543
7. Sørlie T, Perou CM, Tibshirani R, Aas T, Geisler S, Johnsen H, Hastie T, Eisen MB, Van De Rijn M, Jeffrey SS (2001) Gene expression patterns of breast carcinomas distinguish tumor subclasses with clinical implications. Proc Natl Acad Sci U S A 98(19):10869–10874
8. Giltnane JM, Rimm DL (2004) Technology insight: identification of biomarkers with tissue microarray technology. Nat Clin Pract Oncol 1(2):104–111
9. Segal E, Friedman N, Kaminski N, Regev A, Koller D (2005) From signatures to models: understanding cancer using microarrays. Nat Genet 37:S38–S45
10. Potti A, Dressman HK, Bild A, Riedel RF, Chan G, Sayer R, Cragun J, Cottrill H, Kelley MJ, Petersen R (2006) Genomic signatures to guide the use of chemotherapeutics. Nat Med 12(11):1294–1300
11. Huang DW, Sherman BT, Lempicki RA (2009) Bioinformatics enrichment tools: paths toward the comprehensive functional analysis of large gene lists. Nucleic Acids Res 37(1):1–13
12. Khatri P, Draghici S (2005) Ontological analysis of gene expression data: current tools, limitations, and open problems. Bioinformatics 21(18):3587–3595
13. Subramanian A, Tamayo P, Mootha VK, Mukherjee S, Ebert BL, Gillette MA, Paulovich A, Pomeroy SL, Golub TR, Lander ES (2005) Gene set enrichment analysis: a knowledge-based approach for interpreting genome-wide expression profiles. Proc Natl Acad Sci 102(43):15545–15550
14. Glez-Pena D, Gomez-Lopez G, Pisano DG, Fdez-Riverola F (2009) WhichGenes: a web-based tool for gathering, building, storing and exporting gene sets with application in gene set enrichment analysis. Nucleic Acids Res 37(Web Server Issue):W329–W334

15. Medina I, Montaner D, Bonifaci N, Pujana MA, Carbonell J, Tarraga J, Al-Shahrour F, Dopazo J (2009) Gene set-based analysis of polymorphisms: finding pathways or biological processes associated to traits in genome-wide association studies. Nucleic Acids Res 37(Web Server Issue):W340–W344
16. Dennis G Jr, Sherman BT, Hosack DA, Yang J, Gao W, Lane HC, Lempicki RA (2003) DAVID: database for annotation, visualization, and integrated discovery. Genome Biol 4(9):R60
17. Pujana MA, Han JDJ, Starita LM, Stevens KN, Tewari M, Ahn JS, Rennert G, Moreno V, Kirchhoff T, Gold B (2007) Network modeling links breast cancer susceptibility and centrosome dysfunction. Nat Genet 39(11):1338–1349
18. Chuang HY, Lee E, Liu YT, Lee D, Ideker T (2007) Network-based classification of breast cancer metastasis. Mol Syst Biol 3(1):140–149
19. Lupski JR, Reid JG, Gonzaga-Jauregui C, Rio Deiros D, Chen DC, Nazareth L, Bainbridge M, Dinh H, Jing C, Wheeler DA et al (2010) Whole-genome sequencing in a patient with Charcot-Marie-Tooth neuropathy. N Engl J Med 362(13):1181–1191
20. Roach JC, Glusman G, Smit AF, Huff CD, Hubley R, Shannon PT, Rowen L, Pant KP, Goodman N, Bamshad M et al (2010) Analysis of genetic inheritance in a family quartet by whole-genome sequencing. Science 328(5978):636–639
21. Chan D (2006) Clinical proteomics. Clin Proteomics 2(1):1–4
22. Hanash S (2004) Moving forward with clinical proteomics. Clin Proteomics 1(1):3–5
23. Mischak H, Apweiler R, Banks RE, Conaway M, Coon J, Dominiczak A, Ehrich JHH, Fliser D, Girolami M, Hermjakob H et al (2007) Clinical proteomics: a need to define the field and to begin to set adequate standards. Proteomics Clin Appl 1(2):148–156
24. Klampfl CW (2004) Review coupling of capillary electrochromatography to mass spectrometry. J Chromatogr A 1044(1–2):131–144
25. Frohlich T, Arnold GJ (2006) Proteome research based on modern liquid chromatography–tandem mass spectrometry: separation, identification and quantification. J Neural Transm 113(8):973–994
26. Mbeunkui F, Metge BJ, Shevde LA, Pannell LK (2007) Identification of differentially secreted biomarkers using LC-MS/MS in isogenic cell lines representing a progression of breast cancer. J Proteome Res 6(8):2993–3002
27. Needham CJ, Bradford JR, Bulpitt AJ, Westhead DR (2006) Inference in Bayesian networks. Nat Biotechnol 24(1):51–53
28. Lai KC, Chiang HC, Chen WC, Tsai FJ, Jeng LB (2008) Artificial neural network-based study can predict gastric cancer staging. Hepatogastroenterology 55(86–87):1859–1863
29. Amiri Z, Mohammad K, Mahmoudi M, Zeraati H, Fotouhi A (2008) Assessment of gastric cancer survival: using an artificial hierarchical neural network. Pak J Biol Sci 11(8):1076–1084
30. Chi CL, Street WN, Wolberg WH (2007) Application of artificial neural network-based survival analysis on two breast cancer datasets. AMIA Annu Symp Proc 2007:130–134
31. Anagnostopoulos I, Maglogiannis I (2006) Neural network-based diagnostic and prognostic estimations in breast cancer microscopic instances. Med Biol Eng Comput 44(9):773–784
32. Wang HQ, Wong HS, Zhu H, Yip TT (2009) A neural network-based biomarker association information extraction approach for cancer classification. J Biomed Inform 42(4):654–666
33. Meyer D, Leisch F, Hornik K (2003) The support vector machine under test. Neurocomputing 55(1–2):169–186
34. Vapnik VN (1998) Statistical learning theory. Springer, New York
35. Apweiler R, Bairoch A, Wu CH, Barker WC, Boeckmann B, Ferro S, Gasteiger E, Huang H, Lopez R, Magrane M et al (2004) UniProt: the Universal Protein knowledgebase. Nucleic Acids Res 32(Database Issue):D115–D119
36. Chen J, Aronow BJ, Jegga AG (2009) Disease candidate gene identification and prioritization using protein interaction networks. BMC Bioinforma 10:73
37. Kohler S, Bauer S, Horn D, Robinson PN (2008) Walking the interactome for prioritization of candidate disease genes. Am J Hum Genet 82(4):949–958
38. Oti M, Snel B, Huynen MA, Brunner HG (2006) Predicting disease genes using protein-protein interactions. J Med Genet 43(8):691–698

39. Chen JY, Shen C, Sivachenko AY (2006) Mining Alzheimer disease relevant proteins from integrated protein interactome data. Pac Symp Biocomput 2006:367–378
40. Xu J, Li Y (2006) Discovering disease-genes by topological features in human protein-protein interaction network. Bioinformatics 22(22):2800–2805
41. Hamosh A, Scott AF, Amberger JS, Bocchini CA, McKusick VA (2005) Online Mendelian Inheritance in Man (OMIM), a knowledgebase of human genes and genetic disorders. Nucleic Acids Res 33(Database Issue):D514–D517
42. Page L, Brin S, Motwani R, Winograd T (1999) The PageRank citation ranking: bringing order to the web. Technical report, Stanford InfoLab, Nov 1999
43. Kleinberg JM (1999) Authoritative sources in a hyperlinked environment. J ACM 46(5):604–632
44. Wu X, Pandey R, Chen JY (2009) Network topological reordering revealing systemic patterns in yeast protein interaction networks. Conf Proc IEEE Eng Med Biol Soc 2009:6954–6957
45. Huang H, Li J, Chen JY (2009) Disease gene-fishing in molecular interaction networks: a case study in colorectal cancer. Conf Proc IEEE Eng Med Biol Soc 2009:6416–6419
46. Goymer P (2007) Cancer genetics: networks uncover new cancer susceptibility suspect. Nat Rev Genet 8:823
47. Ergün A, Lawrence CA, Kohanski MA, Brennan TA, Collins JJ (2007) A network biology approach to prostate cancer. Mol Syst Biol 3:82
48. Kanehisa M, Goto S, Kawashima S, Okuno Y, Hattori M (2004) The KEGG resource for deciphering the genome. Nucleic Acids Res 32(Database Issue):D277–D280
49. Matthews L, Gopinath G, Gillespie M, Caudy M, Croft D, de Bono B, Garapati P, Hemish J, Hermjakob H, Jassal B et al (2009) Reactome knowledgebase of human biological pathways and processes. Nucleic Acids Res 37(Database Issue):D619–D622
50. Bader GD, Cary MP, Sander C (2006) Pathguide: a pathway resource list. Nucleic Acids Res 34(Database Issue):D504–D506
51. Karp PD, Ouzounis CA, Moore-Kochlacs C, Goldovsky L, Kaipa P, Ahren D, Tsoka S, Darzentas N, Kunin V, Lopez-Bigas N (2005) Expansion of the BioCyc collection of pathway/genome databases to 160 genomes. Nucleic Acids Res 33(19):6083–6089
52. Schaefer CF, Anthony K, Krupa S, Buchoff J, Day M, Hannay T, Buetow KH (2009) PID: the pathway interaction database. Nucleic Acids Res 37(Database Issue):D674–D679
53. Rao PN, Levine E, Myers MO, Prakash V, Watson J, Stolier A, Kopicko JJ, Kissinger P, Raj SG, Raj MH (1999) Elevation of serum riboflavin carrier protein in breast cancer. Cancer Epidemiol Biomarkers Prev 8(11):985–990
54. Dua RS, Isacke CM, Gui GPH (2006) The intraductal approach to breast cancer biomarker discovery. J Clin Oncol 24(7):1209–1216
55. Ou K, Yu K, Kesuma D, Hooi M, Huang N, Chen W, Lee SY, Goh XP, Tan LK, Liu J et al (2008) Novel breast cancer biomarkers identified by integrative proteomic and gene expression mapping. J Proteome Res 7(4):1518–1528
56. Alexander H, Stegner AL, Wagner-Mann C, Du Bois GC, Alexander S, Sauter ER (2004) Proteomic analysis to identify breast cancer biomarkers in nipple aspirate fluid. Clin Cancer Res 10(22):7500–7510
57. Adam PJ, Boyd R, Tyson KL, Fletcher GC, Stamps A, Hudson L, Poyser HR, Redpath N, Griffiths M, Steers G et al (2003) Comprehensive proteomic analysis of breast cancer cell membranes reveals unique proteins with potential roles in clinical cancer. J Biol Chem 278(8):6482–6489
58. Bullinger D, Neubauer H, Fehm T, Laufer S, Gleiter CH, Kammerer B (2007) Metabolic signature of breast cancer cell line MCF-7: profiling of modified nucleosides via LC-IT MS coupling. BMC Biochem 8:25
59. Kulasingam V, Diamandis EP (2007) Proteomics analysis of conditioned media from three breast cancer cell lines: a mine for biomarkers and therapeutic targets. Mol Cell Proteomics 6(11):1997–2011

60. Xiang R, Shi Y, Dillon DA, Negin B, Horvath C, Wilkins JA (2004) 2D LC/MS analysis of membrane proteins from breast cancer cell lines MCF7 and BT474. J Proteome Res 3(6):1278–1283
61. Burdall S, Hanby A, Lansdown M, Speirs V (2003) Breast cancer cell lines: friend or foe? Breast Cancer Res 5(2):89–95
62. Higgs RE, Knierman MD, Gelfanova V, Butler JP, Hale JE (2005) Comprehensive label-free method for the relative quantification of proteins from biological samples. J Proteome Res 4(4):1442–1450
63. Berishaj M, Gao SP, Ahmed S, Leslie K, Al-Ahmadie H, Gerald WL, Bornmann W, Bromberg JF (2007) Stat3 is tyrosine-phosphorylated through the interleukin-6/glycoprotein 130/Janus kinase pathway in breast cancer. Breast Cancer Res 9(3):R32
64. Hu H, Lee HJ, Jiang C, Zhang J, Wang L, Zhao Y, Xiang Q, Lee EO, Kim SH, Lu J (2008) Penta-1,2,3,4,6-O-galloyl-beta-D-glucose induces p53 and inhibits STAT3 in prostate cancer cells in vitro and suppresses prostate xenograft tumor growth in vivo. Mol Cancer Ther 7(9):2681–2691
65. Song H, Jin X, Lin J (2004) Stat3 upregulates MEK5 expression in human breast cancer cells. Oncogene 23(50):8301–8309
66. Nielsen NR, Gronbaek M (2006) Stress and breast cancer: a systematic update on the current knowledge. Nat Clin Pract Oncol 3(11):612–620
67. Zhang F, Chen JY (2010) Discovery of pathway biomarkers from coupled proteomics and systems biology methods. BMC Genomics 11(Suppl 2):S12
68. Ideker T (2004) Systems biology 101: what you need to know. Nat Biotechnol 22(4):473–475
69. Balmain A, Gray J, Ponder B (2003) The genetics and genomics of cancer. Nat Genet 33(3 s):238–244
70. Easton DF, Pooley KA, Dunning AM, Pharoah PD, Thompson D, Ballinger DG, Struewing JP, Morrison J, Field H, Luben R (2007) Genome-wide association study identifies novel breast cancer susceptibility loci. Nature 447(7148):1087–1095
71. Gold B, Kirchhoff T, Stefanov S, Lautenberger J, Viale A, Garber J, Friedman E, Narod S, Olshen AB, Gregersen P (2008) Genome-wide association study provides evidence for a breast cancer risk locus at 6q22. 33. Proc Natl Acad Sci 105(11):4340
72. Barabasi AL, Oltvai ZN (2004) Network biology: understanding the cell's functional organization. Nat Rev Genet 5(2):101–113
73. Goldberger AL, Amaral LAN, Hausdorff JM, Ivanov PC, Peng CK, Stanley HE (2002) Fractal dynamics in physiology: alterations with disease and aging. Proc Natl Acad Sci 99(90001):2466–2472
74. Amaral LAN, Diaz-Guilera A, Moreira AA, Goldberger AL, Lipsitz LA, Kopell NJ (2004) Emergence of complex dynamics in a simple model of signaling networks. Proc Natl Acad Sci U S A 101(44):15551–15555
75. Costa M, Goldberger AL, Peng CK (2005) Broken asymmetry of the human heartbeat: loss of time irreversibility in aging and disease. Phys Rev Lett 95(19):198102–198105
76. Goldberger AL, Moody GB, Peng CK (2006) Techniques, applications and future directions, Heart Rate Viability 2006 Workshop, 20–23 April 2006
77. Tatsumi J, Yamauchi A, Kono Y (1989) Fractal analysis of plant root systems. Ann Bot 64(5):499
78. Palmer MW (1988) Fractal geometry: a tool for describing spatial patterns of plant communities. Plant Ecol 75(1):91–102
79. Peitgen HO, Jugens H, Saupe D (2004) Chaos and fractals: new frontiers of science. Springer, New York
80. Auffray C (2007) Protein subnetwork markers improve prediction of cancer outcome. Mol Syst Biol 3:141–142
81. Nolan GP (2007) What's wrong with drug screening today. Nat Chem Biol 3:187–191
82. McCarthy N (2007) Tumour profiling: networking, protein style. Nat Rev Cancer 7:892–893

83. Morrison JL, Breitling R, Higham DJ, Gilbert DR (2005) GeneRank: using search engine technology for the analysis of microarray experiments. BMC Bioinforma 6(1):233
84. Bar-Joseph Z, Gifford DK, Jaakkola TS (2001) Fast optimal leaf ordering for hierarchical clustering. Bioinformatics 17(Suppl 1):S22–S29
85. Kim SK, Lund J, Kiraly M, Duke K, Jiang M, Stuart JM, Eizinger A, Wylie BN, Davidson GS (2001) A gene expression map for Caenorhabditis elegans. Science 293(5537):2087–2092
86. You Q, Fang S, Chen JY (2008) GeneTerrain: visual exploration of differential gene expression profiles organized in native biomolecular interaction networks. Inf Vis 9(1):1–12. doi:10.1057